新生物学丛书

干细胞的细胞生物学

〔以〕Eran Meshorer 〔美〕Kathrin Plath 著
韩忠朝 李宗金 主译

科学出版社

北京

图字：01-2014-1865

内 容 简 介

本书涉及干细胞生物学的多个方面，范围从胚胎干细胞的基础分子特性到在体内成体干细胞迁移和成体干细胞微环境，最后还讨论了再生和细胞命运的重编程。本书总共十四章，提供了关于干细胞的细胞生物学的新认识，内容丰富详实，对干细胞研究具有指导作用。

本书适合于从事干细胞研究领域的研究生和青年学者阅读使用。

Translation from English language edition: *The Cell Biology of Stem Cells* by Eran Meshorerand Kathrin Plath
Copyright © 2010 Springer New York
Springer New York is a part of Springer+Business Media
All Right Reserved

图书在版编目(CIP)数据

干细胞的细胞生物学/(以)梅肖勒(Meshorer, E.)，(美)普拉思(Plath, K.)著；韩忠朝，李宗金主译. —北京：科学出版社，2014.6
（新生物学丛书）
书名原文：The cell biology of stem cells
ISBN 978-7-03-040867-9

Ⅰ. ①干…　Ⅱ. ①梅…②普…③韩…④李…　Ⅲ. ①干细胞-细胞生物学　Ⅳ. ①Q24

中国版本图书馆 CIP 数据核字（2014）第 118019 号

责任编辑：罗　静　刘　晶 / 责任校对：刘小梅
责任印制：徐晓晨 / 封面设计：美光制版

科学出版社 出版
北京东黄城根北街 16 号
邮政编码：100717
http://www.sciencep.com

北京凌奇印刷有限责任公司 印刷
科学出版社发行　各地新华书店经销
*

2014 年 6 月第 一 版　开本：787×1092　1/16
2020 年 10 月第七次印刷　印张：13 3/4
字数：326 000
定价：86.00 元
（如有印装质量问题，我社负责调换）

《新生物学丛书》专家委员会成员名单

主　任： 蒲慕明

副主任： 吴家睿

专家委员会成员（按姓氏汉语拼音排序）：

昌增益	陈洛南	陈晔光	邓兴旺	高　福
韩忠朝	贺福初	黄大昉	蒋华良	金　力
康　乐	李家洋	林其谁	马克平	孟安明
裴　钢	饶　毅	饶子和	施一公	舒红兵
王　琛	王梅祥	王小宁	吴仲义	徐安龙
许智宏	薛红卫	詹启敏	张先恩	赵国屏
赵立平	钟　扬	周　琪	周忠和	朱　祯

译者名单

主　译　韩忠朝　李宗金

参　译（按姓氏汉语拼音排序）

陈　芳	中国医学科学院血液病医院（血液学研究所）
池　颖	中国医学科学院血液病医院（血液学研究所）
龚　伟	中国医学科学院放射医学研究所
胡　晓	中国医学科学院血液病医院（血液学研究所）
李　雪	中国医学科学院血液病医院（血液学研究所）
梁　璐	细胞产品国家工程研究中心
刘文静	中国医学科学院血液病医院（血液学研究所）
马凤霞	中国医学科学院血液病医院（血液学研究所）
阮　峥	中国医学科学院血液病医院（血液学研究所）
苏位君	南开大学医学院
王丽娜	南开大学医学院
王有为	北京汉氏联合干细胞研究院
周曼倩	南开大学医学院
卓　毅	福建医科大学附属第一医院
卓光生	北京汉氏联合干细胞研究院

《新生物学丛书》丛书序

当前，一场新的生物学革命正在展开。为此，美国国家科学院研究理事会于2009年发布了一份战略研究报告，提出一个"新生物学"（New Biology）时代即将来临。这个"新生物学"，一方面是生物学内部各种分支学科的重组与融合，另一方面是化学、物理、信息科学、材料科学等众多非生命学科与生物学的紧密交叉与整合。

在这样一个全球生命科学发展变革的时代，我国的生命科学研究也正在高速发展，并进入了一个充满机遇和挑战的黄金期。在这个时期，将会产生许多具有影响力、推动力的科研成果。因此，有必要通过系统性集成和出版相关主题的国内外优秀图书，为后人留下一笔宝贵的"新生物学"时代精神财富。

科学出版社联合国内一批有志于推进生命科学发展的专家与学者，联合打造了一个21世纪中国生命科学的传播平台——《新生物学丛书》。希望通过这套丛书的出版，记录生命科学的进步，传递对生物技术发展的梦想。

《新生物学丛书》下设三个子系列：科学风向标，着重收集科学发展战略和态势分析报告，为科学管理者和科研人员展示科学的最新动向；科学百家园，重点收录国内外专家与学者的科研专著，为专业工作者提供新思想和新方法；科学新视窗，主要发表高级科普著作，为不同领域的研究人员和科学爱好者普及生命科学的前沿知识。

如果说科学出版社是一个"支点"，这套丛书就像一根"杠杆"，那么读者就能够借助这根"杠杆"成为撬动"地球"的人。编委会相信，不同类型的读者都能够从这套丛书中得到新的知识信息，获得思考与启迪。

《新生物学丛书》专家委员会
主　任：蒲慕明
副主任：吴家睿
2012年3月

译者前言

近些年来，以干细胞技术为核心的再生医学研究取得了突飞猛进的发展。干细胞独特的自我更新和分化潜能使其成为研究基础生物学问题的模型，而且，现在成体干细胞、胚胎干细胞和诱导多能干细胞均可以用于生产细胞和组织，使其可以服务于再生医学。这一切都使得干细胞技术在再生医学的医疗、科研领域有巨大的应用潜力。掌握更多的干细胞细胞生物学知识将有利于再生医学的进一步发展。

《干细胞的细胞生物学》一书涉及干细胞生物学的多个方面，范围从胚胎干细胞的基础分子特性到成体干细胞的分化和迁移及其微环境，书结尾部分讨论了成体干细胞的重编程。本书共14章，提供了关于干细胞的细胞生物学的新认识，内容丰富翔实，对干细胞研究具有指导作用。本书适合从事干细胞研究领域的研究生和青年学者阅读。

衷心感谢赵钦军、梁璐、王有为等在翻译后校对过程中所付出的辛勤劳作，同时感谢韩之波和郑重在本译作出版过程中所做的贡献。

在译校过程中，虽然力求忠于原作，但限于水平有限，谬误之处在所难免，敬请读者批评指正。

韩忠朝
2014年4月12日

原 书 前 言

近几年来,干细胞获得了大量的关注,干细胞独特的自我更新和分化潜能使其成为研究基础生物学问题引人注目的模型,其涉及的领域包括细胞分裂、复制、转录、细胞命运决定等。由于哺乳动物的胚胎干细胞(ES 细胞)可以发育成不同的细胞类型,并且成体干细胞能够分化成为一个指定谱系的细胞,这使得不同发育阶段的基础问题都可以得到解决。重要的是,成体干细胞和胚胎干细胞为细胞治疗提供了一个极好的手段,使干细胞研究更加切合再生医学。

正如书名"干细胞的细胞生物学"所示,我们的书涉及干细胞生物学的多个方面,范围从它的基础分子特性到在体内成体干细胞迁移和成体干细胞微环境,书结尾部分讨论了再生和细胞命运的重编程。在第 1 章"小鼠早期胚胎细胞的命运决定"中,Amy Ralston 和 Yojiro Yamanaka 描述了支持小鼠胚泡植入前期的早期发育决定的机制,以及目前关于最不成熟的最早期的干细胞类型的来源的理解,其中包括胚胎干(ES)细胞、滋养层干(TS)细胞和胚外内胚层干(XEN)细胞。由此,我们在由 Kelly Morris,Mita Chotalia 和 Ana Pombo 编写的第 2 章"干细胞的核结构"中研究了多能胚胎干细胞的核结构和基因组的组织结构。这一章讨论了 ES 细胞核的三维空间结构和功能,着重分析了这些细胞的染色质、核小体和基因定位的独特特性。Eleni Tomazou 和 Alexander Meissner 编写的第 3 章"细胞多能性的表观遗传学调控"更加深入地分析了胚胎干细胞的表观遗传学。作者阐述了关键染色质修饰的表观遗传的概况,包括 DNA 甲基化和组蛋白修饰,并讨论了这些表观遗传标记的功能性。第 4 章仍然是在 DNA 水平介绍,是 Ichiro Hiratani 和 David Gilbert 编写的"哺乳动物早期发育阶段常染色体复制域的莱昂化",作者通过回顾 50 多年来在这个令人兴奋的领域的研究历程,说明了胚胎干细胞的 DNA 复制的调控和动力学,重提了老观念"常染色体莱昂化"来解释异染色质化的过程。

基因组 DNA 是生命的基本单位,一直处在不断的损伤和修复之中。Peter Stambrook 和 Elisia Tichy 在第 5 章"小鼠胚胎干细胞基因组完整性的保持"一章中,讨论了在胚胎干细胞中的基因突变率、信号通路和 DNA 损伤与修复机制。谈论完 DNA 的包装、复制和损伤,本书进入第 6 章"胚胎干细胞的转录调控",在这章中 Jian-Chien Dominic Heng 和 Huck-Hui Ng 集中在 RNA 水平讨论了转录调控网络,它是多能性状态的核心,并介绍了最近的技术进展,通过这种技术可以系统性地了解 ES 细胞及其分化过程中的转录调控。讨论完转录调控,本书开始介绍 RNA 剪接,David Nelles 和 Gene Yeo 撰写了第 7 章"干细胞自我更新和分化中的选择性剪接",他们综述了有关"剪接"的最新文献,着重介绍了胚胎干细胞中的几个关键的选择性剪接基因,并讨论了最新的全基因组水平的方法,从全局范围来分析剪接和选择性剪接模式。第 8 章由 Collin Melton 和 Robert Blelloch 编写,题为"微小 RNA 调节胚胎干细胞自我更新和分化",这一章阐明了胚胎干细胞的 microRNA 的调控,主要介绍了几个突出的 microRNA,包括 Let-7、Lin-28、miR-134、miR-296 等,它们参与了 ES 细胞的自我更新和/或全能性的调控。第 9 章"成体干细胞和多能胚胎干细胞中的端粒及端粒酶"由 Rosa

Marión 和 Maria Blasco 撰写，这一章概述了在多能性和多能干细胞中的端粒生物学及端粒酶调控，并讨论了从体细胞到多能细胞的核重排过程中使得端粒染色质重塑的潜在机制。在小鼠中，核重排到多能性也需要雌细胞中沉默的 X 染色体的重新激活。第 10 章 "X 染色体失活与胚胎干细胞" 是由 Tahsin Stefan Barakat 和 Joost Gribnau 编写的。这一章讨论诱导分化的雌性胚胎干细胞的 X 染色体失活（XCI）调控，解释了精确的协调整个染色体转录沉默的顺式和反式作用机制，并提出假说阐明了为什么这样一个有趣的过程只发生在雌性细胞中。

讨论完多能胚胎干细胞的细胞核中的分子生物学，接下来的三个章节开始介绍成体干细胞。虽然多能干细胞只是早期胚胎发育的一个短暂阶段，但是成体干细胞群存在于机体的整个生命周期中，直到它们被组织内稳态和/或修复所利用。制约成体干细胞并调节其分化而不是自我更新的信号被认为是细胞和细胞外基质的相互作用控制的，而这就构成了干细胞微环境。第 11 章 "成体干细胞及其细胞龛" 中，Francesca Ferraro、Cristina Celso 和 David Scadden 解释了微环境的概念，讨论了不同哺乳动物微环境的信号通路，并联系了目前关于微环境生物学、癌变和老化的认识。在第 12 章 "成体干细胞的分化和迁移及其对疾病的影响" 中，Ying Zhuge、Zhao-Jun Liu 和 Omaida Velazquez 阐述了造血干细胞（HSC）、间充质干细胞（MSC）和内皮祖细胞（EPC）的迁移，并讨论了在哺乳动物体内控制其迁移调控的机制。Zhuge 等还解释了如何将这些基本过程转化为临床应用。

在接下来的章节 "脊椎动物再生模型对干细胞应用的启示" 中，Christopher Antos 和 Elly Tanaka 讨论了几个脊椎动物模型，如青蛙、鱼和蝾螈的再生机制。他们描述了极大的细胞可塑性，包括了几种组织结构（眼睛、心脏、神经系统及其附属物）的再生，作者总结了一些分子，这些分子促成了指定的组织的转分化和去分化。本书的最后一章 "成体细胞重编程获得多能性" 由 Masato Nakagawa 和 Shinya Yamanaka 撰写。这一章介绍了干细胞生物学最令人兴奋的最新发展，即多能性细胞的重编程。作者简介了 20 世纪 50 年代和 60 年代在青蛙卵母细胞内进行的体细胞核转移实验的历史，介绍了细胞融合实验产生重新编程的细胞、albeit 四倍体，并介绍了他们的实验室在重编程领域的开创性贡献——某些特定转录因子表达使体细胞变成多能干细胞，从而开创了新的蓬勃发展的领域即诱导多能干细胞（iPS）。

现在成体干细胞、胚胎干细胞和诱导多能干细胞可以用于生产细胞和组织，以便应用于细胞治疗。现在应用 iPS 细胞技术能够生成特定患者的多能干细胞，在疾病研究和药物筛选方面有巨大的潜力。为了能够充分利用干细胞的优势和巨大的潜能，我们还需要掌握更多的基本生物学知识。在本书这 14 章中提供了关于干细胞的细胞生物学的新认识，并讨论了许多有待回答的问题。

Eran Meshorer，博士
以色列耶路撒冷希伯来大学生命科学学院遗传学系
Kathrin Plath，博士
美国加州大学洛杉矶分校医学院生物化学系

（梁 璐 译）

原作者简介

  Eran Meshorer，博士，在耶路撒冷希伯来大学遗传学系研究胚胎干细胞和神经干细胞的染色质可塑性。获得希伯来大学分子神经科学博士学位，随后在美国国立卫生研究院（NIH）开展博士后研究。他的实验室主要是应用全基因组和单细胞的技术从染色质的角度来研究多能性、分化和重编程。他是国际干细胞研究学会的成员，是Joseph H. 和Belle R. Braun生命科学学院的高级讲师。

  Kathrin Plath，博士，2004 年以来担任加州大学洛杉矶分校生物化学系助理教授。获德国柏林洪堡大学的博士学位后，她分别在美国加州大学、旧金山和剑桥 Whitehead 研究所、MA 开展博士后研究。Plath 博士的主要研究方向是发育线索如何引起分子水平上染色质结构改变，以及这些改变如何调节细胞命运决定和哺乳动物发育的基因表达。她是国际干细胞研究学会的成员和几家干细胞杂志编委会成员。

原著者名单

Christopher L. Antos
DFG-Center for Regenerative Therapies
Dresden Technische Universität
Dresden
Germany

Tahsin Stefan Barakat
Department of Reproduction
 and Development
Erasmus MC
University Medical Center
Rotterdam
The Netherlands

Maria A. Blasco
Telomeres and Telomerase Group
Molecular Oncology Program
Spanish National Cancer Centre (CNIO)
Madrid
Spain

Robert Blelloch
The Eli and Edythe Broad Center
 of Regeneration Medicine
 and Stem Cell Research
 Center for Reproductive Sciences
Program in Biomedical Science
Department of Urology
University of California San Francisco
San Francisco, California
USA

Cristina Lo Celso
Division of Cell and Molecular Biology
Imperial College London
London, England
UK

Mita Chotalia
Genome Function Group
MRC Clinical Sciences Centre
Imperial College School of Medicine
Hammersmith Hospital Campus
London, England
UK

Francesca Ferraro
Center for Regenerative Medicine
Massachusetts General Hospital and
 Harvard Medical
Department of Stem Cell
 and Regenerative Biology
Harvard University
and
Harvard Stem Cell Institute
Cambridge, Massachusetts
USA

David M. Gilbert
Department of Biological Science
Florida State University
Tallahassee, Florida
USA

Joost Gribnau
Department of Reproduction
 and Development
Erasmus MC
University Medical Center
Rotterdam
The Netherlands

Jian-Chien Dominic Heng
Gene Regulation Laboratory
Genome Institute of Singapore
and
NUS Graduate School for Integrative
 Sciences and Engineering
Singapore

Ichiro Hiratani
Department of Biological Science
Florida State University
Tallahassee, Florida
USA

Zhao-Jun Liu
Division of Vascular Surgery
DeWitt Daughtry Family Department
 of Surgery
and
Sylvester Comprehensive Cancer Center
Miller School of Medicine
University of Miami
Miami, Florida
USA

Rosa M. Marión
Telomeres and Telomerase Group
Molecular Oncology Program
Spanish National Cancer Centre (CNIO)
Madrid
Spain

Alexander Meissner
Department of Stem Cell
 and Regenerative Biology
Harvard University
and
Harvard Stem Cell Institute
and
Broad Institute
Cambridge, Massachusetts
USA

Collin Melton
The Eli and Edythe Broad Center
 of Regeneration Medicine
Stem Cell Research Center
 for Reproductive Sciences
Program in Biomedical Science
 and Department of Urology
University of California San Francisco
San Francisco, California
USA

Eran Meshorer
Department of Genetics
Institute of Life Sciences
The Hebrew University of Jerusalem,
Jerusalem
Israel

Kelly J. Morris
Genome Function Group
MRC Clinical Sciences Centre
Imperial College School of Medicine
Hammersmith Hospital Campus
London, England
UK

Masato Nakagawa
Center for iPS Research and Application
Institute for Integrated Cell-Material
 Sciences
Kyoto University
Kyoto
Japan

David A. Nelles
Department of Cellular and Molecular
 Medicine
Stem Cell Program
University of California, San Diego
La Jolla, California
USA

Huck-Hui Ng
Gene Regulation Laboratory
Genome Institute of Singapore
and
NUS Graduate School for Integrative
 Sciences and Engineering
Singapore

Kathrin Plath
UCLA School of Medicine
Department of Biological Chemistry
Los Angeles, California
USA

Ana Pombo
Genome Function Group
MRC Clinical Sciences Centre
Imperial College School of Medicine
Hammersmith Hospital Campus
London, England
UK

Amy Ralston
Department of Molecular, Cell,
 and Developmental Biology
University of California Santa Cruz
Santa Cruz, California
USA

David Scadden
Department of Stem Cell
 and Regenerative Biology
Harvard University
and
Harvard Stem Cell Institute
Cambridge, Massachusetts
and
Cancer Center, Massachusetts General
 Hospital
Boston, Massachusetts
USA

Peter J. Stambrook
Department of Molecular Genetics,
 Biochemistry and Microbiology
University of Cincinnati College
 of Medicine
Cincinnati, Ohio
USA

Elly M. Tanaka
Max-Planck Institute for Molecular
 Cell Biology and Genetics
Dresden
Germany

Elisia D. Tichy
Department of Molecular Genetics,
 Biochemistry and Microbiology
University of Cincinnati
College of Medicine
Cincinnati, Ohio
USA

Eleni M. Tomazou
Department of Stem Cell Regenerative
 Biology
Harvard University
Cambridge, Massachusetts
USA

Omaida C. Velazquez
Division of Vascular Surgery
DeWitt Daughtry Family Department
 of Surgery
and
Sylvester Comprehensive Cancer Center
Miller School of Medicine
University of Miami
Miami, Florida
USA

Shinya Yamanaka
Center for iPS Research and Application
Institute for Integrated Cell-Material
 Sciences
Kyoto University
Kyoto
Japan

Yojiro Yamanaka
Goodman Cancer Center
Department of Human Genetics
McGill University
Montreal, Quebec
Canada

Gene W. Yeo
Department of Cellular
 and Molecular Medicine
Stem Cell Program
University of California, San Diego
La Jolla, California
USA

Ying Zhuge
Division of Vascular Surgery
DeWitt Daughtry Family Department
 of Surgery
Miller School of Medicine
University of Miami
Miami, Florida
USA

目　　录

《新生物学丛书》丛书序
译者前言
原书前言
原作者简介
原著者名单

第1章　小鼠早期胚胎细胞的命运决定 · 1
　引言 · 1
　胚层的建立和前干细胞程序：囊胚的形成 · 2
　胚系的维持和干细胞程序：囊胚以外 · 5
　第二胚层决定：ICM 的细分 · 5
　细胞信号调节 PE/EPI 形成 · 6
　EPI 胚层多能性的建立和调节 · 8
　结论 · 9
　参考文献 · 10

第2章　干细胞的核结构 · 13
　引言 · 13
　胚胎干细胞核的功能分区 · 13
　干细胞其他核质亚区的特点 · 17
　胚胎干细胞核特有的染色质特征 · 18
　结论 · 19
　参考文献 · 20

第3章　细胞多能性的表观遗传学调控 · 23
　引言 · 23
　表观遗传调控 · 24
　胚胎干细胞的表观遗传组学 · 28
　结论 · 32
　参考文献 · 33

第4章　哺乳动物早期发育阶段常染色体复制域的莱昂化 · 37
　引言 · 37
　复制时序编程：对基因组结构的初步测定 · 38
　一种在进化上相对保守的表观遗传学印记 · 43
　复制时序作为染色体三维结构的定量指标 · 44
　复制时序反映表观遗传特征的改变：常染色体在外胚层阶段的莱昂化 · 46
　复制时序及细胞重编程：常染色体莱昂化的进一步证明 · 47
　复制时序程序的维持和改变及其潜在作用 · 48

结论	49
参考文献	50

第5章 小鼠胚胎干细胞基因组完整性的保持 … 54
引言和历史观点 … 54
体细胞的突变频率 … 56
小鼠 ES 细胞基因组的保护 … 57
结论 … 64
参考文献 … 65

第6章 胚胎干细胞的转录调控 … 68
引言 … 68
胚胎干细胞可作为研究转录调控的模型细胞 … 69
转录因子决定胚胎干细胞的多能性 … 69
转录调控网络 … 72
转录调控网络分析技术 … 72
转录调控网络核心：Oct4、Sox2 和 Nanog … 73
扩大化的转录调控网络 … 75
增强体：转录因子复合体 … 76
信号通路参与转录网络 … 77
转录调控与表观遗传学调控的相互作用 … 78
结论 … 79
参考文献 … 79

第7章 干细胞自我更新和分化中的选择性剪接 … 82
引言 … 82
选择性剪接的概述 … 82
干细胞干性维持及分化中涉及的选择性剪接基因 … 84
基因组学方法鉴别、检测选择性剪接 … 86
RNA 结合蛋白对选择性剪接的调控 … 87
结论和展望 … 90
参考文献 … 90

第8章 微小 RNA 调节胚胎干细胞自我更新和分化 … 93
引言：自我更新程序 … 93
胚胎干细胞 … 94
微小 RNA 的生成和功能 … 94
促进自我更新的 ESCC miRNA … 95
ESC 分化过程中诱导的 miRNA 阻止自我更新程序 … 97
控制 miRNA 表达的调节网络 … 99
miRNA 能促进或抑制 IPS 细胞分化 … 99
成体干细胞中的 miRNA … 100
肿瘤细胞中的 miRNA … 100
结论 … 101

参考文献 ··· 101

第9章 成体干细胞和多能胚胎干细胞中的端粒及端粒酶 104
引言 ··· 104
端粒及端粒酶的调节 ··· 106
端粒和端粒酶在成体干细胞中的作用 ·· 107
在体细胞核移植中端粒及端粒酶的调控 ··· 108
在iPS产生过程中端粒及端粒酶的调控 ·· 109
端粒酶活性对于产生"高"质量的iPS细胞是必需的 ··· 110
端粒重编程调控 ·· 110
结论 ··· 111
参考文献 ··· 111

第10章 X染色体失活与胚胎干细胞 115
引言 ··· 115
X染色体失活中的顺势作用因子 ··· 117
X染色体失活中的反式作用因子 ··· 118
数量及选择 ·· 119
沉默与沉默的维持 ·· 123
X染色体失活与人ES细胞 ··· 126
结论 ··· 129
参考文献 ··· 130

第11章 成体干细胞及其细胞龛 137
"龛"的概念、定义与历史 ··· 137
干细胞龛成分 ·· 138
与龛功能相关的分子通路 ··· 140
细胞外基质与细胞-细胞间相互作用 ·· 141
干细胞龛动态性 ·· 142
干细胞龛衰老 ·· 143
恶性干细胞龛 ·· 143
结论 ··· 144
参考文献 ··· 145

第12章 成体干细胞的分化和迁移及其对疾病的影响 149
分化 ··· 149
间充质干细胞 ·· 152
迁移 ··· 153
结论 ··· 157
参考文献 ··· 158

第13章 脊椎动物再生模型对干细胞应用的启示 162
脊椎动物再生模型的属性 ··· 162
成熟组织再生的机制 ··· 163
结论 ··· 180

参考文献 ………………………………………………………………………… 181
第 14 章　成体细胞重编程获得多能性 …………………………………………… 189
　引言 ……………………………………………………………………………… 189
　青蛙成体细胞核重编程研究 …………………………………………………… 189
　克隆羊"多莉"的诞生 ………………………………………………………… 190
　改变细胞命运的因子 MyoD（成肌分化抗原）………………………………… 190
　通过细胞融合来重编程体细胞 ………………………………………………… 190
　转染 Sox2、Oct3/4、Klf4 和 c-Myc 从而产生诱导多能干细胞 ……………… 191
　iPS 细胞诱导的方法 …………………………………………………………… 192
　iPS 细胞产生的分子机制 ……………………………………………………… 193
　直接重编程：来源于胰腺细胞的 β 细胞 ……………………………………… 193
　直接重编程：来源于成纤维细胞的神经元细胞 ……………………………… 194
　可用于临床的疾病 iPS 细胞系 ………………………………………………… 194
　结论 ……………………………………………………………………………… 195
　参考文献 ………………………………………………………………………… 195

第1章 小鼠早期胚胎细胞的命运决定

Yojiro Yamanaka*，Amy Ralston*

摘要：发育过程中，最初的胚胎全能干细胞特异性地形成独立的组织胚层。小鼠中形成的第一胚系是胚外组织。同时，没有成为胚外层的细胞仍保留多能干性，它们可以形成胚胎的所有胚层。多能干细胞系来源于胚胎的几个发育阶段。有趣的是，在同一时期胚外层已经有多能干细胞系了。因此，检测早期胚胎细胞命运决定的调控是研究干细胞系建立的一个难得的机会。以往的研究为前三个胚层的形成提出了深刻的见解，而现代分子成像技术的运用进一步推动了这一领域的发展。本章我们将介绍目前所发现的小鼠发育过程早期三个胚层建立和维持的多样的分子机制。

引言

在小鼠发育的初期，最初全能细胞的发育潜力是受到限制的，仅发育成小鼠第一胚层。而在非哺乳动物物种中，第一胚层的建立则可能涉及主体轴的形成，哺乳动物的不同在于其首先要完成着床。因此，几天内胚胎和胚外组织之间的差异包括了前两个胚层决定（图1-1）及之前胚层（外胚层、中胚层、内胚层）和胚系的建立。这种独特的哺乳动物发育模式涉及一种独特的细胞类型，可以从中分离出来并扩增为稳定的细胞系。因此，认识胚外组织的起源有助于我们了解干细胞的形成与分化。以往的研究对前三个胚层的形成提出了颇具洞察力的见解，而利用现代分子生物学和成像技术将进一步推进这一领域的发展。

受精后第三天，小鼠胚胎（囊胚）包含三个组织系：外胚层（EPI）、滋养层（TE）和原始内胚层（PE）。对来源于这些胚层的干细胞系的分离和研究加深了我们对早期胚胎细胞命运决定的认识。已经从囊胚中分离得到三类干细胞系：胚胎干细胞、滋养层干细胞和胚外内胚层干细胞（ES细胞、TS细胞和XEN细胞）。它们均具有干细胞自我更新及分化为各类成熟细胞等的特性。同时，每种干细胞系仍保有与其来源相关的特性，包括组织特异性的发育潜力、形态、转录因子表达和生长因子需求[1]。这些干细胞系的提出不仅为那些需要大量初始原料的研究提供了可扩增的纯细胞群，也是了解干细胞起源的一个契机。

从胚胎干细胞的研究中我们可以对细胞命运选择基因进行更深入的分子水平分析。

---

* Corresponding Authors：Yojiro Yamanaka—Goodman Cancer Center, Department of Human Genetics, Faculty of Medicine, McGill University, Montreal, QC H3A1A3. Email：yojiro.yamanaka@mcgill.ca. Amy Ralston—Department of Molecular, Cell and Developmental Biology, University of California Santa Cruz, California, USA. Email：ralston@biology.ucsc.edu

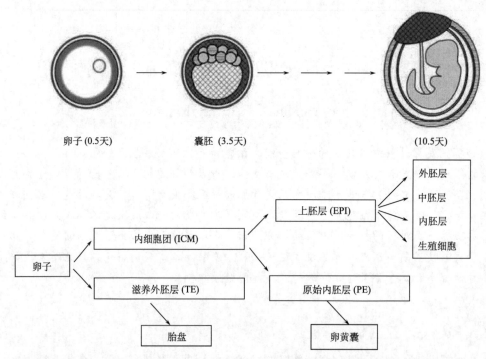

图1-1 小鼠发育过程中的前两个胚系决定的概述。起始全能性的受精卵发育成囊胚,其中包含三个胚层:EPI(蓝色)、TE(红色,网格)和PE(黄色,直线)。这些胚层会发育成为胎儿、胎盘和发育晚期卵黄囊的一部分。此图的彩版请见 www.landesbioscience.com/curie

操控某一特定种系调节基因可以引起干细胞发生相应的命运改变。例如,一个滋养细胞转录因子 Cdx2 就足够使 ES 细胞转变成 TS-样的细胞[2]。这类研究表明 ES 细胞具有很强的可塑性,以及作为胚层决定因子 Cdx2 基因的核心作用。胚胎干细胞也为研究胚层决定基因之间的分子相互作用提供了一个机会,从而成为认识胚胎的细胞命运选择的模式。然而,在胚层决定基因的研究中发现其发挥作用主要在相对晚期,因此便提出胚胎早期是如何决定三个胚层的疑问。

参与第一胚层形成的机制可能有很多种,包括细胞的位置、形状、极化、信号和分裂面。在新提出的模式中,早期的前干细胞程序使组织胚层特异性形成囊胚。而在着床及以后,细胞命运则由干细胞系中的一个活化程序所掌控(图1-2)。

胚层的建立和前干细胞程序:囊胚的形成

这里我们将讨论胚层形成的第一个阶段:TE 和内细胞团(ICM)的形成共同组成囊胚。TE 将分化为胎盘,而 ICM 是胚胎和原始内胚层祖细胞的混合物。在囊胚中,TE 围绕 ICM 和中空的囊胚腔,胚层示踪实验表明 TE 和 ICM 细胞群是从胚胎的内外细胞群开始发育[3]。受精卵细胞分裂为 2、4、8 和 16 细胞,这群少量的细胞被外细胞包围。通过连续分裂,内外细胞的数量增加,TE 上皮化和囊胚腔扩张,形成囊胚结构。

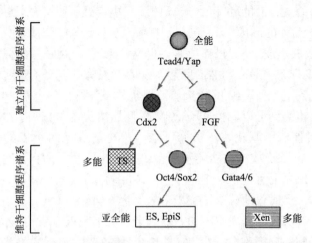

图1-2 小鼠早期发育过程中形成和维持细胞命运的分子间相互作用的概述。Tead4/Yap复合物从最初的全能干细胞（灰色）中选择TE（红色，网格）。不能成为TE细胞则发育为EPI（蓝色）和PE（黄色，直线）。胚层内的信号有利于EPI和PE命运的走向。胚层特异性转录因子参与各个胚层的成熟过程。此图的彩版请见www.landesbioscience.com/curie

拓扑结构如何与细胞的命运相连系的机制尚不清楚，目前已经提出了几种模型。例如，细胞的命运可能是细胞位置的结果（图1-3A）；或者预先确定的细胞的命运可以驱使细胞到适当的拓扑学位置（图1-3B）。随后的机制预测了在内外细胞群形成之前就可以检测到前内细胞和前外细胞。然而，尽管人们在这一领域付出了相当多的努力，但目前还没有找到支持这个预先确定机制的证据。

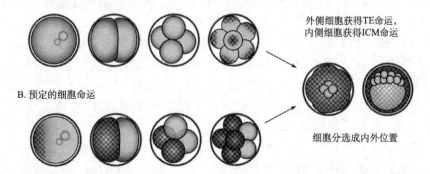

图1-3 TE形成的两种可能模式。A. 细胞位置决定细胞的命运，外层细胞或细胞外部分选择TE细胞的命运（红色，网格）。B. TE的命运是预先确定的，细胞的一个特定子集继承了TE的命运决定分子。此图的彩版请见www.landesbioscience.com/curie

胚层示踪法与分子分析法作为两个主要的研究策略，被运用于寻找关于囊胚期之前的细胞存在预定理论的证据。在胚层示踪方面，对两种细胞共存阶段的细胞研究发现，两细胞之间有偏差的发育潜能和 TE/ICM 的胚层决定是不相关的，这是因为这两个细胞均参与了 TE 和 ICM 的形成[4~13]。同样，4 细胞阶段、8 细胞阶段的所有细胞也参与了 TE 和 ICM 胚层的形成[14,15]。虽然有报道认为 4 细胞阶段胚层的发育潜能受到了限制[7]，因而胚外层没有完全形成，然而胚层示踪实验没有找到证据证明在内外细胞群形成之前细胞被预先确定为 TE 或 ICM 细胞。至于分子分析方面，在 16 细胞阶段之前的细胞亚群中，没有检测到可以调控 TE/ICM 的胚层决定的蛋白质。在 4 细胞阶段的卵裂球中存在一种组蛋白甲基化水平分布不均，这与嵌合体的小鼠的生育能力降低有关[16]，而其在 TE/ICM 形成中的功能重要性仍然有待阐明。所以，也没有分子证据表明在内外细胞群形成之前有前 TE 细胞或前 ICM 细胞的存在。相反，当内外细胞群确定了其在胚胎中的位置时，便确立了它未来的发育方向。

如果是细胞位置决定细胞发育方向，那一定存在细胞感应胚胎内位置的机制。长期的研究结果表明细胞在 8 细胞阶段极化[17]，这个理论认为内/外轴在细胞水平上存在差异。被保守的极性蛋白如非典型蛋白激酶 C（aPKC）、Par3 和 Par6 极化是维持细胞位置所必需的[8]，细胞接触则是细胞极化所必需的[17]。但还没有分子水平上的证据来确认位置、极化和细胞命运之间的联系。利用常规的基因敲除技术来研究这个领域是具有挑战性的。与细胞位置、细胞接触相关的许多蛋白质是大的基因家族的成员，如 aPKC，这表明单基因敲除可能会被遗传冗余所掩盖。此外，早期发育阶段部分是通过母体提供的蛋白质调控的，因此需要敲除生殖细胞的基因来确定表型。最后，许多这类蛋白质参与了细胞基本过程如细胞分裂，这使得在发育的过程中研究其功能变得困难。再者，显性抑制或 siRNA 的过表达会导致短期或局部功能丧失，这也将阻碍表型的确立。

最终需要一个不同的定位转录因子来将内/外差异转换成基因表达的变化。确定早期胚层发育相关的转录因子有以下几个策略：利用芯片分析着床前的转录表达来确定候选的转录因子，随后通过原位杂交筛选囊胚中表达受限的候选因子[18]；或者通过对囊胚来源的干细胞系进行基因芯片比较来筛选候选因子[19]。还有就是通过敲除来偶然发现早期致死表型，Cdx2 和 Tead4[20~22]便是这样发现的。

虽然 TE 的发育需要 Cdx2，但是 Cdx2 可能在 TE 的形成中没有起到决定性的作用[23,24]。研究表明 Cdx2 mRNA[25]定位在 8 细胞阶段的细胞外表面，但 Cdx2 蛋白[24,26]却没有。在没有 Cdx2 的胚胎中，TE 的标记 Gata3 仍持续表达[19]，这说明无论是形态学上[23,24]还是分子水平上，Cdx2 都不是 TE 所必需的，很难相信定位的 Cdx2 mRNA 的表达会在胚层确立中起决定性作用。最近一个与 Tead4 及辅助因子相关的新的通路被证明在第一胚层决定中发挥了决定性作用。在 Cdx2 表达的激活中，转录共激活子 Yap 和相关蛋白 Taz，表现出对细胞位置敏感的变化[27]。在囊胚期之前，Yap/Taz 定位在外细胞的细胞核和内细胞的细胞质，这一定位是由 Hippo 信号通路成员 Lats1/2 的磷酸化来调节的。此外，改变细胞位置会导致 Yap 定位的相应变化：外细胞嵌入内部，细胞聚集并失去核 Yap，而内细胞脱离了周围的外细胞并获得核 Yap。Tead4 是一种 DNA 结合蛋白，Yap/Taz 与 Tead4 的直接相互作用是 Cdx2[21,22]和其他滋养层标记[19]表达所必需的。虽然 Yap/Taz 调节信号感知细胞位置的特性或者本质仍然未知，但 Hippo 信号通路以及与细胞接触有关的蛋白质如钙黏蛋白可能参与了这一过程，这无疑将是一个令人兴奋的研究领域。

目前已经清楚了 Yap/Tead4 的上游通路，而对它的下游通路还不完全清楚。tead4 是 Cdx2 表达所必需的，缺失 Tead4 的胚胎不能形成囊胚，而 Cdx2 敲除的胚胎在囊胚形成后死亡。Tead4 在 ICM 中不是必需的[21,22]，所以一定还有其他的基因与 TE 中的 Cdx2 同时作用。目前已经开始研究一些基因，如 Gata3[19]，今后寻找参与促进外细胞增殖和构建囊胚的 Tead4 的靶点将是重要的研究方向。

胚系的维持和干细胞程序：囊胚以外

在囊胚中，早期阶段的胚系决定转录子之间相互作用加强了 TE 和 ICM 的命运的建立。在这个阶段起中心作用的是 *Oct4*（*Pou5f1*）和 *Cdx2*。*Oct4* 是 ICM 成熟所必需的[28]，*Cdx2* 是 TE 成熟所必需的[23]。初步推测这两个因子之间的相互制约导致了第一胚层的命运决定。在囊胚的 TE 中，*Cdx2* 抑制 *Oct4* 和其他 ICM 的基因的表达[23]。但在 *Cdx2* 缺失的胚胎中，TE 仍然可以形成，其他 TE 的标记也仍然表达[19]。同样，在着床后囊胚形成 1 天后，ICM 中的 *Oct4* 抑制 *Cdx2* 的表达[19]。因此，在没有 *Oct4* 或 *Cdx2* 时胚层发育的最初是正常的，但胚胎不能正常表达胚层基因。*Cdx2* 缺失的胚胎尽管表达 ICM 基因，却不完全遵照 ICM 的命运发育。在 *Cdx2* 缺失的胚胎 TE 中，TE 的标记 Gata3 仍表达[19]，但此胚胎的凋亡水平比野生型高[23]。因此可以肯定的是，*Cdx2* 维持了那些将发育成 TE 的细胞的存活和增殖。与此一致的是，*Cdx2* 在后期的滋养层细胞增殖区持续表达[29]。*Oct4* 表达缺失的胚胎不能存活，其原因目前尚未明确。

Oct4 和 *Cdx2* 之间的制约关系在来源于囊胚的干细胞中得到了证实。*Oct4* 缺失的胚胎中没有 ES 细胞，*Cdx2* 缺失的胚胎没有 TS 细胞[23,28]。已有的 ES 细胞缺失 *Oct4* 后会导致 *Cdx2* 的上调，并在 TS 细胞培养基中形成 TS-样细胞[30]。同样，在 ES 细胞中过表达 *Cdx2* 会导致 *Oct4* 表达抑制并形成 TS-样细胞[2]。在 ES 细胞中其他滋养层细胞因子如 Eomes 和 Gata3 也可以诱导滋养层细胞相关的基因表达[2,19]。这些因子在滋养层细胞成熟过程的晚期发挥作用，却与定位无关[23,31,32]。囊胚一旦形成，TE/ICM 中的干细胞将启动某一基因程序维持两者间的相互制约作用，这使得干细胞的获得需要跨越囊胚期这一假设显得合理。而对 ICM 发育的深入了解则需基于对第二胚层的进一步了解，这将在下一节进行讨论。

第二胚层决定：ICM 的细分

受精后第 3 天，囊胚中的 ICM 包含两种类型的细胞：外胚层（EPI）和原始内胚层（PE）。只有 EPI 发育成胚胎，而 PE 是一个胚层外系，它将发育成于卵黄囊（图 1-1）[33~36]。PE 胚层在刚着床后有两个重要的角色：其一是为胚胎提供营养；其二，它是一个信号中心，协助原肠胚的前后极性的形成[37]。TE 胚层是一个来源于 PE 胚层的特殊的干细胞系。多能的干细胞系也被称为 XEN 细胞，已确认其来源于 PE 胚层[38]（图 1-2）。此外，当过表达 PE 转录因子如 Gata4 和 Gata6 时，ES 细胞可被诱导成 PE

状细胞[39]。然而 Gata4/6 在 PE 发育的后期发挥作用[40,41],这说明,作为 TE 胚层,PE 是由在干细胞基因上游的作用机制所决定的。对 PE 胚层决定的研究揭示了一个独特的、基于细胞信号的机制。

异质性和祖细胞分选

受精后第 4 天,囊胚着床。在这个阶段,PE 是 ICM 囊胚腔表面的一个独特的单细胞层。由此,最初的假设认为 PE 是在囊胚着床时期由直接面向囊胚腔的 ICM 细胞产生的,并推测在这个阶段面向囊胚腔的细胞和深层细胞之间的微环境的差异参与了胚层决定。然而,最近的研究表明,在着床的一天前可以在囊胚腔检测到 EPI 祖细胞和 PE 祖细胞[36,42,43]。在这一阶段,ICM 是一个 EPI 祖细胞和 PE 祖细胞的混合细胞群,并表达胚层特异性转录因子。在这个阶段之前,ICM 中的所有细胞均表达 Nanog 和 Gata6;在囊胚扩张过程中,Nanog 和 Gata6 的表达逐渐出现相互排斥,从而以不依赖于位置的方式使两种祖细胞发生分化[36,44]。值得注意的是,在 ICM 中没有定型的模式决定两种祖细胞的分布,它们像盐和胡椒一样随机散布在整个 ICM。

这些结果表明,着床后两种随机分布的胚层祖细胞形成两个形态不同的胚层。利用活体成像技术对表达荧光胚层标记的转基因小鼠中的囊泡扩张进行实时观察支持了这一理论模型。在 PdgfraH2B-GFP 小鼠的 PE 中表达了组蛋白 H2B-GFP,这表明两个胚层的分离包括了细胞凋亡和细胞迁移的过程[36]。在日益扩增的 ICM 中细胞不断地被重新排列[36,45],一旦 PE 祖细胞来到 ICM 的表面,它们便停留在那里。与此相一致的是,PE 的成熟需要逐步进行,而这与在 ICM 中的位置密切相关[46]。其中一个突出的问题是,PE 细胞的排列是由于细胞定向运动还是随机运动和位置识别相组合形成的?

几个突变体的发育表现为一个整体的 PE 层形成的缺陷[47~51]。在这些突变体中,表达 Gata4 的细胞被认为是 PE 细胞,它们在 ICM 的中间聚集,这表明 PE 祖细胞开始形成但未能形成一个形态不同的表层细胞。与此相反,在 TE 中胚层定位(位置)要先于胚层形成。而在 PE 中,胚层形成要先于胚层定位。因此,了解在 ICM 中 PE 的命运的选择是了解 PE/EPI 胚层选择的关键。

细胞信号调节 PE/EPI 形成

ICM 中早期的异质性表明,PE 和 EPI 胚层的形成是通过非位置依赖机制调节的。现在已经证实 FGF 信号是 PE 在体内和体外形成所必需的[52~54]。但是细胞外信号通路如 FGF 信号通路如何参与 PE 和 EPI 在 CMI 中的随机分布目前尚不清楚。例如,胚胎内的某些前 PE 细胞可能倾向于对信号应答,或者细胞可以随机接收信号从而成为 PE 祖细胞。

有两种模型概括了这些可能:来源依赖模型和信号依赖模型[17,55](图 1-4A、B)。来源依赖模型是基于对在卵裂阶段内细胞生成过程的了解[56]。桑椹胚的内细胞将发育成为囊胚的 ICM,它们是通过 8~16 细胞和 16~32 细胞阶段两轮不对称分裂产生的。

根据来源依赖模型,单个 ICM 细胞发育的来源决定它们的命运。也就是说,内细

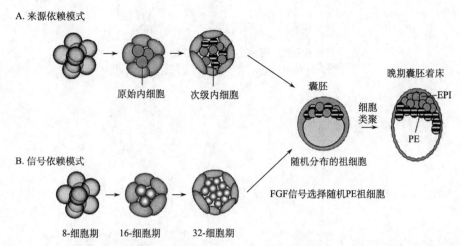

图 1-4 小鼠胚胎中 PE/ EPI 形成的两个模型。A. 来源依赖模型，ICM 细胞的发育起源调节 EPI/PE 的形成。8 细胞阶段后通过两轮不对称分裂生成 ICM 细胞。原始内细胞（蓝色）产生 EPI 胚层，次级内细胞（黄色，直线）产生 PE 胚层。B. 信号依赖模型，原始和次级内细胞之间的胚层潜能没有差别。每个内细胞对 FGF 信号随机应答。应答的细胞形成 PE 胚层，无应答的细胞形成 EPI 胚层。PE/ EPI 胚系决定后，EPI 和 PE 祖细胞表达胚层特异性转录因子 Nanog 或 Gata6，这些祖细胞在囊胚的 ICM 中随机分布。在着床 4.5 天后，这两个祖细胞形成了 EPI 和 PE 的两个不同的胚层。此图的彩版请见 www.landesbioscience.com/curie

胞（原始内细胞）是在第一轮分裂后产生的，它会优先选择 EPI，而在第二轮分裂后产生的细胞（次级内细胞）将优先成为 PE[42,57]（图 1-4A）。由于次级内细胞长期处于外部位置，而且 TE 细胞也在外部，次级内细胞将倾向于成为胚外层[17]。为了验证来源依赖模型，首次直接观察活胚胎中内细胞的生成，然后分析在后期它们的后代对 EPI 和 PE 胚层形成的贡献[44]。由于原始和次级内细胞的后代都参与了 EPI 和 PE 胚层发育，并且没有明显的偏倚，所以没有检测到原始和次级内细胞之间胚层潜能的差异。因此，这些发现说明来源依赖模型是不成立的。

第二个模型是信号依赖模型，即单个 ICM 细胞对一定水平的 FGF 信号随机应答，从而选择 EPI 或 PE 的命运（图 1-4B）。如上所述，FGF 信号是胚胎中 PE 形成所必需的[52~54]。当使用化学抑制剂或基因敲除将 FGF 信号阻断后，所有的 ICM 细胞选择了 EPI 的命运[42,58]。有趣的是，高剂量的外源性 FGF4 可诱导相反的表型：所有的 ICM 细胞选择了 PE 的命运[44]。这表明，所有早期的 ICM 细胞具有对 FGF 信号应答并发育成 PE 的潜能。然而，在正常的发育过程中，有限的内源性 FGF 会限制对 FGF 应答的 ICM 细胞的比例（图 1-5）。单个的 ICM 细胞对有限的 FGF 是否应答是由细胞-细胞的变化随机决定的，而对细胞-细胞变化的敏感性是由细胞依赖机制还是非细胞依赖机制决定的，目前仍然不清楚[59]。内源性 FGF 水平由发育遗传程序控制，从而生成比例大致相等的可再生的 EPI/PE 胚层，而不是通过确定性的发育机制。

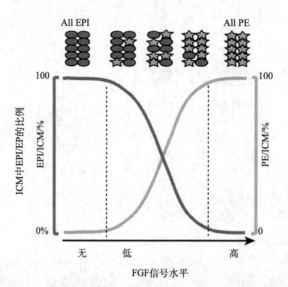

图 1-5 依赖 FGF 信号的 PE 和 EPI 胚层形成的原理模型。x 轴表示 FGF 信号的激活水平。y 轴表示 EPI（蓝色）和 PE（黄色，星）在 ICM 的比例。当信号低于阈值时，所有的 ICM 细胞选择 EPI 的命运；然而，当信号水平高时，所有的 ICM 细胞选择 PE 的命运。在激活水平适中时，单个 ICM 细胞对 FGF 信号随机应答。在这种模式下，FGF 信号的水平控制 ICM 中两个胚层的比例，但不参与其分布。此图的彩版请见 www.landesbioscience.com/curie

EPI 胚层多能性的建立和调节

经过两轮胚层形成后，即先是 TE 形成，再是 PE 形成，建立的 EPI 是一个多能干细胞胚层。多能性的一般定义是具有形成来自所有三个胚层的组织的能力，最近对多个多能干细胞系的鉴定表明全能性显然不是一个单一的状态。相反，胚胎的全能性可能包括一系列与发育相关的状态[60]。小鼠胚胎的全能性至少有两个状态，有两种类型的多能干细胞：胚胎干细胞（ES）和外胚层来源的干细胞（EpiSC）[58]。这些细胞系来源于 EPI 胚层，但代表了两种截然不同的胚胎阶段：ES 细胞相当于胚胎刚着床后的 EPI 细胞[19,58]，而 EpiSC 相当于胚胎着床后原肠胚前阶段的 EPI 细胞[61,62]。虽然把 EpiSC 注射到囊胚中，EpiSC 不能形成胚胎，这可能是由于未能整合到宿主 ICM，但是它们可以生成具有所有三个胚层的畸胎瘤。多能性的基因如 Oct4 和 Sox2 在早期和晚期阶段均表达，但在着床过程中它的几个特性与 EPI 细胞不同。例如，从一个无组织的细胞团到上皮单层的细胞形态发生了改变。此外，在转型期某些基因的表达发生了显著改变，如 Rex1 下调而 FGF5 上调[63]。转型后，晚期 EPI 细胞有能力接收感应信号并产生三个胚层。在第一、第二和随后的胚层建立中如何保证多能干细胞的状态是一个研究热点。

Nanog 可能参与了多能干细胞状态的维持。最初发现 Nanog 是一个维持 ES 细胞多能性的转录因子[64,65]，但随后的研究发现，它的作用相当于一个守门员。ES 细胞中 Nanog 的水平波动依赖于 FGF 信号[66,67]，当 Nanog 的水平低时，ES 细胞更易于分化。下调 Nanog 虽然不能启动分化，但是允许了分化的发生。与此一致的是，在着床过程中内源性 Nanog 的表达是短暂下调的。Nanog 缺失的囊胚的形态是正常的，但囊胚期后 ICM 细胞很快就退化。Nanog 缺失的 ICM 细胞滞留在了前多能的状态，既不是 EPI 也不是 PE 细胞，而是一个无活力的不确定状态[66]。

整个胚胎期中调节 EPI 组织转变的机制目前仍知之甚少。干细胞的体外研究提出了一些见解。在 EpiSC 培养条件中培养 ES 细胞时发现 ES 细胞易成为 EpiSC，条件培养基中包括 FGF2 和激活素 A[68]。有趣的是，Fgf4 是 ES 细胞分化所必需的[69]。这些结果表明，FGF 信号的质量或数量可能参与了从 ES 细胞到 EpiSC 的转变，进而可能参与了从 ICM 到 EPI 的转变。

有趣的是，两个多能干细胞状态之间的转变也有可能发生逆转。KLF4 是一个原始的重编程因子[70]，过表达 KLF4 后 EpiSC 可以变成 ES-样的细胞，尽管这种逆转发生的频率很低[68]。然而，当在传统的 ES 细胞培养条件下培养 EpiSC 或来自原肠期胚胎的外胚层细胞 10～20 天后，出现了重新编程的 ES 细胞样细胞（rES 细胞）[71]。虽然逆转过程比从 ES 细胞发育成 EpiSC 需要更多时间，但 rES 细胞已完全重建了早期的多能干细胞状态。

结论

这一章中我们描述了目前研究得最广泛的哺乳动物胚胎即小鼠胚胎的胚系形成。其中一个最有趣的发现是在小鼠胚胎中存在多潜能状态，即 ES 细胞和 EpiSC 的存在。迄今为止，大鼠 ES 细胞是已发现的唯一一个与小鼠 ES 细胞相似的哺乳动物 ES 细胞株。相比之下，在形态、基因表达和生长因子的依赖方面，人胚胎干细胞（hES）更类似于小鼠的 EpiSC 而不是小鼠 ES 细胞[61,62,72]。目前尚不清楚人 ES 细胞是否同样具有多潜能细胞状态存在。有趣的是，成人细胞和小鼠细胞是由来自各自 ES 细胞的相同的因子重新编程的。也就是说，人 iPS 细胞类似 hES 细胞[73,74]，而小鼠 iPS 细胞类似小鼠 ES 细胞[70]，而不是小鼠 EpiSC。也许在多潜能细胞状态稳定性方面存在种属特异性的差异。即使所有的哺乳动物是由一个囊胚发育而来，着床的发育时间和早期着床后胚胎的形态也具有很大差异。对胚系形成和其他哺乳动物干细胞的进一步分析可能会对这些问题有启示。

（梁　璐　译）

参 考 文 献

1. Ralston A, Rossant J. Genetic regulation of stem cell origins in the mouse embryo. Clin Genet 2005; 68(2):106-112.
2. Niwa H, Toyooka T, Shimosato D et al. Interaction between Oct3/4 and Cdx2 determines trophectoderm differentiation. Cell 2005; 123(5):917-929.
3. Fleming T. A quantitative analysis of cell allocation to trophectoderm and inner cell mass in the mouse blastocyst. Dev Biol 1987; 119(2):520-531.
4. Gardner R. The early blastocyst is bilaterally symmetrical and its axis of symmetry is aligned with the animal-vegetal axis of the zygote in the mouse. Development 1997; 124(2):289-301.
5. Gardner R. Specification of embryonic axes begins before cleavage in normal mouse development. Development 2001; 128(6):839-847.
6. Piotrowska K, Wianny F, Pedersen R et al. Blastomeres arising from the first cleavage division have distinguishable fates in normal mouse development. Development 2001; 128(19):3739-3748.
7. Fujimori T, Kurotaki Y, Miyazaki J et al. Analysis of cell lineage in two- and four-cell mouse embryos. Development 2003; 130(21):5113-5122.
8. Plusa B, Frankenberg S, Chalmers A et al. Downregulation of Par3 and aPKC function directs cells towards the ICM in the preimplantation mouse embryo. J Cell Sci 2005; 118(Pt 3):505-515.
9. Kurotaki Y, Hatta K, Nakao K et al. Blastocyst axis is specified independently of early cell lineage but aligns with the ZP shape. Science 2007; 316(5825):719-723.
10. Alarcón V, Marikawa Y. Deviation of the blastocyst axis from the first cleavage plane does not affect the quality of mouse postimplantation development. Biol Reprod 2003; 69(4):1208-1212.
11. Chróścicka A, Komorowski S, Maleszewski M. Both blastomeres of the mouse 2-cell embryo contribute to the embryonic portion of the blastocyst. Mol Reprod Dev 2004; 68(3):308-312.
12. Motosugi N, Bauer T, Polanski Z et al. Polarity of the mouse embryo is established at blastocyst and is not prepatterned. Genes Dev 2005; 19(9):1081-1092.
13. Bischoff M, Parfitt D, Zernicka-Goetz M. Formation of the embryonic-abembryonic axis of the mouse blastocyst: relationships between orientation of early cleavage divisions and pattern of symmetric/asymmetric divisions. Development 2008; 135(5):953-962.
14. Piotrowska-Nitsche K, Perea-Gomez A, Haraguchi S et al. Four-cell stage mouse blastomeres have different developmental properties. Development 2005; 132(3):479-490.
15. Johnson M, McConnell J. Lineage allocation and cell polarity during mouse embryogenesis. Semin Cell Dev Biol 2004; 15(5):583-597.
16. Torres-Padilla M, Parfitt D, Kouzarides T et al. Histone arginine methylation regulates pluripotency in the early mouse embryo. Nature 2007; 445(7124):214-218.
17. Yamanaka Y, Ralston A, Stephenson R et al. Cell and molecular regulation of the mouse blastocyst. Dev Dyn 2006; 235(9):2301-2314.
18. Yoshikawa T, Piao Y, Zhong J et al. High-throughput screen for genes predominantly expressed in the ICM of mouse blastocysts by whole mount in situ hybridization. Gene Expr Patterns 2006; 6(2):213-224.
19. Ralston A, Cox B, Nishioka N et al. Gata3 regulates trophoblast development downstream of Tead4 and in parallel to Cdx2. Development 2010; 137(3):395-403.
20. Chawengsaksophak K, James R, Hammond V et al. Homeosis and intestinal tumours in Cdx2 mutant mice. Nature 1997; 386(6620):84-87.
21. Nishioka N, Yamamoto S, Kiyonari H et al. Tead4 is required for specification of trophectoderm in pre-implantation mouse embryos. Mech Dev 125(3-4):270-283.
22. Yagi R, Kohn M, Karavanova I et al. Transcription factor TEAD4 specifies the trophectoderm lineage at the beginning of mammalian development. Development 2007; 134(21):3827-3836.
23. Strumpf D, Mao CA, Yamanaka Y et al. Cdx2 is required for correct cell fate specification and differentiation of trophectoderm in the mouse blastocyst. Development 2005; 132(9):2093-2102.
24. Ralston A, Rossant J. Cdx2 acts downstream of cell polarization to cell-autonomously promote trophectoderm fate in the early mouse embryo. Dev Biol 2008; 313(2):614-629.
25. Jedrusik A, Parfitt D, Guo G et al. Role of Cdx2 and cell polarity in cell allocation and specification of trophectoderm and inner cell mass in the mouse embryo. Genes Dev 2008; 22(19):2692-2706.
26. Dietrich J, Hiiragi T. Stochastic patterning in the mouse pre-implantation embryo. Development 2007; 134(23):4219-4231.
27. Nishioka N, Inoue K, Adachi K et al. The Hippo signaling pathway components Lats and Yap pattern Tead4 activity to distinguish mouse trophectoderm from inner cell mass. Dev Cell 2009; 16(3):398-410.
28. Nichols J, Zevnik B, Anastassiadis K et al. Formation of pluripotent stem cells in the mammalian embryo depends on the POU transcription factor Oct4. Cell 1998; 95(3):379-391.
29. Beck F, Erler T, Russell A et al. Expression of Cdx-2 in the mouse embryo and placenta: possible role in patterning of the extra-embryonic membranes. Dev Dyn 1995; 204(3):219-227.

30. Niwa H, Miyazaki J, Smith A. Quantitative expression of Oct-3/4 defines differentiation, dedifferentiation or self-renewal of ES cells. Nat Genet 2000; 24(4):372-376.
31. Russ A, Wattler S, Colledge W et al. Eomesodermin is required for mouse trophoblast development and mesoderm formation. Nature 2000; 404(6773):95-99.
32. Ma G, Roth M, Groskopf J et al. GATA-2 and GATA-3 regulate trophoblast-specific gene expression in vivo. Development 1997; 124(4):907-914.
33. Gardner RL, Rossant J. Investigation of the fate of 4-5 day postcoitum mouse inner cell mass cells by blastocyst injection. J Embryol Exp Morphol 1979; 52:141-152.
34. Gardner RL. Investigation of cell lineage and differentiation in the extraembryonic endoderm of the mouse embryo. J Embryol Exp Morphol 1982; 68:175-198.
35. Gardner RL. An in situ cell marker for clonal analysis of development of the extraembryonic endoderm in the mouse. J Embryol Exp Morphol 1984; 80:251-288.
36. Plusa B, Piliszek A, Frankenberg S et al. Distinct sequential cell behaviours direct primitive endoderm formation in the mouse blastocyst. Development 2008; 135(18):3081-3091.
37. Rossant J, Tam PP. Blastocyst lineage formation, early embryonic asymmetries and axis patterning in the mouse. Development 2009; 136(5):701-713.
38. Kunath T, Arnaud D, Uy G et al. Imprinted X-inactivation in extra-embryonic endoderm cell lines from mouse blastocysts. Development 2005; 132(7):1649-1661.
39. Fujikura J, Yamato E, Yonemura S et al. Differentiation of embryonic stem cells is induced by GATA factors. Genes Dev 2002; 16(7):784-789.
40. Morrisey E, Ip H, Lu M et al. GATA-6: a zinc finger transcription factor that is expressed in multiple cell lineages derived from lateral mesoderm. Dev Biol 1996; 177(1):309-322.
41. Koutsourakis M, Keijzer R, Visser P et al. Branching and differentiation defects in pulmonary epithelium with elevated Gata6 expression. Mech Dev 2001; 105(1-2):105-114.
42. Chazaud C, Yamanaka Y, Pawson T et al. Early lineage segregation between epiblast and primitive endoderm in mouse blastocysts through the Grb2-MAPK pathway. Dev Cell 2006; 10(5):615-624.
43. Kurimoto K, Yabuta Y, Ohinata Y et al. An improved single-cell cDNA amplification method for efficient high-density oligonucleotide microarray analysis. Nucleic Acids Res 2006; 34(5):e42.
44. Yamanaka Y, Lanner F, Rossant J. FGF signal-dependent segregation of primitive endoderm and epiblast in the mouse blastocyst. Development 2010; 137(5):715-724.
45. Meilhac S, Adams R, Morris S et al. Active cell movements coupled to positional induction are involved in lineage segregation in the mouse blastocyst. Dev Biol 2009; 331(2):210-221.
46. Gerbe F, Cox B, Rossant J et al. Dynamic expression of Lrp2 pathway members reveals progressive epithelial differentiation of primitive endoderm in mouse blastocyst. Dev Biol 2008; 313(2):594-602.
47. Morris S, Tallquist M, Rock C et al. Dual roles for the Dab2 adaptor protein in embryonic development and kidney transport. EMBO J 2002; 21(7):1555-1564.
48. Yang D, Smith E, Roland I et al. Disabled-2 is essential for endodermal cell positioning and structure formation during mouse embryogenesis. Dev Biol 2002; 251(1):27-44.
49. Gao F, Shi H, Daughty C et al. Maspin plays an essential role in early embryonic development. Development 2004; 131(7):1479-1489.
50. Smyth N, Vatansever H, Murray P et al. Absence of basement membranes after targeting the LAMC1 gene results in embryonic lethality due to failure of endoderm differentiation. J Cell Biol 1999; 144(1):151-160.
51. Stephens L, Sutherland A, Klimanskaya I et al. Deletion of beta 1 integrins in mice results in inner cell mass failure and peri-implantation lethality. Genes Dev 1995; 9(15):1883-1895.
52. Arman E, Haffner-Krausz R, Chen Y et al. Targeted disruption of fibroblast growth factor (FGF) receptor 2 suggests a role for FGF signaling in pregastrulation mammalian development. Proc Natl Acad Sci USA 1998; 95(9):5082-5087.
53. Feldman B, Poueymirou W, Papaioannou V et al. Requirement of FGF-4 for postimplantation mouse development. Science 1995; 267(5195):246-249.
54. Cheng A, Saxton T, Sakai R et al. Mammalian Grb2 regulates multiple steps in embryonic development and malignant transformation. Cell 1998; 95(6):793-803.
55. Rossant J, Chazaud C, Yamanaka Y. Lineage allocation and asymmetries in the early mouse embryo. Philos Trans R Soc Lond B Biol Sci 2003; 358(1436):1341-1348; discussion 1349.
56. Johnson MH, Chisholm JC, Fleming TP et al. A role for cytoplasmic determinants in the development of the mouse early embryo? J Embryol Exp Morphol 1986; 97 Suppl:97-121.
57. Chisholm JC, Houliston E. Cytokeratin filament assembly in the preimplantation mouse embryo. Development 1987; 101(3):565-582.
58. Nichols J, Silva J, Roode M et al. Suppression of Erk signalling promotes ground state pluripotency in the mouse embryo. Development 2009; 136(19):3215-3222.

59. Huang S. Non-genetic heterogeneity of cells in development: more than just noise. Development 2009; 136(23):3853-3862.
60. Rossant J. Stem cells and early lineage development. Cell 2008; 132(4):527-531.
61. Brons I, Smithers L, Trotter M et al. Derivation of pluripotent epiblast stem cells from mammalian embryos. Nature 2007; 448(7150):191-195.
62. Tesar P, Chenoweth J, Brook F et al. New cell lines from mouse epiblast share defining features with human embryonic stem cells. Nature 2007; 448(7150):196-199.
63. Pelton T, Sharma S, Schulz T et al. Transient pluripotent cell populations during primitive ectoderm formation: correlation of in vivo and in vitro pluripotent cell development. J Cell Sci 2002; 115(Pt 2):329-339.
64. Mitsui K, Tokuzawa Y, Itoh H et al. The homeoprotein Nanog is required for maintenance of pluripotency in mouse epiblast and ES cells. Cell 2003; 113(5):631-642.
65. Chambers I, Colby D, Robertson M et al. Functional expression cloning of Nanog, a pluripotency sustaining factor in embryonic stem cells. Cell 2003; 113(5):643-655.
66. Silva J, Nichols J, Theunissen T et al. Nanog is the gateway to the pluripotent ground state. Cell 2009; 138(4):722-737.
67. Lanner F, Lee K, Sohl M et al. Heparan Sulfation Dependent FGF Signalling Maintains Embryonic Stem Cells Primed for Differentiation in a Heterogeneous State. Stem Cells 2009.
68. Guo G, Yang J, Nichols J et al. Klf4 reverts developmentally programmed restriction of ground state pluripotency. Development 2009; 136(7):1063-1069.
69. Kunath T, Saba-El-Leil M, Almousailleakh M et al. FGF stimulation of the Erk1/2 signalling cascade triggers transition of pluripotent embryonic stem cells from self-renewal to lineage commitment. Development 2007; 134(16):2895-2902.
70. Takahashi K, Yamanaka S. Induction of pluripotent stem cells from mouse embryonic and adult fibroblast cultures by defined factors. Cell 2006; 126(4):663-676.
71. Bao S, Tang F, Li X et al. Epigenetic reversion of post-implantation epiblast to pluripotent embryonic stem cells. Nature 2009; 461(7268):1292-1295.
72. Thomson J, Itskovitz-Eldor J, Shapiro S et al. Embryonic stem cell lines derived from human blastocysts. Science 1998; 282(5391):1145-1147.
73. Takahashi K, Tanabe K, Ohnuki M et al. Induction of pluripotent stem cells from adult human fibroblasts by defined factors. Cell 2007; 131(5):861-872.
74. Yu J, Vodyanik M, Smuga-Otto K et al. Induced pluripotent stem cell lines derived from human somatic cells. Science 2007; 318(5858):1917-1920.

第2章 干细胞的核结构

Kelly J. Morris[§], Mita Chotalia[§], Ana Pombo[*]

摘要：基因组调控的基本特点有赖于线性DNA序列，以及局部调控基因表达的DNA和染色质相关蛋白的细胞类型特异性修饰。三维核空间中基因组组织的结构特点确定哪些与核亚区相关、有特定生化活性的基因占据优先位点，从而影响表达的状态。干细胞核中基因组的结构和时序组织，加上特定的表观遗传特点和转录调控，是影响多能性和分化的关键因素[1,2]。

引言

干细胞一个显著的潜能就是在早期的生命和生长过程中，能发育成多种细胞类型。对干细胞的研究，主要集中在啮齿类和人类的两类干细胞：多潜能胚胎干细胞和多能组织特异性干细胞。来自着床前囊胚阶段胚胎内细胞团的胚胎干细胞，其特征是培养时具有无限自我更新能力，同时显示出巨大的可塑性和向三个胚层（外胚层、内胚层、中胚层）所有细胞分化的能力。

无论是多潜能阶段，还是已定向分化为某一细胞谱系，胚胎干细胞的自我更新过程伴随着对发育相关的一些重要基因的对立的、抑制性调节。对这些基因的抑制是表观遗传学层面的，当细胞分化时，这些基因会高效地激活。自我更新过程需要胚胎干细胞基因组一分为二，且不影响其多能性；而分化过程引起错综复杂的基因表达变化，使胚胎干细胞进入特定的细胞分化路径。弄清胚胎干细胞可塑性的分子机制对细胞治疗应用是非常必要的，我们需要了解其多能性和发育的基因调控情况。

胚胎干细胞核的功能分区

真核生物细胞核包含染色体，每个染色体是由一条DNA、组蛋白和形成染色质的其他蛋白质构成的复合物。人细胞核里包含22对常染色体和2条性染色体。染色体通过核膜和细胞质隔离，蛋白质和RNA通过核膜上分布的核孔进出细胞核。核内膜提供了一个功能性界标，参与细胞核内染色体的组织；核内膜上还存在一些结构蛋白，构成核纤层。进一步在核内部进行结构和功能性分区，通过形成高阶的复合物，凝聚成一些

[§] Both authors contributed equally to this work

[*] Corresponding Author: Ana Pombo-Genome Function Group, MRC Clinical Sciences Centre, Imperial College School of Medicine, Hammersmith Hospital Campus, Du Cane Road, London W12 0NN, UK. Email: ana.pombo@csc.mrc.ac.uk

大的结构域,如核仁和剪接斑(splicing speckle);或一些较小的核小体,如卡哈尔体和多梳体(polycomb body)。由局部相互作用介导的,或与功能域关联的核内部染色质间的相互作用,被认为对基因表达的程序既重要又依赖。

核内染色体和单基因的组织

在间期细胞核形成期间,细胞刚退出有丝分裂,染色体部分去凝聚,分别占据一定的核区域,在边界处相互混杂[3],形成大范围的染色体间相互作用[4]。核空间里染色体领域(chromosome territory,CT)的定位,以及该领域基因的定位,影响了基因的表达和基因组的稳定[5]。

对人胚胎干细胞的研究显示,染色体也形成离散的CT,和在体细胞中观察到的分布一样[6,7]。例如,通过全染色体探针的荧光原位杂交(fluorescence in situ hybridization,FISH)技术,发现人富含基因的19号染色体和基因缺乏的18号染色体的核定位,在人胚胎干细胞和淋巴母细胞非常类似,分别占据核内部和核周缘[6]。在人胚胎干细胞分化前和分化早期,其他的染色体(10号、12号、15号、17号和19号)也存在类似相关性,这些数据表明存在一定程度的保守结构,这种结构与细胞分化无关,在发育早期已经存在[7]。

对人胚胎干细胞包含多潜能基因的CT区,包括含有Nanog基因的12p染色体的分析揭示,和淋巴母细胞相比,人胚胎干细胞的这些区域向心分布[6]。Nanog基因周边环绕着其他一些与多潜能相关的基因。很多有特殊功能的体细胞,向心分布是其转录活动的一个特点[8]。至于位于6p染色体、包含在活化的主要组织相容性复合体基因簇中的Oct4,其在胚胎干细胞和淋巴母细胞中皆定位于核内部,这可能是由于局部活性染色质彼此靠近的缘故。对特定基因座相对于其CT定位的深入研究显示,Oct4基因座呈解螺旋状态,在胚胎干细胞位于其CT外部,在淋巴母细胞处于内部[6]。

染色质转录活性区域的"成环"现象在诱导分化和维甲酸诱导表达 *Hox* 基因的胚胎干细胞中也能观察到[9]。*Hox d* 和 *Hox b* 基因座包含与发育调节相关的 *Hox* 基因簇,这些基因对发育中的胚胎早期和后期至关重要,在胚胎干细胞中处于沉默状态,在体外分化过程中呈现表达。在胚胎干细胞中,*Hox* 基因簇位于CT内部;当分化时,*Hox b* 去凝聚并和侧翼的基因一起重新定位于CT的外部[9,10]。*Hox b* 基因簇重新定位于CT外部,和与RNA聚合酶II(RNA polymerase II,RNAPII)伸长形式共定位的活化等位基因的时序表达相符合[10]。分化时,去凝聚和 *Hox* 基因座定位于其CT的外部伴随侧翼基因的成环,但侧翼基因的表达并没有增加[10]。在CT的外部,与 *Hoxb* 基因簇毗邻的活化侧翼基因的转录活性稍稍高一些,但在CT内部也能检测到新的转录本。

Hox b 基因座在可检测到的表达前,已经能够观察到活化的组蛋白修饰,因此基因的调控可能是两步模式。首先,基因座发生表观遗传学的改变;然后,去凝聚呈许可状态,基因座随后成环并时序表达。非常重要的是,体外增加组蛋白的乙酰化水平并不足以引起基因座的重塑,因此,染色质的高级结构不是简单的组蛋白修饰。包含 *Hox* 基因簇染色质的重新组织可能促进染色质的移动,使其更易于接近位于CT内部或外部的转录复合物,增加其机会[9]。

在 X 染色体失活（X chromosome inactivation，XCI）过程中，也能观察到高级结构的变化，该过程起始于雌性胚胎干细胞的分化。XCI 现象是雌性哺乳动物细胞的一条 X 染色体沉默，以保证雄性和雌性细胞中 X 染色体上基因表达量相当。XCI 的起始受 X 失活中心（X-inactivation centre，Xic）控制，这是一个特定的染色组区域，包含多个非编码基因，如 *Xist*、*Tsix* 和 *Xite*，这些基因是调节 XCI 的[11,12]。当 XCI 发生时，再将要失活的那条 X 染色体上的 *Xist* 表达上调。*Xist* 是一个很大的非编码 RNA 分子，通过包裹染色体上顺式作用元件起始 XCI，并通过调节染色质结构触发转录沉默[11]。随机的 XCI 的机制并没有完全澄清，但最近的一些分析认为，在早期分化过程中 Xics 的定位至关重要。

在 XCI 起始前，Xics 在核内短暂共定位[13~15]，这种相互作用通过募集染色质绝缘子 CTCF 介导，并依赖 *Tsix* 和 *Xite* 的共转录活性[16]。同源染色体配对是胚胎干细胞 X 染色体独有的结构特征，而经典的染色体配对发生在间期核的异源染色体间[17]。在存在结合因子的情况下，通过计算机模拟聚合物折叠发现，未分化细胞 Xics 间在空间上的接近度可能依赖于结合亲和力和限制因子浓度，如 CTCF[16]，其将优先结合某一 Xic[18]。

在 XCI 发生时进一步的分析显示，染色质去凝聚和 XCI 调节区域间广泛的染色质相互作用是按发育调控和性别特异性模式调节的[19]。对女性细胞 X 染色体的三维形态学分析显示，和失活的 X 染色体相比，未失活的 X 染色体更长，而失活的 X 染色体呈球状。分析人类 7 号染色体形态，其与 X 染色体包含相同的 DNA 容量，结果表明，其在形态和体积上与未失活的 X 染色体类似。因此，活化的染色体呈现扁平、瘦长的形态[20]，这种形态可能更易于和邻近的染色体接触，增强了转录活性[17]。关于染色体的组织，更深层次的方面是被 *Xist* 包裹的 X 染色体如何维持其转录沉默状态。*Xist* 在失活 X 染色体上的积聚形成抑制的核区域，抑制转录因子和 RNAPII 的作用。很有意思的是，逃脱沉默的基因定位于失活 X 染色体 CT 的外部，而沉默基因定位于内部，不能靠近转录复合物[21]。

除了染色体和多能标记物及发育调节基因定位存在差异外，胚胎干细胞染色质高级结构也存在不同。胚胎干细胞着丝粒成簇形成，和分化细胞相比，定位于核内部，这可能是由于胚胎干细胞细胞周期更快[6,7,22]。当人胚胎干细胞分化时，着丝粒（通过对着丝粒异染色质标记物 α-satellite/CENP-A 染色标记）重新组织并定位于核周缘[7,22]。这些剧烈变化可能导致细胞定向期间大规模的染色质重塑，协同基因表达的改变，但详细的机制还不是很清楚。

核纤层和核周缘

核纤层蛋白是核内膜结构的主要组成部分，在许多核活动过程中发挥重要作用，包括有丝分裂后核的组装、染色质组织和基因表达。核纤层是由中间丝组成的网状结构，镶嵌核纤层相关蛋白，包括 B 型核纤层蛋白 B1 和 B2，这些蛋白质对细胞存活至关重要并在所有细胞类型中都表达。A 型核纤层蛋白来源于单一基因（LMNA）的不同剪接体，其在特定的组织表达。核纤层被认为在核膜、异染色质和组蛋白之间起重要的连接作用，在小鼠成纤维细胞中敲除后，会导致异染色质从核周缘脱落[23]。和核纤层关联的一些染色质曾被用来下调某些基因的表达,但并非所有相关基因;而且在分化的哺乳动物细胞中,兼性异染色质常会在这些区域富集[24~30]。

人类和小鼠的胚胎干细胞在核周缘缺少核纤层蛋白 A/C(图 2-1)[22,31,32],但表达核纤层蛋白 B1 和 B2。核纤层蛋白 A/C 出现在分化的早期[22,31],在多潜能标记 Oct4 下调之前[31]。胚胎干细胞中核纤层蛋白 A/C 的缺乏意味着核保持一定的刚性[32]和形态[22]。缺少核纤层蛋白 A/C 很可能有助于胚胎干细胞的可塑性和核中染色质的移动[32],而其在分化过程中的表达可能是促进和/或维持分化表型的重要因素[31]。

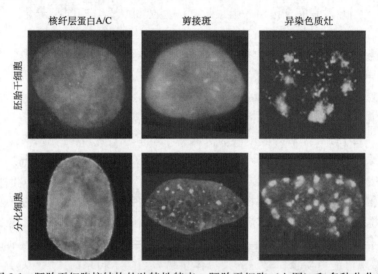

图 2-1 胚胎干细胞核结构的独特性特点。胚胎干细胞(上图)和多种分化细胞(下图)的核质亚区。核纤层,核纤层蛋白 A/C 在未分化的人胚胎干细胞中缺乏,在分化时表达。图像由 Butler JT 和 Lawrence JB 惠赠。剪接斑,当细胞分化时,SC35 结构域形成。在未分化的人胚胎干细胞,SC35 呈弥散状态,随着分化的起始,SC35 结构域开始形成。图像由 Butler JT 和 Lawrence JB 惠赠。异染色质灶(heterochromatin foci),和神经祖细胞相比,胚胎干细胞中异染色质结合蛋白 HP1α 染色阳性区较少,但比较大。转自 Meshorer 等,Hyperdynamic plasticity of chromatin proteins in pluripotent embryonic stem cells. Dev Cell 2006;10 (1):105-16

对胚胎干细胞相关核周缘基因的表观遗传及核分布的定量、细胞学分析显示,相对于分化细胞(胚胎源 NIH-3T3 成纤维细胞和神经祖细胞),胚胎干细胞组蛋白 H3 第 27 位赖氨酸三甲基化(H3K27me3:一种定位于被抑制基因启动子的表观遗传学标记)富集。此外,10% 活化标记的染色质(H3K4me3)和数百个转录位点定位于胚胎干细胞的核周缘,分化细胞具有类似的分布[33]。

与核周缘相关的基因定位的变化在发育相关基因中也能看到,如 Mash1(Ascl1),它是一种必不可少的神经前体细胞分化和特异的神经定向分化的转录因子,其定位的变化与分化和基因表达变化相关联[28]。在胚胎干细胞,Mash1 启动子与 H3K27me3 相关联并定位于核周缘附近。有趣的是,其定位不受组蛋白甲基转移酶 Ezh2/Eed 敲除的影响。当向神经谱系诱导时，Mash1 转录增加超过 100 倍，同时基因座向内部重新定位，Mash1 基因减少 H3K27me3，增加组蛋白 H3 第 9 位赖氨酸乙酰化（H3K9ac：一种活化的表观遗传标记），其邻近基因并无这些变化。基因座向内部定位，但不影响侧翼基因的表达，这显示有另外的调节机制，当核环境改变时，这些机制以基因特异的方式维持其表达状态。这些结果提出了一个有趣的可能性，即核周缘对全能胚胎干细胞基因组的表观遗传状态有所帮助，尽管其分子机制尚未阐明。

干细胞其他核质亚区的特点

转录、RNA 加工、DNA 复制和修复等核加工过程中形成的大分子复合物在核中的动态定位被认为是自组织的，能够形成发挥功能的位点[34,35]。最近的研究结果表明，这些结构在胚胎干细胞和体细胞中是有差别的。

剪接斑和卡哈尔体

剪接斑是富含剪接组分的核区域，如 SC35、polyA$^+$ RNA 和许多 mRNA 代谢因子[36~38]。人胚胎干细胞核免疫荧光法显示，SC35 散在分布于整个核质，分化时形成更集中的独立斑点（图 2-1）[22]。同样的发现也可以用来描述卡哈尔体，它是一种涉及转录和剪接复合物的装配或修饰的核结构[39]。人胚胎干细胞核不包含卡哈尔体和 Coilin。Coilin 是卡哈尔体的一种标记物，在整个核质中表现为一种弱的、弥散的信号分布。卡哈尔体更精细的定义和定位，是在分化后的干细胞群边缘[22]。在一些细胞体系中，与剪接斑和卡哈尔体有关的基因与它们的表达相关联，并可能与转录本加工的特定方面有关[38,39]。

早幼粒细胞白血病核小体

早幼粒细胞白血病核小体由几个调节蛋白组成，与多个细胞过程有关，包括染色质组织、DNA 复制和转录调控。在体细胞中，早幼粒细胞白血病核小体表现为统一的球形结构。在人胚胎干细胞培养物中，仅一小部分细胞包含几个球形早幼粒细胞白血病核小体，而相当大一部分细胞含有早幼粒细胞白血病核小体样的结构。这些结构有确定的形态，被归为两类：长线性"杆状物"；大的（>2μm）"玫瑰花样物"[22]。人胚胎干细胞中早幼粒细胞白血病核小体不仅形态不同，蛋白质组成也不同。那些定义体细胞早幼粒细胞白血病核小体的标记物包括 SUMO、Sp100 和 Daxx，由于含量不丰富而不能用作标记。这些独特的早幼粒细胞白血病核小体结构在细胞定向的早期阶段短暂出现，在分化后消失。它们还常和核膜有关联，体细胞中未观察到这一特点[22]。

多梳体

在许多细胞类型中，多梳蛋白对基因抑制起重要作用[40]。在胚胎干细胞，它们对抑制发育调节基因至关重要[41]。多梳抑制复合物 1 包含一个亚单位，由 CBX 蛋白家族组成，在体外可与 H3K27me3 结合[42]。荧光成像显示在未分化小鼠胚胎干细胞中，不同的 CBX 家族成员表现为不同的亚核分布。大部分 CBX 蛋白聚集形成多梳体，而当分化时，随着 CBX 蛋白消失，多梳体消失[43]。

胚胎干细胞核特有的染色质特征

真核基因组的结构和组织是多层次的。局部 DNA 缠绕在组蛋白八聚体上形成核小体并进一步压缩，压缩水平取决于连接 DNA 的长度和组蛋白 H1。在大部分组织细胞中，异染色质（闭合的）或常染色质（开放的）分别定位于核周缘或核中心，通过控制蛋白与染色质的可及性赋予其调控的潜能[44]。除了与染色质稳定结合的核心组蛋白外，哺乳动物细胞核中高度动态结合的染色质因子，决定了异染色质和常染色质的状态[45,46]。在某些细胞类型中，细胞分化过程中基本转录/修复因子 IIH (TFIIH) 与染色质的动态结合表明，胚胎干细胞基因组已准备好谱系特异性程序，该过程涉及一组调控因子[47]。

结构染色质蛋白的高流动性和异染色质的形成

由异染色质蛋白 1 族 (HP1) 和维持压缩的连接组蛋白 H1 组成的结构染色质蛋白，通常抑制异染色质域[48~50]。出乎意料的是，HP1、H1 和核心组蛋白的活细胞成像揭示，和谱系限定及分化细胞相比，胚胎干细胞具有独特的染色质状态[51]。在胚胎干细胞中，几个结构染色质蛋白以高动态、松散结合或可溶片段的方式存在。有趣的是，蛋白移动性的降低与多能性的丧失相关，而不是胚胎干细胞分化本身。胚胎干细胞染色质蛋白的高流动性被认为不仅有助于维护胚胎干细胞的多能性，同时对胚胎干细胞基因组在整个核中结构的重新组织至关重要，尤其是异染色质的组织[51]。

比较胚胎干细胞和分化细胞的异染色质域，在分化的早期阶段，异染色质经历了急剧的空间重排[51~53]。直接观察异染色质（使用针对主要卫星重复序列的 DNA 探针）和检测异染色质结合蛋白 (HP1α)，结果显示，相对于 NPC（图 2-1），胚胎干细胞异染色质部分去凝聚。胚胎干细胞异染色质散在分布，而 NPC 异染色质域更紧凑且集中分布于不同的灶，非常类似在其他体细胞类型中观察到的那样[51]。在几个胚胎干细胞系中证实，异染色质灶存在类似的差异[54,55]。核纤层蛋白 A/C 和核纤层相关蛋白可以直接与染色质和组蛋白进行相互作用，从而限制了染色质的移动性，并影响染色质在整个核内的组织[23]。胚胎干细胞核纤层蛋白 A/C 的缺失连同染色质去凝聚可能有助于胚胎干细胞的可塑性。

分化过程中不同异染色质灶的形成伴随着组蛋白 H3 第 9 位赖氨酸三甲基化 (H3K9me3：异染色体一种表观遗传标记) 的增加，以及 H3、H4 乙酰化降低（转录活化常染色质的表观遗传标记）和 DNA 甲基化的增加。这些表观遗传标记的变化是与胚胎干细胞染色质比谱系限定及分化细胞类型有更开放构象的想法相符合的[51,56]。参与维持这种开放染色质状态的分子之一是染色质改造因子 Chd1，因为缺乏 Chd1 的鼠胚胎干细胞异染色质和分化细胞中观察到的特点类似。此外，Chd1 缺乏细胞多能性的减弱，意味着 Chd1 在建立或维持胚胎干细胞多能性及异染色质形成方面起作用[57]。

胚胎干细胞的高转录和 DNA 复制

基因表达的全基因组分析和 Br-UTP（5-溴尿苷 5′-三磷酸）标记转录位点显示，胚胎干细胞基因组可能比其分化后代转录更活跃[51,55,58]。胚胎干细胞染色质的特点是活化组蛋白标记（例如，组蛋白 H3 第 36 位赖氨酸二甲基化：H3K36me2[55]；终末端延长的 RNAPⅡ）的增加。全基因组覆瓦式阵列芯片分析显示鼠胚胎干细胞蛋白编码和非编码区的高转录活性，而当细胞沿着神经路径分化时，转录地标越来越明确。与 NPC 相比，胚胎干细胞中一半的标记基因转录水平增高。由于高流动性染色质会更容易接近转录因子和转录复合物[55]，观察到的高转录可能只是独特的开放染色质结构的副产品[51]。深入检查过表达基因发现，大多数是编码常见转录因子和染色质改造蛋白的基因，因此染色质改造因子转录水平的提高对建立和维持胚胎干细胞独特的开放染色质构象是非常重要的。尽管高转录和开放的染色质结构间的关系还不清楚，胚胎干细胞高活力的转录被认为是维持胚胎干细胞多潜能的关键机制，这有助于基因组的可塑性。

DNA 复制的时机是整个核内染色质状态的另一个指标[59~62]。染色质结构和 DNA 复制之间的相互关系源于对红系细胞 β-珠蛋白位点、分化中雌性胚胎干细胞 XCI[63] 及分化中胚胎干细胞其干性关联基因[60,61]的相关研究。这些研究表明，复制时机的变化是染色质结构和压缩程度变化的反映。鼠胚胎干细胞复制时机的高分辨率资料揭示，胚胎干细胞和 NPC 相比，尽管复制时机和总的转录趋势变化一致，但出现的许多例外情况说明，两者间没有直接关系。当沿着神经路径分化时，复制域时间上的重组反映了染色体空间上的重组[61]。

发育调节基因的沉默机制

胚胎干细胞基因组的一个独特特征是，约 2500 个沉默的发育调节基因含有二价的组蛋白修饰，修饰使染色质呈活化（H3K4me3）和抑制（H3K27me3；组蛋白 H2A 第 119 位赖氨酸单泛素化；H2Aub1）状态[64~66]，后者由多梳阻遏复合物 PRC1 和 PRC2 介导。出人意料的是，发育调节基因也与 RNAPⅡ 的非寻常形式有关，这种形式能够在 C 端结构域缺乏第 2 位丝氨酸磷酸化的情况下通过编码区进行转录[66]。对含有二价组蛋白修饰的胚胎干细胞和分化细胞的全基因组比较研究显示，发育调节基因的二价结构域是多能干细胞的特征，决定胚胎干细胞分化时基因的活化或抑制，这与基因表达时相关改变一致[67]。

二价结构域不局限于胚胎干细胞，在谱系限定的多能干细胞[68]，以及分化细胞（包括祖细胞、终末分化神经元[69]和结肠直肠癌细胞[70,71]）中也能确认。有趣的是，对淋巴细胞的分析鉴定了诱导初级反应基因的亚群含有二价染色质，对适当的胞外信号能迅速反应，这表明这种许可染色质状态对多种细胞类型的基因表达调控有重要的作用[72,73]。

结论

从局部染色质折叠的复杂细节到细胞核内染色体总的组织，很明显，基因组结构和功能紧密相连。最近对胚胎干细胞独特而又高度可塑基因组的研究让我们首次了解多能状态下核结构和染色质组织的特征，以及随之在建立谱系特异性程序中的变化。尽管对胚胎干细胞核亚区的分析非常有限，但有一个共同的发现，即胚胎干细胞的核结构比分化细胞简单。在某些情况下，尽管组装形成功能亚区的组分已经存在于未分化细胞中，但它们以看似无序的方式散在分布，随着分化开始，无序状态消失。在另一些情况中，如核纤层蛋白 A/C，诱导其表达足以促进分化。少数调查胚胎干细胞核的功能性组织的研究已将多能性和高动态的基因组相互关联，这反映了其独特的核结构。

致谢

非常感谢 J. T. Butler 和 J. B. Lawrence 向我们提供尚未发表的图片，同时对医学研究理事会（英国）的支持也表示感谢。

（龚　伟　译）

参 考 文 献

1. Keenen B, de la Serna IL. Chromatin remodeling in embryonic stem cells: regulating the balance between pluripotency and differentiation. J Cell Physiol 2009; 219(1):1-7.
2. Meshorer E, Misteli T. Chromatin in pluripotent embryonic stem cells and differentiation. Nat Rev Mol Cell Biol 2006; 7(7):540-546.
3. Branco MR, Pombo A. Intermingling of chromosome territories in interphase suggests role in translocations and transcription-dependent associations. PLoS Biol 2006; 4(5):e138.
4. Lieberman-Aiden E, van Berkum NL, Williams L et al. Comprehensive mapping of long-range interactions reveals folding principles of the human genome. Science 2009; 326(5950):289-293.
5. Parada L, Misteli T. Chromosome positioning in the interphase nucleus. Trends Cell Biol 2002; 12(9):425-432.
6. Wiblin AE, Cui W, Clark AJ et al. Distinctive nuclear organisation of centromeres and regions involved in pluripotency in human embryonic stem cells. J Cell Sci 2005; 118(Pt 17):3861-3868.
7. Bartova E, Galiova G, Krejci J et al. Epigenome and chromatin structure in human embryonic stem cells undergoing differentiation. Dev Dyn 2008; 237(12):3690-3702.
8. Solovei I, Kreysing M, Lanctot C et al. Nuclear architecture of rod photoreceptor cells adapts to vision in mammalian evolution. Cell 2009; 137(2):356-368.
9. Chambeyron S, Bickmore WA. Chromatin decondensation and nuclear reorganization of the HoxB locus upon induction of transcription. Genes Dev 2004; 18(10):1119-1130.
10. Morey C, Kress C, Bickmore WA. Lack of bystander activation shows that localization exterior to chromosome territories is not sufficient to up-regulate gene expression. Genome Res 2009; 19(7):1184-1194.
11. Senner CE, Brockdorff N. Xist gene regulation at the onset of X inactivation. Curr Opin Genet Dev 2009; 19(2):122-126.
12. Lee JT. Regulation of X-chromosome counting by Tsix and Xite sequences. Science 2005; 309(5735):768-771.
13. Xu N, Tsai CL, Lee JT. Transient homologous chromosome pairing marks the onset of X inactivation. Science 2006; 311(5764):1149-1152.
14. Bacher CP, Guggiari M, Brors B et al. Transient colocalization of X-inactivation centres accompanies the initiation of X inactivation. Nat Cell Biol 2006; 8(3):293-299.
15. Augui S, Filion GJ, Huart S et al. Sensing X chromosome pairs before X inactivation via a novel X-pairing region of the Xic. Science 2007; 318(5856):1632-1636.
16. Xu N, Donohoe ME, Silva SS et al. Evidence that homologous X-chromosome pairing requires transcription and Ctcf protein. Nat Genet 2007; 39(11):1390-1396.
17. Khalil A, Grant JL, Caddle LB et al. Chromosome territories have a highly nonspherical morphology and nonrandom positioning. Chromosome Res 2007; 15(7):899-916.
18. Scialdone A, Nicodemi M. Mechanics and dynamics of X-chromosome pairing at X inactivation. PLoS Comput Biol 2008; 4(12):e1000244.
19. Tsai CL, Rowntree RK, Cohen DE et al. Higher order chromatin structure at the X-inactivation center via looping DNA. Dev Biol 2008; 319(2):416-425.
20. Eils R, Dietzel S, Bertin E et al. Three-dimensional reconstruction of painted human interphase chromosomes: active and inactive X chromosome territories have similar volumes but differ in shape and surface structure. J Cell Biol 1996; 135(6 Pt 1):1427-1440.

21. Chaumeil J, Le Baccon P, Wutz A et al. A novel role for Xist RNA in the formation of a repressive nuclear compartment into which genes are recruited when silenced. Genes Dev 2006; 20(16):2223-2237.
22. Butler JT, Hall LL, Smith KP et al. Changing nuclear landscape and unique PML structures during early epigenetic transitions of human embryonic stem cells. J Cell Biochem 2009; 107(4):609-621.
23. Gruenbaum Y, Margalit A, Goldman RD et al. The nuclear lamina comes of age. Nat Rev Mol Cell Biol 2005; 6(1):21-31.
24. Francastel C, Schubeler D, Martin DI et al. Nuclear compartmentalization and gene activity. Nat Rev Mol Cell Biol 2000; 1(2):137-143.
25. Akhtar A, Gasser SM. The nuclear envelope and transcriptional control. Nat Rev Genet 2007; 8(7):507-517.
26. Finlan LE, Sproul D, Thomson I et al. Recruitment to the nuclear periphery can alter expression of genes in human cells. PLoS Genet 2008; 4(3):e1000039.
27. Ragoczy T, Bender MA, Telling A et al. The locus control region is required for association of the murine beta-globin locus with engaged transcription factories during erythroid maturation. Genes Dev 2006; 20(11):1447-1457.
28. Williams RR, Azuara V, Perry P et al. Neural induction promotes large-scale chromatin reorganisation of the Mash1 locus. J Cell Sci 2006; 119(Pt 1):132-140.
29. Kosak ST, Skok JA, Medina KL et al. Subnuclear compartmentalization of immunoglobulin loci during lymphocyte development. Science 2002; 296(5565):158-162.
30. Hewitt SL, High FA, Reiner SL et al. Nuclear repositioning marks the selective exclusion of lineage-inappropriate transcription factor loci during T helper cell differentiation. Eur J Immunol 2004; 34(12):3604-3613.
31. Constantinescu D, Gray HL, Sammak PJ et al. Lamin A/C expression is a marker of mouse and human embryonic stem cell differentiation. Stem Cells 2006; 24(1):177-185.
32. Pajerowski JD, Dahl KN, Zhong FL et al. Physical plasticity of the nucleus in stem cell differentiation. Proc Natl Acad Sci USA 2007; 104(40):15619-15624.
33. Luo L, Gassman KL, Petell LM et al. The nuclear periphery of embryonic stem cells is a transcriptionally permissive and repressive compartment. J Cell Sci 2009; 122(Pt 20):3729-3737.
34. Kaiser TE, Intine RV, Dundr M. De novo formation of a subnuclear body. Science 2008; 322(5908):1713-1717.
35. Misteli T. Self-organization in the genome. Proc Natl Acad Sci USA 2009; 106(17):6885-6886.
36. Hall LL, Smith KP, Byron M et al. Molecular anatomy of a speckle. Anat Rec A Discov Mol Cell Evol Biol 2006; 288(7):664-675.
37. Xie SQ, Martin S, Guillot PV et al. Splicing speckles are not reservoirs of RNA polymerase II, but contain an inactive form, phosphorylated on serine2 residues of the C-terminal domain. Mol Biol Cell 2006; 17(4):1723-1733.
38. Lawrence JB, Clemson CM. Gene associations: true romance or chance meeting in a nuclear neighborhood? J Cell Biol 2008; 182(6):1035-1038.
39. Morris GE. The Cajal body. Biochim Biophys Acta 2008; 1783(11):2108-2115.
40. Schwartz YB, Pirrotta V. Polycomb complexes and epigenetic states. Curr Opin Cell Biol 2008; 20(3):266-273.
41. Brookes E, Pombo A. Modifications of RNA polymerase II are pivotal in regulating gene expression states. EMBO Rep 2009; 10(11):1213-1219.
42. Bernstein E, Duncan EM, Masui O et al. Mouse polycomb proteins bind differentially to methylated histone H3 and RNA and are enriched in facultative heterochromatin. Mol Cell Biol 2006; 26(7):2560-2569.
43. Ren X, Vincenz C, Kerppola TK. Changes in the distributions and dynamics of polycomb repressive complexes during embryonic stem cell differentiation. Mol Cell Biol 2008; 28(9):2884-2895.
44. Bancaud A, Huet S, Daigle N et al. Molecular crowding affects diffusion and binding of nuclear proteins in heterochromatin and reveals the fractal organization of chromatin. EMBO J 2009; 28(24):3785-3798.
45. Kimura H, Cook PR. Kinetics of core histones in living human cells: little exchange of H3 and H4 and some rapid exchange of H2B. J Cell Biol 2001; 153(7):1341-1353.
46. Phair RD, Scaffidi P, Elbi C et al. Global nature of dynamic protein-chromatin interactions in vivo: three-dimensional genome scanning and dynamic interaction networks of chromatin proteins. Mol Cell Biol 2004; 24(14):6393-6402.
47. Giglia-Mari G, Theil AF, Mari PO et al. Differentiation driven changes in the dynamic organization of Basal transcription initiation. PLoS Biol 2009; 7(10):e1000220.
48. Cheutin T, McNairn AJ, Jenuwein T et al. Maintenance of stable heterochromatin domains by dynamic HP1 binding. Science 2003; 299(5607):721-725.
49. Festenstein R, Pagakis SN, Hiragami K et al. Modulation of heterochromatin protein 1 dynamics in primary Mammalian cells. Science 2003; 299(5607):719-721.
50. Misteli T, Gunjan A, Hock R et al. Dynamic binding of histone H1 to chromatin in living cells. Nature 2000; 408(6814):877-881.

51. Meshorer E, Yellajoshula D, George E et al. Hyperdynamic plasticity of chromatin proteins in pluripotent embryonic stem cells. Dev Cell 2006; 10(1):105-116.
52. Aoto T, Saitoh N, Ichimura T et al. Nuclear and chromatin reorganization in the MHC-Oct3/4 locus at developmental phases of embryonic stem cell differentiation. Dev Biol 2006; 298(2):354-367.
53. Kobayakawa S, Miike K, Nakao M et al. Dynamic changes in the epigenomic state and nuclear organization of differentiating mouse embryonic stem cells. Genes Cells 2007; 12(4):447-460.
54. Cammas F, Oulad-Abdelghani M, Vonesch JL et al. Cell differentiation induces TIF1beta association with centromeric heterochromatin via an HP1 interaction. J Cell Sci 2002; 115(Pt 17):3439-3448.
55. Efroni S, Duttagupta R, Cheng J et al. Global transcription in pluripotent embryonic stem cells. Cell Stem Cell 2008; 2(5):437-447.
56. Bibikova M, Chudin E, Wu B et al. Human embryonic stem cells have a unique epigenetic signature. Genome Res 2006; 16(9):1075-1083.
57. Gaspar-Maia A, Alajem A, Polesso F et al. Chd1 regulates open chromatin and pluripotency of embryonic stem cells. Nature 2009; 460(7257):863-868.
58. Faro-Trindade I, Cook PR. A conserved organization of transcription during embryonic stem cell differentiation and in cells with high C value. Mol Biol Cell 2006; 17(7):2910-2920.
59. Donaldson AD. Shaping time: chromatin structure and the DNA replication programme. Trends Genet 2005; 21(8):444-449.
60. Perry P, Sauer S, Billon N et al. A dynamic switch in the replication timing of key regulator genes in embryonic stem cells upon neural induction. Cell Cycle 2004; 3(12):1645-1650.
61. Hiratani I, Ryba T, Itoh M et al. Global reorganization of replication domains during embryonic stem cell differentiation. PLoS Biol 2008; 6(10):e245.
62. Hiratani I, Leskovar A, Gilbert DM. Differentiation-induced replication-timing changes are restricted to AT-rich/long interspersed nuclear element (LINE)-rich isochores. Proc Natl Acad Sci USA 2004; 101(48):16861-16866.
63. Goren A, Cedar H. Replicating by the clock. Nat Rev Mol Cell Biol 2003; 4(1):25-32.
64. Azuara V, Perry P, Sauer S et al. Chromatin signatures of pluripotent cell lines. Nat Cell Biol 2006; 8(5):532-538.
65. Bernstein BE, Mikkelsen TS, Xie X et al. A bivalent chromatin structure marks key developmental genes in embryonic stem cells. Cell 2006; 125(2):315-326.
66. Stock JK, Giadrossi S, Casanova M et al. Ring1-mediated ubiquitination of H2A restrains poised RNA polymerase II at bivalent genes in mouse ES cells. Nat Cell Biol 2007; 9(12):1428-1435.
67. Mikkelsen TS, Ku M, Jaffe DB et al. Genome-wide maps of chromatin state in pluripotent and lineage-committed cells. Nature 2007; 448(7153):553-560.
68. Cui K, Zang C, Roh TY et al. Chromatin signatures in multipotent human hematopoietic stem cells indicate the fate of bivalent genes during differentiation. Cell Stem Cell 2009; 4(1):80-93.
69. Mohn F, Weber M, Rebhan M et al. Lineage-specific polycomb targets and de novo DNA methylation define restriction and potential of neuronal progenitors. Mol Cell 2008; 30(6):755-766.
70. Rodriguez J, Munoz M, Vives L et al. Bivalent domains enforce transcriptional memory of DNA methylated genes in cancer cells. Proc Natl Acad Sci USA 2008; 105(50):19809-19814.
71. McGarvey KM, Van Neste L, Cope L et al. Defining a chromatin pattern that characterizes DNA-hypermethylated genes in colon cancer cells. Cancer Res 2008; 68(14):5753-5759.
72. Lim PS, Hardy K, Bunting KL et al. Defining the chromatin signature of inducible genes in T-cells. Genome Biol 2009; 10(10):R107.
73. Araki Y, Wang Z, Zang C et al. Genome-wide analysis of histone methylation reveals chromatin state-based regulation of gene transcription and function of memory CD8+ T-cells. Immunity 2009; 30(6):912-925.

第3章 细胞多能性的表观遗传学调控

Eleni M. Tomazou, Alexander Meissner*

摘要：表观遗传学调控是指在不改变DNA序列的情况下引起基因表达改变的一种调控模式。目前了解比较清楚的表观遗传学标记包括组蛋白的翻译后修饰和DNA甲基化。这两者在生物发育和疾病发生过程中都扮演了重要的角色。多能干细胞为再生医学带来了曙光，目前在这方面研究的重点为调控细胞多能性的分子网络。本章概括介绍了胚胎干细胞的表观遗传学调控，同时简单介绍了表观遗传学机制及其在多能干细胞中扮演的角色。

引言

Conrad Waddington于1942年首先提出了表观遗传学（epigenetic）这一概念[1]。它曾经被用来描述基因通过与环境的相互作用来影响表型。现在，表观遗传学的概念指在没有发生DNA序列变化的情况下产生的可以通过有丝分裂或减数分裂遗传的基因表达改变[2]。在希腊语中，词缀epi表示基因组之外的改变，在表观遗传学这个词中恰如其分。表观遗传学改变可以通过影响DNA的可接近性、蛋白分子的募集和染色质结构来调控基因功能。所以，表观遗传学在界定、维持、延续细胞所处状态方面扮演了极为重要的角色[3]。表观遗传学不但与细胞当前的状态有关，还与细胞将要到达的新状态相关[4,5]，藉此调控胚胎干细胞的发育可塑性[6]。

胚胎干细胞可由着床前胚胎的内细胞团分离得到（图3-1）[7~10]，并且具有两个重要特性：自我更新（self-renewal）和多向分化（pluripotency）。自我更新代表胚胎干细胞能无限分裂产生与其相同的细胞；多向分化代表在体内或体外，在不同发育信号的影响下胚胎干细胞能产生不同体细胞及生殖细胞[11]。胚胎干细胞可用来研究哺乳动物的发育，并为再生医学提供细胞来源[12]。从1998年第一次分离出5株人类胚胎干细胞开始，过去10年，已经有数百株人类胚胎干细胞分离出来[13,14]。此外，利用特定培养条件，也可以从发育晚期的胚胎中分离出多能细胞，如外胚层干细胞（epiblast stem cell, EpiSC）、返祖胚胎干细胞（reverted embryonic stem cell, rES cell）[15]和胚胎生殖细胞（embryonic germ cell, EG）等[16]。最近，诱导性多能干细胞（induced pluripotent stem cell, iPS）技术通过过表达4个转录因子，可以使大多数体细胞类型产生胚胎干细胞样细胞[17]。

---

*Corresponding Author: Alexander Meissner—Department of Stem Cell and Regenerative Biology, Harvard University, 7 Divinity Avenue, Cambridge, Massachusettes 02138, USA. Email: alexander_meissner@harvard.edu

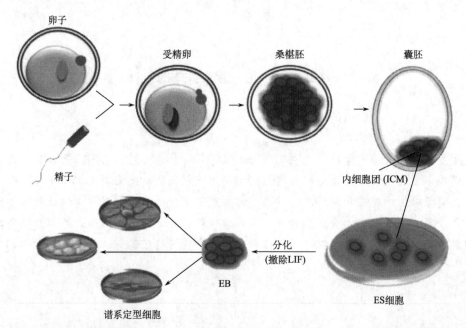

图 3-1 小鼠胚胎干细胞（mouse ES cell，mES）的起源与分化。mES 细胞来源于小鼠发育早期（第 3.5 天）囊胚期胚胎的内细胞团（inner cell mass，ICM）。mES 在白细胞抑制因子（leukemia inhibitory factor，LIF）存在的情况下可进行持久的离体培养。撤除 LIF 后，mES 接种于非吸附性培养皿可形成类胚体（embryoid body，EB）。在不同培养条件下，胚体细胞可以分化为各胚层，包括内胚层（endoderm）、中胚层（mesoderm）、外胚层（ectoderm）

从分子水平理解 ES 细胞状态调控及多能性调控是目前分子生物学研究的重点。过去的研究发现，转录因子如 OCT4、NANOG、SOX2，对于维持多能干细胞的非分化状态很重要[18]。这一调控网络使控制干细胞存活和增殖的转录因子激活，并抑制控制分化的转录因子[19]。转录因子与表观遗传标记的相互作用很可能对稳定这一网络具有重要作用[20]。在接下来的部分我们将具体讨论表观遗传学与多能干细胞的关系。

表观遗传调控

在真核生物中，DNA 被包装成染色质（chromatin），见图 3-2。染色质的基本单位为核小体（nucleosome），是由约 147 个碱基缠绕组蛋白八聚体核心（H3、H4、H2A 和 H2B 各两个分子）两圈组成[21]。相邻核小体由接头 DNA（linker DNA）连在一起，接头 DNA 结合组蛋白 H1。通过改变调控元件的可及性，以及转录因子与其靶区域的结合，染色质结构可以影响基因的表达[22]。能够影响染色质结构的重要因素包括组蛋白翻译后修饰（posttranslational histone modification）和基因组 DNA 甲基化（methylation of genomic DNA）[23,24]，此外，ATP 依赖的染色质重塑因子（ATP-dependent chromatin remodeling factor）也会影响染色质结构[25]（图 3-2）。

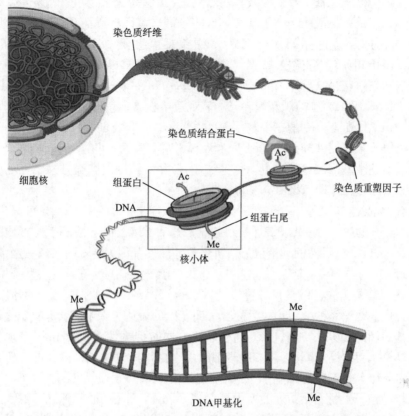

图 3-2　表观遗传调控。细胞核中，DNA 编码和结构信息都位于染色质中。染色质的结构基本单位为核小体，包含 147bp DNA 及其缠绕的组蛋白核心。核小体进一步形成更高级的结构，其包装程度可以影响包括基因表达在内的所有 DNA 相关的生物学过程。一系列事件，包括组蛋白翻译后修饰、DNA 甲基化和染色质改构因子的结合都会影响核小体的定位及染色质构象

染色质结构的调控因子

ATP 依赖染色质重塑因子是一类能够通过水解 ATP 破坏 DNA 与组蛋白之间连接的酶类。它们可以改变核小体构象及其定位，进而改变染色质高级结构[26,27]。这些因子可以增加 DNA 的可及性，使调控因子及转录因子能够结合在调控区域。这些因子分为 4 个家族：SWI/SNF 家族、ISWI 家族、NuRDMi-2/CHD 家族和 INO80 家族。每个家族都参与了各种生命过程的调控，包括 DNA 修复、细胞周期检验点调控、DNA 复制、端粒维持及染色体分离，这些表明它们对维持遗传稳定性至关重要[25]。

组蛋白修饰

核心组蛋白是核小体的主要组成部分（图 3-2），在进化上较为保守。组蛋白的氨基端从核小体伸出，而羧基端球状结构域组成核小体的骨架。氨基端组蛋白可作为各种

翻译后修饰的受体,包括:赖氨酸残基的乙酰化、甲基化和泛素化,丝氨酸和苏氨酸残基的磷酸化,以及精氨酸残基的甲基化。这些修饰与基因的激活和抑制密切相关[28]。能够激活转录或打开染色体的表观遗传修饰包括:组蛋白 H3K9 和 K14 的乙酰化(H3K9ac、H3K14ac),组蛋白 H4K16 的乙酰化(H4K16ac),组蛋白 H3K4 的单甲基、双甲基和三甲基化(H3K4me1、H3K4me2、H3K4me3),以及组蛋白 H3 K36 的三甲基化(H3K36me3)。抑制基因转录的表观遗传学修饰包括组蛋白 H3K27 的三甲基化(H3K27me3)和 H3K9 的三甲基化(H3K9me3)。有观点认为,发生于一个基因的多种组蛋白修饰不仅反映当前的基因表达状态,而且还决定了在发育后期或者在后代细胞中细胞激活这个基因表达的能力[4]。组蛋白的各种修饰的组合会带来不同的生物学效应,可称其为组蛋白编码(histone code)[5]。

在理解多能性与发育潜能相关性过程中,抑制性 H3K27me3 修饰和激活性 H3K4me3 修饰的共存提供了某种可能性。这种特殊组合被称为二价染色体域(bivalent domain)[29]。目前的理论认为二价染色体域使基因处在随时可以被激活的状态。有相当一部分对早期发育至关重要的基因,其激活依赖于分化过程中抑制性 H3K27me3 修饰的丢失。但是,二价染色体域也可以是一个双向开关,不但能激活基因的表达,也能在谱系变化定型之后完全关闭基因的表达。在细胞命运决定之后,非诱导性双向基因倾向于失去 H3K4me3 活化状态,同时抑制性 H3K27me3 表观遗传学标记得到恢复[30,31]。虽然一直认为仅多能干细胞才有二价染色体域,但最近的研究也在一些具有非常有限分化潜能的细胞中鉴定出了这些结构[30]。

除了核心组蛋白之外,在 DNA 中也发现了 H2A 和 H3 的其他几种形式,并且最近的研究也显示其具有重要的功能。它们以核心组蛋白的非等位形式存在,并以非细胞周期依赖形式进入染色体。它们以替换的形式取代核心组蛋白之后会改变局部染色体结构,进而改变转录状态[32]。

已知有许多酶可以改变和维持组蛋白的表观遗传学修饰。这些酶可以对组蛋白进行乙酰化(acetylation)[33]、甲基化(methylation)[34]、磷酸化(phosphorylation)[35]、泛素化(ubiquitination)[36]、泛素样蛋白修饰[sumoylation, small ubiquitin-like modifier (SUMO)与目标蛋白的赖氨酸结合][37]、ADP 糖基化(ADP-ribosylation)[38]、精氨酸瓜化(deimination)[39]及脯氨酸异构化修饰(proline isomerization)[40]。其中的大多数修饰是可逆的,如组蛋白去甲基化酶(demethylases)和去乙酰化酶(deacetylaces)可以去除上述修饰,这使得通过组蛋白修饰来灵活调控转录成为可能。

第一个被发现同时也是研究得最深入的表观遗传修饰酶为多梳蛋白家族(polycomb group,PcG)[41]。这些发育调控因子形成多梳抑制复合物(polycomb-repressive complex,PRC)。PcG 蛋白催化两种组蛋白修饰:H3K27 三甲基化修饰和 H2A 单泛素化修饰。H3K27 三甲基化修饰由 Zeste 基因增强子同源物 2(Ezh2)催化完成,而 Ezh2 可以与胚胎外胚层发育蛋白(Eed)和 zeste12 抑制因子(Suz12)组成 PRC2 复合体。H3K27me3 可以募集 PRC1 复合物介导染色体紧缩。PRC1 通过 Ringb 介导 H2A 的泛素化,这一催化活性在 Bmi1 和 Mel18 存在的情况下被激活[42]。在哺乳动物细胞中 PRC2 复合物与 H3K27me3 共定位,而 PRC1 却没有这种共定位,这说明 PRC1 在 PcG

介导的基因沉默中具有其他稳定作用[43]。

DNA 甲基化

DNA 甲基化是研究最多的一种表观遗传学修饰，并且是唯一一种发生在 DNA 分子上的修饰。从生物化学水平讲，DNA 甲基化是一种发生在胞嘧啶 5 号碳上的共价修饰，以甲基取代氢原子（图 3-3）。在哺乳动物中，DNA 甲基化修饰主要发生在 CpG 岛[44]，也有报道称在某些细胞系中甲基化修饰可以发生在非 CpG 岛（图 3-3）[45~49]。短暂的去甲基化发生在生殖细胞中及胚胎着床前[50]。DNA 甲基化与染色质局部空间重构有关（形成更为紧密的染色质结构），还与转录因子与启动子区域结合的改变及高稳定性基因沉默相关[51]。

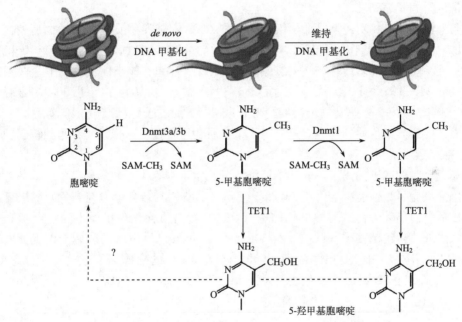

图 3-3　DNA 甲基化机制。DNA 甲基转移酶（DNMT 1，3a，3b）催化 5-甲基胞嘧啶，将 S-腺苷甲硫氨酸（S-adenosylmethionine，SAM）的一个甲基转移到胞嘧啶 5 碳位置。最近有报道称 TET1 催化 5-甲基胞嘧啶转化为 5-羟甲基胞嘧啶，后者在鼠胚胎干细胞中被检测到[105]。目前推测 5-羟甲基胞嘧啶是去甲基化的一个中间产物，但是还有待验证

有两类 DNA 甲基转移酶（DNA methyltransferase）催化 DNA 甲基化。DNA 甲基转移酶 1（Dnmt1）被认为是一种维持甲基化状态的甲基转移酶，因为它催化细胞分裂后处于半甲基化状态 DNA 双链中新生链的 CpG 岛的甲基化，从而维持细胞在多次分裂后的甲基化状态不发生变化[52,53]。两个从头甲基化酶，即 Dnmt3a 和 Dnmt3b 负责在胚胎发育和分化过程中建立新的甲基化位点[54]。它们在胚胎干细胞中高表达，在分化细胞中低表达。Dnmt3l 与 Dnmt3a 及 Dnmt3b 有很高的同源性[55]，但是缺少甲基转移酶活性，可作为 Dnmt3a、Dnmt3b 的调节因子。当第 4 个赖氨酸缺少甲基化修饰时，

Dnmt3l 可以与组蛋白 H3 的氮端结合[56]，后者是 Dnmt31 促进生殖细胞核小体 DNA 重头甲基化的一种可能机制。

胚胎干细胞的表观遗传组学

染色质结构及其动态调控

鼠胚胎干细胞拥有异乎寻常的开放染色质结构，富含松散的常染色质（euchromatin）[57~59]和散开的异染色质（heterochromatin）[60]。开放的染色质结构导致转录因子和"转录单位"较易接近，造成了整个染色质的高转录（hyper-transcription），这为胚胎干细胞的可塑性（plasticity）提供了可能；谱系的定向分化（lineage commitment）常伴有染色质的紧缩和转录的失活[61]。

与染色质状态相对应，胚胎干细胞中几种 ATP 依赖染色质重构因子也高表达。例如，Brg1（Smarca4）作为 SWI/SNF 复合物中的 ATP 水解酶，通过与 Nanog 相互作用调控胚胎干细胞的多能性[62]。与之类似，NuRD 复合物（核小体重构及组蛋白去乙酰化）成分 Mbd3 对分化至关重要[63]，而 Chd1 失活细胞不再具有多能性。Cdh1 含有一个 ATP 水解酶 SNF-2 样解旋酶结构域，与活化基因的启动子区域结合。有趣的是，Chd1 的低表达导致异染色质的形成，这支持了染色质的可及性对于保持干细胞独特属性是必需的这一观点[64]。

染色质重构因子的高水平表达不但维持了胚胎干细胞染色质的松散状态，还对胚胎干细胞的细胞可塑性有影响[57]。在胚胎干细胞中，染色质结合蛋白与染色质的结合不紧密，因此具有高动力性。随着分化的进行，这些蛋白质会在染色质上固定以抑制多能性。缺少核小体组装因子（nucleosome assembly factor）HirA 的胚胎干细胞拥有较高水平的未结合组蛋白，拟胚体（embryoid body，EB）的形成速度也较快。与此相反，具有受限 H1 交换的胚胎干细胞表现为分化终止（differentiation arrest）[60]。因此，染色质的高动力性是多能性的一个重要标志。

组蛋白修饰

位于活跃基因组区域的组蛋白修饰

胚胎干细胞除了具有开放染色质结构外，活跃染色质区域位于整个胚胎干细胞。这些区域通常具有组蛋白 H3 和 H4 乙酰化修饰、H3K4 二甲基化或三甲基化修饰及 H3K9/14 乙酰化修饰[30,65]。许多酶可以识别或者催化这些修饰进而成为多能性的调控因子。例如，Rbp2（Jarid1a）特异性催化 H3K4me2/3 去甲基化，通过结合分化相关基因的启动子区域以调控其表观遗传学状态[66]。P300 是一种组蛋白乙酰转移酶，可以与延伸状态的 DNA 聚合酶 II（PolII）结合，定位于 Nanog、Oct4、Sox2[67]等转录因子结合区域，同时 Tip60-p400 复合物可作为组蛋白乙酰基转移酶和染色质重构因子，通过调控 Nanog 结合区域维持胚胎干细胞的多能性[68]。活化的启动子也含有 H3K4me2 修饰，而 H3K36me3 和 H3K79me2 定位于转录活跃区域（图 3-4）[65]。

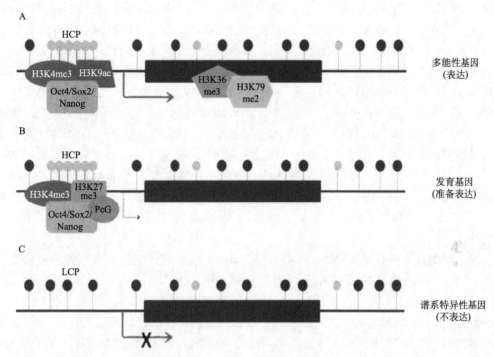

图3-4 胚胎干细胞表观遗传学调控机制简图。表观遗传学调控模式可分为三类：A. 表达的多能性相关基因；B. 即将表达的发育相关基因就绪表达；C. 需要分化诱导才表达的基因（谱系特异性基因）。根据CpG岛的密度不同，基因启动子区域被分为HCP和LCP。CpG岛由棒糖状图标表示。白色的代表未甲基化CpG岛，黑色的代表甲基化CpG岛。位于基因内及LCP中的CpG岛多被甲基化，而位于HCP中的CpG岛未被甲基化。多能性相关基因的启动子区具有活化状态的组蛋白修饰（H3K4me3、H3K9ac），而在其基因内则具有H3K36me3、H3K79me2表观遗传学修饰。二价染色质域是发育相关基因的启动子区的特征。PcG与启动子的相互作用也被显示在图中。在特定条件下，它们显示了转录起始过程（用浅色的小箭头表示）

除了组蛋白赖氨酸修饰外，组蛋白H3 R17和R26位点的精氨酸甲基化可作为新的表观遗传学机制以调控多能性[69,70]。Oct4和Sox2的启动子区含有H3R16和H3R26甲基化，并可检测到辅活化因子相关的精氨酸甲基转移酶1（co-activator-associated-protein-arginine mathyltransferase 1，CARM1）。在胚胎干细胞中，抑制CARM1导致分化；而过表达导致Nanog启动子区组蛋白H3精氨酸甲基化，延迟分化[70]。

在人类胚胎干细胞中，增强子（enhancer）区域存在较多H3K4me1、H3K27ac修饰[71]。增强子是指可以活化相关启动子的DNA序列，通过独立于方向和位置的模式调控基因表达[72]。

即将表达基因启动子中的组蛋白修饰

研究显示，在人类胚胎干细胞中有3/4基因启动子区的组蛋白被活化修饰，并通过DNA聚合酶Ⅱ起始转录。但是只有其中的一半能够生成可检测到的转录本。例如，

H3K4me3 出现在胚胎干细胞大约 80% 的启动子区[65]。其中的许多区域并不能产生目前技术可检测到的全长转录本。在许多情况下这些启动子区也会有抑制性表观遗传标记 H3K27me3，表现为二价性。染色质二价域多与胚胎干细胞中的分化相关基因关联，在上文中我们已经讨论过这一点（图 3-4）。有趣的是，大约一半的二价染色质域都有三个转录因子 Oct4、Nanog、Sox2 中至少一个转录因子的结合位点[19,73]。二价染色质域多有 PcG 结合。然而，尽管胚胎干细胞中 PRC1 和 PCR2 会同时出现在许多启动子区[74,75]，也有一些启动子只有一个复合物。在胚胎干细胞中，根据 PRC1、PRC2 同时出现或只有 PRC2 出现，可将二价染色质域分为两类[43]。有趣的是，出现两种复合物的启动子区可以在分化过程中有效地保持 PcG 介导的染色质结构。非典型性 PRC2 同样也出现在人类胚胎干细胞中。这种复合物中包含 EZH1——一种 EZH2 的同源物。含有 EZH1 的复合物选择性调控发育相关关键基因，抑制这些基因的表达可以阻止胚胎干细胞的分化[76]。

基因功能丧失实验证实 PRC2（和 H3K27me3）在胚胎干细胞中发挥着至关重要的作用。PRC2 成分的缺失可以导致胚胎干细胞分化缺陷[76~78]。然而，纯合删除 Eed、Suz12 或 Ezh2 后，小鼠囊胚中仍可分离出胚胎干细胞并能够在体外传代[76~78]。最近对 Jarid2（与 Jarid1 家族密切相关）的研究显示 Jarid2 能调控 PRC2 的催化活性及其定位，从而可以精细调控 H3K27me3 水平[79~82]。已知 Jarid2 在胚胎干细胞和诱导多能干细胞[83]中高表达并对分化至关重要，但似乎与胚胎干细胞的自我更新无关[79,80]。

最近，有表观遗传学研究试图鉴定 H3K4me3、H3K27me3 之外的二价染色质域。目前已经在胚胎干细胞编码发育调控因子的二价启动子区域中鉴定出了 H3K9me3[84]。但截止到目前，还没有能够成功鉴定出这种三价染色质结构[30,85]。同样，有报道称组蛋白异构体 H2A.Z 可与 PcG 蛋白共定位，并与 PCR2 成分 Suz12 相互作用[86]。但使用人类细胞（U2OS）的另一实验却证明没有被转录的常染色质基因的启动子区不存在 H2A.Z[87]。

在人类胚胎干细胞中，除了 H3K4me3 组蛋白修饰会同时出现在活化或失活基因启动子区外，H3K56ac 也有类似的分布[88]。有趣的是，与 H3K4me3 相比，NANOG、SOX2 和 OCT4 与 H3K56ac 结合更加紧密，说明 H3K56ac 参与多能性相关核心调控网络[88]。当然，这还需要进一步功能研究予以证实。

基因沉默相关的组蛋白表观遗传学标记

包括转座子和重复元件在内的基因沉默相关区域通常具有抑制性组蛋白修饰，包括 H3K9me3 和 H4K20me3[30,89]。H3K64me3（H3K64 位于 H3 组蛋白的球状结构域）是一个新鉴定出的抑制性修饰，多位于着丝点周围异染色质。有趣的是，与分化细胞相比，H3K64me3 在小鼠胚胎干细胞中水平较高[90]，这说明分化过程中重复序列的表观遗传学修饰与胚胎干细胞的可塑性相关[89,90]。

功能性研究进一步证实了这些组蛋白修饰在胚胎干细胞中的重要性。一方面，失去 H3K9 甲基转移酶 Suv39h 的小鼠胚胎干细胞中会有大量的重复序列相关转录本产生[89]。另一方面，H3K9 去甲基化酶 Jmjd1a 和 Jmjd2c 被认为是 Oct4 的靶基因[91]。在胚胎干胞中抑制 Jmjd1a/2c 的表达会导致细胞分化及 H3K9me2/3 的升高，说明这两个去甲基化酶及 H3K9 甲基化水平与多能性的维持相关[91]。

DNA 甲基化

胚胎干细胞中 DNA 甲基化的功能关联

胚胎干细胞的 DNA 甲基化修饰与体细胞不同。一方面，从头甲基转移酶——特别是 Dnmt3a2 和 Dnmt3b1（分别为 Dnmt3a 和 Dnmt3b 的异构体）在胚胎干细胞中高表达；另一方面，基因调控元件（gene-regulatory elements）中 DNA 甲基化缺失。此外，胚胎干细胞可以忍受 DNA 甲基化的完全缺失[92,93]。Dnmt1[93]、Dnmt3a/Dnmt3b[92,93] 或 Dnmt1/3a/3b 缺失小鼠胚胎干细胞可以存活，尽管完全失去了 DNA 甲基化，但依然维持它们自我更新的能力。虽然保留了多能性，但只有恢复 Dnmts 表达之后，这些胚胎干细胞才能继续分化[92,93]。与此相反，在体细胞中通过抑制 Dnmts 活性降低 CpG 岛甲基化水平可以导致生长抑制、细胞死亡、逆转录转座子的激活及基因组不稳定性[94~97]，说明 CpG 甲基化对于维持基本细胞功能具有重要作用。

胚胎干细胞同样也高表达 Dnmt3l，但目前我们对其功能还没有很多了解[98]。Dnmt3l 具有结合组蛋白 H3 未甲基化尾和激活 DNA 甲基转移酶的双重功能[56]，这表明在胚胎干细胞中具有高密度 CpG 岛启动子区域可以通过 H3K4 甲基化标记避免 DNA 甲基化。

为了研究在胚胎干细胞中 DNA 甲基化如何调控基因表达，有研究者对缺少三种 Dnmts 的小鼠胚胎干细胞（TKO 细胞[49]）的全基因表达进行了分析[99]。在低甲基化情况下高表达的基因多为组织特异性基因，包括转录因子和信号分子；此外，睾丸和卵母细胞特异性基因在 TKO 细胞中表达也较高。有趣的是，在去甲基化胚胎干细胞中上调基因仅有 5% 与 PcG 结合，只有 1.7% 的基因结合有 Nanog/Oct4/Sox2，这表明 DNA 甲基化与 Oct4、Sox2 和 Nanog 控制的转录网络在调控胚胎干细胞多能性方面具有不同的作用。

胚胎干细胞中 DNA 的甲基化模式

技术的进步使在单个碱基对水平对人类 DNA 甲基化组学的研究成为可能[48]。在人类胚胎干细胞中 76% 的 CpG 岛被甲基化。与先前在小鼠中的研究结果类似[100]，在野生型胚胎干细胞中 CpG 岛的甲基化表现为两面性，大多数基因组区域表现为高度非甲基化或高度甲基化。CpG 岛的甲基化状态与周围 CpG 岛的密度相关。位于高密度 CpG 岛区域（>7%超过 300 个碱基）的 CpG 易为非甲基化状态，CpG 岛低密度区域（<5%）的 CpG 易发生甲基化（图 3-4）[100]。值得注意的是，高密度 CpG 岛启动子（high-CpG-density promoters，HCP）一般与两类基因相关：管家基因和发育相关关键基因。

另外，低 CpG 岛密度启动子（low-CpG-density promoters，LCP）通常与组织特异性基因相关[101]。在胚胎干细胞中，位于 LCP 中的 CpG 岛大多会被甲基化，但有 H3K4me3 和 H3K4me2 修饰的 LCP 区域则没有发生甲基化。位于远距离调控区域的 CpG 岛 [如增强子、沉默子（silencers）、边界元件（boundary elements）] 表现出这种负相关，只在缺乏 H3K4me1/2 修饰的区域发生 DNA 甲基化。这些数据表明组蛋白甲基化模式比只参考 CpG 岛密度能更为准确地预测 DNA 甲基化。这与之前非全基因组研究结果相一致[99,102]，表明 DNA 甲基化与组蛋白修饰有内在关联[103]。

目前，人们对小鼠胚胎干细胞的基因组重复区域 DNA 甲基化模式也有研究。位于长末端重复序列（long terminal repeat，LTR）和长散在元件（long interspersed element，LINE）中的 CpG 岛一般高度甲基化，即使是在 CpG 岛密度很高的情况下也是如此。相反，与非重复序列的情况类似，位于短散在元件（short interspersed element，SINE）中的 CpG 岛的甲基化程度与 CpG 岛密度相关[100]。与去 H3K9 甲基化相比，DNA 去甲基化导致重复元件转录本水平略微升高（参见上文）。

胚胎干细胞中的非 CpG 岛甲基化

以上所讨论的人类全基因组甲基化分析[48]显示，在人类胚胎干细胞中存在非 CpG 岛甲基化现象，这与之前在小鼠胚胎干细胞[46,49]和胚胎[104]中的研究结果相符。非 CpG 岛甲基化多出现在基因区域，而在蛋白结合区和增强子区不存在。有趣的是，非 CpG 岛甲基化在诱导分化的胚胎干细胞中消失，而在诱导多能干细胞中恢复[48]。非 CpG 岛甲基化只在高表达 Dnmt 哺乳动物胚胎干细胞中出现。对于非 CpG 岛甲基化的功能还需进一步的研究。

除 CpG 岛甲基化之外，最近的研究显示，包括小鼠胚胎干细胞在内的某些细胞中存在 5-羟甲基胞嘧啶（5-hydroxymethylcytosine，5hmc）[105]。5hmc 由 10～11 易位 1（ten-eleven translocation 1，TET1）蛋白催化产生（图 3-3）。现在认为通过 DNA 损伤修复 5hmc 可重新转化为非甲基化胞嘧啶，因此认为 5hmc 可能是 DNA 去甲基化过程中的中间产物（图 3-3）。值得注意的是，目前 DNA 甲基化检测技术并不能区分 5mc 和 5hmc，所以研究 5hmc 的功能还是有必要的。

结论

目前，全基因组技术的发展使解析胚胎干细胞的表观遗传学模式成为可能（图 3-4），同时表观遗传学调控因子的功能研究也在启动。此外，胚胎干细胞的表观遗传学标记也有待深入研究。

缺少包括 Dnmts、组蛋白甲基化酶/去甲基化酶、染色质重构因子和 PcG 蛋白[106]等表观遗传学关键基因的胚胎干细胞仍能存活并保留其自我更新能力，这说明表观遗传学调控对维持胚胎干细胞性状并非必要。此外，二价染色质域并非只存在于胚胎干细胞，这些都提示表观遗传学机制可能不只对胚胎干细胞多能性的维持较为重要[107]。与此相应，目前认为表观遗传学能缓冲转录可变性[108]，这可能是对胚胎干细胞中超转录（hyper-transcription）所伴随的噪声的一种控制[61]。但是，具有表观遗传学因子缺陷的胚胎干细胞通常表现出有缺陷的分化过程，反映出这些因子可能参与分化及之后谱系状态决定[106]。

这些事实说明表观遗传学的功能在发育过程中可能是动态的。分化过程中表观遗传修饰模式和相关因子的表达改变可以反映表观遗传学的"意向",从多能干细胞中的缓冲地带转变到分化细胞中的"看门人"[3]。

为了更好地了解表观遗传学在多能性和谱系决定中的作用,把本章中介绍的胚胎干细胞表观遗传学研究与哺乳动物早期发育的调节机制进行综合考虑是很有必要的。新的表观遗传学研究方法已经可以对少量细胞进行研究[109,110],这为该类研究提供了保障。研究发育过程中不同细胞状态下的表观遗传学调控无疑将帮助我们进行更加深入的干细胞和再生医学研究。

致谢

感谢 Sigrid Hard(马萨诸塞州,剑桥,Broad 研究所)为我们设计了图 3-2。感谢 Meissner 实验室成员,特别是 Natalie Jager 和 Casey Gifford 对文章提出的意见。Alexander Meissner 来自于 NIH、马萨诸塞州生命科学中心和 Pew Charitable 基金会。

(苏位君　李宗金　译)

参 考 文 献

1. Waddington C. The epigenotype. Endeavour 1942; 1:18-20.
2. Allis CD, Jenuwein T, Reinberg D. Epigenetics, x, 502 p. (Cold Spring Harbor Laboratory Press, Cold Spring Harbor, N.Y., 2007).
3. Hemberger M, Dean W, Reik W. Epigenetic dynamics of stem cells and cell lineage commitment: digging Waddington's canal. Nat Rev Mol Cell Biol 2009; 10:526-37.
4. Turner BM. Cellular memory and the histone code. Cell 2002; 111:285-91.
5. Turner BM. Defining an epigenetic code. Nat Cell Biol 2007; 9:2-6.
6. Spivakov M, Fisher AG. Epigenetic signatures of stem-cell identity. Nat Rev Genet 2007; 8:263-71.
7. Evans MJ, Kaufman MH. Establishment in culture of pluripotential cells from mouse embryos. Nature 1981; 292:154-6.
8. Martin GR. Isolation of a pluripotent cell line from early mouse embryos cultured in medium conditioned by teratocarcinoma stem cells. Proc Natl Acad Sci USA 1981; 78:7634-8.
9. Stojkovic M, Lako M, Stojkovic P et al. Derivation of human embryonic stem cells from day-8 blastocysts recovered after three-step in vitro culture. Stem Cells 2004; 22:790-7.
10. Thomson JA, Itskovitz-Eldor J, Shapiro SS et al. Embryonic stem cell lines derived from human blastocysts. Science 1998; 282:1145-7.
11. Yu J, Thomson JA. Pluripotent stem cell lines. Genes Dev 2008; 22:1987-97.
12. Murry CE, Keller G. Differentiation of embryonic stem cells to clinically relevant populations: lessons from embryonic development. Cell 2008; 132:661-80.
13. Cowan CA, Klimanskaya I, McMahon J et al. Derivation of embryonic stem-cell lines from human blastocysts. N Engl J Med 2004; 350:1353-6.
14. Sidhu KS, Ryan JP, Tuch BE. Derivation of a new human embryonic stem cell line, endeavour-1 and its clonal propagation. Stem Cells Dev 2008; 17:41-51.
15. Bao S et al. Epigenetic reversion of post-implantation epiblast to pluripotent embryonic stem cells. Nature 2009; 461:1292-5.
16. Matsui Y, Zsebo K, Hogan BL. Derivation of pluripotential embryonic stem cells from murine primordial germ cells in culture. Cell 1992; 70:841-7.
17. Takahashi K, Yamanaka S. Induction of pluripotent stem cells from mouse embryonic and adult fibroblast cultures by defined factors. Cell 2006; 126:663-76.
18. Jaenisch R, Young R. Stem cells, the molecular circuitry of pluripotency and nuclear reprogramming. Cell 2008; 132:567-82.
19. Boyer LA, Lee TI, Cole MF et al. Core transcriptional regulatory circuitry in human embryonic stem cells. Cell 2005; 122:947-56.
20. Wang J, Rao S, Chu J et al. A protein interaction network for pluripotency of embryonic stem cells. Nature 2006; 444:364-8.
21. Luger K, Mader AW, Richmond RK et al. Crystal structure of the nucleosome core particle at 2.8 A resolution. Nature 1997; 389:251-60.

22. Fraser P, Bickmore W. Nuclear organization of the genome and the potential for gene regulation. Nature 2007; 447:413-7.
23. Campos EI, Reinberg D. Histones: annotating chromatin. Annu Rev Genet 2009; 43:559-99.
24. Bernstein BE, Meissner A, Lander ES. The mammalian epigenome. Cell 2007; 128:669-81.
25. Morrison AJ, Shen X. Chromatin remodelling beyond transcription: the INO80 and SWR1 complexes. Nat Rev Mol Cell Biol 2009; 10:373-84.
26. Li B, Carey M, Workman JL. The role of chromatin during transcription. Cell 2007; 128:707-19.
27. Gutierrez J et al. Chromatin remodeling by SWI/SNF results in nucleosome mobilization to preferential positions in the rat osteocalcin gene promoter. J Biol Chem 2007; 282:9445-57.
28. Kouzarides T. Chromatin modifications and their function. Cell 2007; 128:693-705.
29. Bernstein BE, Mikkelsen TS, Xie X et al. A bivalent chromatin structure marks key developmental genes in embryonic stem cells. Cell 2006; 125:315-26.
30. Mikkelsen TS, Ku M, Jaffe DB et al. Genome-wide maps of chromatin state in pluripotent and lineage-committed cells. Nature 2007; 448:553-60.
31. Pietersen AM, van Lohuizen M. Stem cell regulation by polycomb repressors: postponing commitment. Curr Opin Cell Biol 2008; 20:201-7.
32. Henikoff S. Nucleosome destabilization in the epigenetic regulation of gene expression. Nat Rev Genet 2008; 9:15-26.
33. Sterner DE, Berger SL. Acetylation of histones and transcription-related factors. Microbiol Mol Biol Rev 2000; 64:435-59.
34. Sims RJ, 3rd, Nishioka K, Reinberg D. Histone lysine methylation: a signature for chromatin function. Trends Genet 2003; 19:629-39.
35. Nowak SJ, Corces VG. Phosphorylation of histone H3: a balancing act between chromosome condensation and transcriptional activation. Trends Genet 2004; 20:214-20.
36. Shilatifard A. Chromatin modifications by methylation and ubiquitination: implications in the regulation of gene expression. Annu Rev Biochem 2006; 75:243-69.
37. Nathan D, Ingvarsdottir K, Sterner DE et al. Histone sumoylation is a negative regulator in Saccharomyces cerevisiae and shows dynamic interplay with positive-acting histone modifications. Genes Dev 2006; 20:966-76.
38. Hassa PO, Haenni SS, Elser M et al. Nuclear ADP-ribosylation reactions in mammalian cells: where are we today and where are we going? Microbiol Mol Biol Rev 2006; 70:789-829.
39. Cuthbert GL, Daujat S, Snowden AW et al. Histone deimination antagonizes arginine methylation. Cell 2004; 118:545-53.
40. Nelson CJ, Santos-Rosa H, Kouzarides T. Proline isomerization of histone H3 regulates lysine methylation and gene expression. Cell 2006; 126:905-16.
41. Schuettengruber B, Chourrout D, Vervoort M et al. Genome regulation by polycomb and trithorax proteins. Cell 2007; 128:735-45.
42. Elderkin S, Maertens GN, Endoh M et al. A phosphorylated form of Mel-18 targets the Ring1B histone H2A ubiquitin ligase to chromatin. Mol Cell 2007; 28:107-20.
43. Ku M, Koche RP, Rheinbay E et al. Genomewide analysis of PRC1 and PRC2 occupancy identifies two classes of bivalent domains. PLoS Genet 2008; 4:e1000242.
44. Bird A. DNA methylation patterns and epigenetic memory. Genes Dev 2002; 16:6-21.
45. Grandjean V, Yaman R, Cuzin F et al. Inheritance of an epigenetic mark: the CpG DNA methyltransferase 1 is required for de novo establishment of a complex pattern of nonCpG methylation. PLoS One 2007; 2:e1136.
46. Ramsahoye BH, Biniszkiewicz D, Lyko F, Clark V et al. Non-CpG methylation is prevalent in embryonic stem cells and may be mediated by DNA methyltransferase 3a. Proc Natl Acad Sci USA 2000; 97:5237-42.
47. Finnegan EJ, Peacock WJ, Dennis ES. DNA methylation, a key regulator of plant development and other processes. Curr Opin Genet Dev 2000; 10:217-23.
48. Lister R, Pelizzola M, Dowen RH et al. Human DNA methylomes at base resolution show widespread epigenomic differences. Nature 2009; 462:315-22.
49. Meissner A, Gnirke A, Bell GW et al. Reduced representation bisulfite sequencing for comparative high-resolution DNA methylation analysis. Nucleic Acids Res 2005; 33:5868-77.
50. Morgan HD, Santos F, Green K et al. Epigenetic reprogramming in mammals. Hum Mol Genet 2005; 14 Spec No 1:R47-58.
51. Klose RJ, Bird AP. Genomic DNA methylation: the mark and its mediators. Trends Biochem Sci 2006; 31:89-97.
52. Bestor TH. The DNA methyltransferases of mammals. Hum Mol Genet 2000; 9:2395-402.
53. Goll MG, Bestor TH. Eukaryotic cytosine methyltransferases. Annu Rev Biochem 2005; 74:481-514.
54. Okano M, Bell DW, Haber DA et al. DNA methyltransferases Dnmt3a and Dnmt3b are essential for de novo methylation and mammalian development. Cell 1999; 99:247-57.
55. Aapola U, Kawasaki K, Scott HS et al. Isolation and initial characterization of a novel zinc finger gene, DNMT3L, on 21q22.3, related to the cytosine-5-methyltransferase 3 gene family. Genomics 2000; 65:293-8.

56. Ooi SK, Qiu C, Bernstein E et al. DNMT3L connects unmethylated lysine 4 of histone H3 to de novo methylation of DNA. Nature 2007; 448:714-7.
57. Meshorer E, Misteli T. Chromatin in pluripotent embryonic stem cells and differentiation. Nat Rev Mol Cell Biol 2006; 7:540-6.
58. Francastel C, Schubeler D, Martin DI et al. Nuclear compartmentalization and gene activity. Nat Rev Mol Cell Biol 2000; 1:137-43.
59. Arney KL, Fisher AG. Epigenetic aspects of differentiation. J Cell Sci 2004; 117:4355-63.
60. Meshorer E, Yellajoshula D, George E et al. Hyperdynamic plasticity of chromatin proteins in pluripotent embryonic stem cells. Dev Cell 2006; 10:105-16.
61. Efroni S, Duttagupta R, Cheng J et al. Global transcription in pluripotent embryonic stem cells. Cell Stem Cell 2008; 2:437-47.
62. Liang J, Wan M, Zhang Y, et al. Nanog and Oct4 associate with unique transcriptional repression complexes in embryonic stem cells. Nat Cell Biol 2008; 10:731-9.
63. Kaji K, Caballero IM, MacLeod R et al. The NuRD component Mbd3 is required for pluripotency of embryonic stem cells. Nat Cell Biol 2006; 8:285-92.
64. Gaspar-Maia A, Alajem A, Polesso F et al. Chd1 regulates open chromatin and pluripotency of embryonic stem cells. Nature 2009; 460:863-8.
65. Guenther MG, Levine SS, Boyer LA et al. A chromatin landmark and transcription initiation at most promoters in human cells. Cell 2007; 130:77-88.
66. Pasini D, Hansen KH, Christensen J et al. Coordinated regulation of transcriptional repression by the RBP2 H3K4 demethylase and Polycomb-Repressive Complex 2. Genes Dev 2008; 22:1345-55.
67. Chen X, Xu H, Yuan P et al. Integration of external signaling pathways with the core transcriptional network in embryonic stem cells. Cell 2008; 133:1106-17.
68. Fazzio TG, Huff JT, Panning B. Chromatin regulation Tip(60)s the balance in embryonic stem cell self-renewal. Cell Cycle 2008; 7:3302-6.
69. Torres-Padilla ME, Parfitt DE, Kouzarides T et al. Histone arginine methylation regulates pluripotency in the early mouse embryo. Nature 2007; 445:214-8.
70. Wu Q, Bruce AW, Jedrusik A et al. CARM1 is required in embryonic stem cells to maintain pluripotency and resist differentiation. Stem Cells 2009; 27:2637-45.
71. Heintzman ND, Hon GC, Hawkins RD et al. Histone modifications at human enhancers reflect global cell-type-specific gene expression. Nature 2009; 459:108-12.
72. Heintzman ND, Ren B. Finding distal regulatory elements in the human genome. Curr Opin Genet Dev 2009; 19:541-9.
73. Boiani M, Scholer HR. Regulatory networks in embryo-derived pluripotent stem cells. Nat Rev Mol Cell Biol 2005; 6:872-84.
74. Boyer LA, Plath K, Zeitlinger J et al. Polycomb complexes repress developmental regulators in murine embryonic stem cells. Nature 2006; 441:349-53.
75. Lee TI, Jenner RG, Boyer LA et al. Control of developmental regulators by Polycomb in human embryonic stem cells. Cell 2006; 125:301-13.
76. Shen X, Liu Y, Hsu YJ et al. EZH1 mediates methylation on histone H3 lysine 27 and complements EZH2 in maintaining stem cell identity and executing pluripotency. Mol Cell 2008; 32:491-502.
77. Chamberlain SJ, Yee D, Magnuson T. Polycomb repressive complex 2 is dispensable for maintenance of embryonic stem cell pluripotency. Stem Cells 2008; 26:1496-505.
78. Pasini D, Bracken AP, Hansen JB et al. The polycomb group protein Suz12 is required for embryonic stem cell differentiation. Mol Cell Biol 2007; 27:3769-79.
79. Peng JC, Valouev A, Swigut T et al. Jarid2/Jumonji coordinates control of PRC2 enzymatic activity and target gene occupancy in pluripotent cells. Cell 2009; 139:1290-302.
80. Shen X, Kim W, Fujiwara Y et al. Jumonji modulates polycomb activity and self-renewal versus differentiation of stem cells. Cell 2009; 139:1303-14.
81. Li G, Margueron R, Ku M et al. Jarid2 and PRC2, partners in regulating gene expression. Genes Dev. 2010 Feb 15;24(4):368-80.
82. Pasini D, Cloos PA, Walfridsson J et al. JARID2 regulates binding of the Polycomb repressive complex 2 to target genes in ES cells. Nature. 2010 Jan 14. [Epub ahead of print].
83. Mikkelsen TS, Hanna J, Zhang X et al. Dissecting direct reprogramming through integrative genomic analysis. Nature 2008; 454:49-55.
84. Bilodeau S, Kagey MH, Frampton GM et al. SetDB1 contributes to repression of genes encoding developmental regulators and maintenance of ES cell state. Genes Dev 2009; 23:2484-9.
85. Squazzo SL, O'Geen H, Komashko VM et al. Suz12 binds to silenced regions of the genome in a cell-type-specific manner. Genome Res 2006; 16:890-900.
86. Creyghton MP, Markoulaki S, Levine SS et al. H2AZ is enriched at polycomb complex target genes in ES cells and is necessary for lineage commitment. Cell 2008; 135:649-61.
87. Hardy S, Jacques PE, Gévry N et al. The euchromatic and heterochromatic landscapes are shaped by antagonizing effects of transcription on H2A.Z deposition. PLoS Genet 2009; 5:e1000687. .

88. Xie W, Song C, Young NL et al. Histone h3 lysine 56 acetylation is linked to the core transcriptional network in human embryonic stem cells. Mol Cell 2009; 33:417-27.
89. Martens JH, O'Sullivan RJ, Braunschweig U et al. The profile of repeat-associated histone lysine methylation states in the mouse epigenome. EMBO J 2005; 24:800-12.
90. Daujat S, Weiss T, Mohn F et al. H3K64 trimethylation marks heterochromatin and is dynamically remodeled during developmental reprogramming. Nat Struct Mol Biol 2009; 16:777-81.
91. Loh YH, Zhang W, Chen X et al. Jmjd1a and Jmjd2c histone H3 Lys 9 demethylases regulate self-renewal in embryonic stem cells. Genes Dev 2007; 21:2545-57.
92. Jackson M, Krassowska A, Gilbert N et al. Severe global DNA hypomethylation blocks differentiation and induces histone hyperacetylation in embryonic stem cells. Mol Cell Biol 2004; 24:8862-71.
93. Tsumura A, Hayakawa T, Kumaki Y et al. Maintenance of self-renewal ability of mouse embryonic stem cells in the absence of DNA methyltransferases Dnmt1, Dnmt3a and Dnmt3b. Genes Cells 2006; 11:805-14.
94. Jackson-Grusby L, Beard C, Possemato R et al. Loss of genomic methylation causes p53-dependent apoptosis and epigenetic deregulation. Nat Genet 2001; 27:31-9.
95. Gaudet F, Hodgson JG, Eden A et al. Induction of tumors in mice by genomic hypomethylation. Science 2003; 300:489-92.
96. Bourc'his D, Bestor TH. Meiotic catastrophe and retrotransposon reactivation in male germ cells lacking Dnmt3L. Nature 2004; 431:96-9.
97. Dodge JE, Okano M, Dick F et al. Inactivation of Dnmt3b in mouse embryonic fibroblasts results in DNA hypomethylation, chromosomal instability and spontaneous immortalization. J Biol Chem 2005; 280:17986-91. .
98. Nimura K, Ishida C, Koriyama H et al. Dnmt3a2 targets endogenous Dnmt3L to ES cell chromatin and induces regional DNA methylation. Genes Cells 2006; 11:1225-37.
99. Fouse SD, Shen Y, Pellegrini M et al. Promoter CpG methylation contributes to ES cell gene regulation in parallel with Oct4/Nanog, PcG complex and histone H3 K4/K27 trimethylation. Cell Stem Cell 2008; 2:160-9.
100. Meissner A, Mikkelsen TS, Gu H et al. Genome-scale DNA methylation maps of pluripotent and differentiated cells. Nature 2008; 454:766-70.
101. Weber M, Hellmann I, Stadler MB et al. Distribution, silencing potential and evolutionary impact of promoter DNA methylation in the human genome. Nat Genet 2007; 39:457-66.
102. Mohn F, Weber M, Rebhan M et al. Lineage-specific polycomb targets and de novo DNA methylation define restriction and potential of neuronal progenitors. Mol Cell 2008; 30:755-66.
103. Cedar H, Bergman Y. Linking DNA methylation and histone modification: patterns and paradigms. Nat Rev Genet 2009; 10:295-304.
104. Haines TR, Rodenhiser DI, Ainsworth PJ. Allele-specific nonCpG methylation of the Nf1 gene during early mouse development. Dev Biol 2001; 240:585-98.
105. Tahiliani M, Koh KP, Shen Y et al. Conversion of 5-methylcytosine to 5-hydroxymethylcytosine in mammalian DNA by MLL partner TET1. Science 2009; 324:930-5.
106. Niwa H. How is pluripotency determined and maintained? Development 2007; 134:635-46.
107. Silva J, Smith A. Capturing pluripotency. Cell 2008; 132:532-6.
108. Chi AS, Bernstein BE. Developmental biology. Pluripotent chromatin state. Science 2009; 323:220-1.
109. Gu H, Bock C, Mikkelsen TS et al. Genome-scale DNA methylation mapping of clinical samples at single-nucleotide resolution. Nat Methods. 2010 Feb;7(2):133-6.
110. Goren A, Ozsolak F, Shoresh N et al. Chromatin profiling by directly sequencing small quantities of immunoprecipitated DNA. Nat Methods 7:47-9.

第4章 哺乳动物早期发育阶段常染色体复制域的莱昂化

Ichiro Hiratani, David M. Gilbert*

摘要：真核生物基因组在细胞周期 S 期的复制遵循确定时间顺序的这一现象发现已有五十年。尽管目前对复制时序程序的机制尚不完全清楚，但这一过程与染色体动态和静态特性的相关性仍使它成为一个探索令人费解的染色体高级组织的研究平台。事实上，对于 DNA 复制的研究为我们提供了一个简单而直接的途径以了解基因组在染色体线性和细胞核内三维空间环境的物理结构。在本章中，我们总结了 50 年来以哺乳动物细胞为主进行的有关复制时序程序及其调控发育过程的研究历史。我们首先介绍复制时序程序过程的发现，讨论了发育过程中对常染色体和女性 X 染色体失活过程的复制时序的调控，最后介绍了近年来通过全基因组分析对复制时序程序调控的新发现，尤其是在小鼠胚胎干细胞分化过程中这一程序发生的变化。我们尝试揭示这些新发现所传达的信息及其与胚胎发育的潜在联系。在此过程中，我们重新使用了"常染色体莱昂化作用"这个旧有概念来描述"兼性异染色质化"。而在常染色体上发生的单个复制域的不可逆失活，与小鼠外胚层植入后阶段发生的 X 染色体失活相似。

引言

大自然包罗万象，而最主要的两个问题便是：现象背后的潜在机制（如何发生？）和它的生物学意义（为什么发生？）。所有真核细胞的 DNA 都是按特定的时间顺序进行复制的，但是有关"复制时序程序"的机制和它的生物学意义至今仍是未解之谜。理解"复制时序程序"的机制也许可以使我们通过人为操纵复制时序来探求其背后的生物学意义。但即便借助了芽殖酵母和裂殖酵母中有力的分子遗传学研究工具，我们仍未能深刻理解"复制时序程序"的机制[1]。目前许多研究结果认为，对于复制时序的调控发生在大的染色体区域或亚核空间水平。虽然传统分子与生物化学研究手段对于这一水平的研究力所不及，但是对复制时序的研究打开了我们认识染色体更高级结构和功能组织的一扇大门。与此同时，理解"复制时序研究"的生物意义也可为我们进一步了解其背后的机理提供线索。一些研究通过变异率和/或抑制染色体重组提示了复制时序可能的进化作用[2,3]，而其他研究则提示复制时序具有有利于利用有限的代谢前体协同大基因组复制的最基本的生物作用[4]。长期以来，对于早期复制时序和转录能力之间的联系的提法也在近期的全基因组研究的工作中得到证据

---

* Corresponding Author: David M. Gilbert—Department of Biological Science, Florida State University, Tallahassee, Florida, USA. Email: gilbert@bio.fsu.edu

的支持[1,5]。但究竟是转录驱动早期复制还是早期复制驱动转录，目前它们两者之间的关系尚不明确。

复制时序如果与转录能力有关系，这一过程就应该受发育调节。在这一章中，我们总结了哺乳动物复制时序发育调控研究的有关证据。从最早期作为一个与女性X染色体失活过程有关的染色体特性[6]，到最近研究发现的常染色体上由上千万碱基形成的染色体"复制域"及在这种"复制域"水平发生的常染色体复制时序变化[7,8]，在此过程中，我们重新使用了"常染色体莱昂化作用"这个旧有概念来描述"兼性异染色质化"。而在常染色体上发生的单个复制域的不可逆失活让人们联想起发生在与小鼠外胚层植入后阶段同等时期的X染色体失活[8,9]。此外，比较生物学研究揭示：与复制起始位置或染色体等基因组区的GC组成比例相比，复制时序编程及其在发育过程中的变化在进化中具有保守性。这些发现提示复制程序的形成并不仅是简单地将基因组任意分割成几个临时存在的片段，而是与染色体结构及功能紧密相关的生物学过程。

复制时序编程：对基因组结构的初步测定

早期实验

染色体组织的基本概念源自早期有关DNA复制的研究。1953年，在人们刚认识DNA结构时，Dr. J. Herbert Taylor利用合成的氚标记胸腺嘧啶核苷，使人类第一次观察到活细胞中DNA的合成。与此同时，Dr. Taylor的胸腺嘧啶脱氧核苷系列标记实验不仅第一次证明了DNA半保留复制和DNA反向平行的性质（此发现至少比Drs. Messelson和Stahl早一年[12]）[13]，而且显示了在植物[14]和动物[15]细胞中染色体的DNA复制都遵循一种特定的时序。尤其是通过用脉冲标记中国仓鼠细胞（其染色体极易通过大小来区分）并在脉冲后不同时间检查分裂中期染色体，Dr. Taylor发现不同的染色体片段会在S期特定时间阶段发生复制（图4-1）。通过这项研究[15]，Dr. Taylor断定"部分染色体的复制时序受遗传的控制，这一调控可能有一定的生物功能意义。"然而即便是在50年后的今天，这个功能的生物学意义仍完全是一个谜。

19世纪70年代，科学家们发现这些染色体的等位标记片段在大小和出现频率方面，与用吉姆萨染色获得的染色体条带相似。科学家发现在DNA中掺入BrdU（以替换氚标记的胸腺嘧啶核苷）会抑制Hoechst染料的荧光[16]，于是建立了一种新的染色体显带方法（"复制显带"），这一方法可以避免使用放射性及长时间的放射显影曝光（图4-1）。虽然这一判断标准并不绝对，但通常来说，转录活跃、GC含量丰富，R带通常位于复制早期；而转录不活跃、AT含量丰富，G带则位于复制晚期[17,18,19]。这些结果支持了最早由Lima-de-Faria在研究蝗虫性染色质时提出异染色质会发生延迟复制的假说[20]。研究发现，染色体片段复制时序与其转录活动息息相关，提示协同复制的染色体片段可能不仅为染色体的结构单位，还可能是功能单位。这是一个非常吸引人的理论。但如果复制仅与转录相关，则在不同的细胞类型中应该会发现不同的染色体条带模式。不幸的是，科学家对不同细胞中复制条带的分析并没有发现这种差别[21]。当然，受

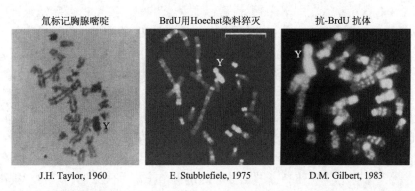

图 4-1 DNA 合成进行短时脉冲标记显示了大小在数百万碱基的"复制域"。对中国仓鼠细胞在 S 期采用不同试剂进行短时标记后，加入未标记的培养基使细胞进入有丝分裂期观察。左图，S 期晚期用氚标记胸腺嘧啶核苷标记 10min，用放射自显影计数分析细胞分裂中期分布；右图，采用 BrdU 进行同样标记，用抗-BrdU 抗体进行免疫荧光检验观察细胞分裂中期分布；中图，除在细胞 S 期晚期中有 1h 用胸腺嘧啶脱氧核苷标记外，其余时间用 BrdU 持续标记，BrdU 用 Hoechst 染料淬灭后进行观察。这些细胞遗传学方法发现：大小在数百万碱基的巨大基因组片段在很短的时间里被标记，并产生了染色体特异的带型，这些带型在 S 期会发生变化，其中代表常染色体的 R 带复制较早，而代表异染色质的 G 带则延迟复制。值得注意的是，异染色质化的 Y 染色体在 S 期晚期 10min 内几乎完全被标记。由于 DNA 复制叉点移动的速率一定，只有当整条染色体上的复制起点几乎同步启动复制才能实现上述结果。本图数据分别采自 J. H. Taylor[15]（左图 © Taylor，1960，原文发表在 The Journal of Biophysical and Biochemical Cytology. 7：455-463）、E. Stubblefield[18]（中图）和 D. M. Gilbert[102]（右图），分别获得了 Journal of Biophysical and Biochemical Cytology[15]、Chromosoma[18] 的版权许可

限于分析方法的精度，上述研究的结果并不能排除局部改变的存在[10]。此外，对于蛙胚及哺乳动物 X 染色体失活（下面我们会讨论这一点）的研究表明，复制时序会在发育阶段发生变化，而这种变化是与转录活动息息相关的。

X 染色体失活的启示

Taylor 在其 1960 年的研究中发现，雌性细胞中两条 X 染色体的复制是不同步的。巧合的是，Mary Lyon 在 1961 年提出了她著名的假说：巴氏小体[23]作为 X 染色体失活的细胞学表现，其出现与早期发育期间 X 染色体失活是同步发生的。这种结构和功能的替换会在雌性细胞的两条 X 染色体间发生同源随机选择，并会在其后代体细胞中稳定维持[24]。Taylor 随后证明了巴氏小体在雌性人类细胞中会延迟复制，并证明了在拥有几条 X 染色体的细胞中所有的巴氏小体都会延迟复制[6]。这些发现引入了一个完全崭新而神秘的概念：两个基因完全相同的 X 染色体可有不同的生物学表现，这意味着同源染色质在同一个细胞既可以是异染色质状态也可以是常染色质状态，从而引出了一个必然的结论——复制时序不仅由序列本身决定，更受到表观遗传的影响。

在这之后的 20 年中，在早期胚胎阶段延迟复制的 Xi 被当作 X 染色体失活的一个可靠的细胞学标记，因为它是 X 染色体失活最稳定的特征之一[25]。事实上，虽然在有胎

盘哺乳动物（真哺乳亚纲）中 Xi 的复制延迟现象，在有袋类动物和部分卵生的鸭嘴兽（单孔类动物）中也可以观察到[26,27]，然而 *Xist* 基因表达、"抑制性"组蛋白修饰的富集和启动子 DNA 甲基化等真核哺乳动物亚纲 X 染色体失活的特征在单孔目或有袋类动物中要么还未发现，要么至今尚未被报道[26,28]。在小鼠胚体发生随机的 X 染色体失活时，植入后胚层阶段的 Xi 会发生延迟复制[29]，这一过程先于 Xi 染色体上基因启动子的 DNA 甲基化[30]。随后建立的胚胎干细胞（ESC）的体外分化体系，使我们可以对发育过程中各种事件的时序做更为准确的划分。对 ESC 分化的研究提示：X 染色体随机失活过程中，*Xist* 基因产物包被失活的 X 染色体是最早发生的事件，之后 RNA 聚合酶 II 从 Xi 上被剔除，接下来标志染色体转录"活性"的组蛋白修饰丢失并同时获得"抑制性"组蛋白标记[31]。Xi 延迟复制与组蛋白修饰改变同时发生或稍后于组蛋白修饰的改变（这些结果基于细胞分裂中期分布分析[32,33]），而启动子重新甲基化则发生在较晚时期[32]。虽然推测早期发生的事件导致了 Xi 的复制延迟是一个非常吸引人的论点，但这个观点与在缺失 Xist 和"抑制性"组蛋白修饰丢失的有袋类动物[28]中同样发生 Xi 染色体复制延迟的事实是冲突的[26]。到目前为止，还没有关于染色质修饰在调节 Xi 复制时序中的因果作用的确实报道。有趣的是，利用可诱导 *Xist* 转基因的小鼠 ESC 分化模型，Wutz 等证明了 X 染色体失活的时间点是独立于 X 染色体转录下调的，但在时序上却几乎与整个 X 染色体向延迟复制的 Xi 转变紧密相关[32,34]。

相比胚体细胞中复制延迟的 Xi，胚外谱系的细胞就不同了。在胚外组织中，父源的 X 染色体（Xp）以印记的方式处于失活状态。在滋养外胚层和原始内胚层中，Xp 的复制先呈现一个短期的提前复制过程，其复制比任何常染色体都要早，发生在 S 期的极早期[35]。X 染色体从同步复制时序到非同步复制时序的转变就发生在胚外组织分化之后，滋养外胚层细胞发生在 E3.5 天，原始内胚层细胞发生在 E4.0～4.5 天[35]。之后在 E6.0～6.4 天，这个两个胚层的 Xp 染色体（在这个时期，原始内胚层变为了内脏内胚层，滋养外胚层变成了额外的胚胎外胚层）转变为延迟复制，并在此之后稳定维持[29]。这种几乎全染色体的复制时序的突然转变可能仅发生在一个细胞周期内[36]。这些观察表明胚外谱系细胞内 Xp 不寻常的提前复制是短暂的，然而胚胎组织及胚外组织细胞的失活 Xi 的复制延迟转变却是保守的，而且在发育过程中稳定维持。有趣的是，在 E3.5 天出现的滋养外胚层及 E4.0 天出现的原始内胚层代表了第一个从胚体中分离出胚外组织细胞的细胞谱系分化，因而早期胚胎发生过程中两类主要的细胞命运的分化伴随了 Xi 染色体的复制时序的转变。然而在 E6.0～6.4 天时胚外谱系细胞中发生的复制延迟转变所代表的生物学意义目前还不得而知[29]。胚外组织和母体组织之间的相互作用有可能参与了这些过程。

常染色体的复制时序谱

复制时序改变是 X 染色体所特有的现象还是常染色体在分化期间也有类似的复制时序改变但却未能由细胞学检测发现？是否常染色体上的单个复制子或复制子群中也具有类似的"莱昂化"程序作用（即兼性染色体异质化），只是由于它们的片段太小而无法通过显微镜观察？在 20 世纪 70 年代，Carl Schildkraut 和 Walt Fangman 最先使用分子手段分别在哺乳动物细胞和出芽酵母[38]中研究复制时序顺序。他们的研究共同证实了复制时序精确调节程序在所有真核细胞中的保守性。20世纪80年代，科学家在不同细胞系中测定了几十个特定基因的复制时序[39~42]，并清楚地区分常染色体复制时序在不同细胞间的差异。当时总结的经验法则认为：如果一个基因转录活跃则将较早复制，而较晚复制的基因常常处于失活状态。非洲爪蟾（Xenopus）中处于转录活性状态与失活状态的5S rDNA基因簇竞争相同的转录因子，对它们的复制时序研究揭示：活性基因簇明显复制得更早。这启发我们创建如下模型：提前复制为基因提供了一个在复制叉区域优先接受数量有限的转录活性因子作用的优势（"先到先得原则"）[43]。这个模型目前尚未被直接证明或推翻。

染色体的复制时序在发育过程中的变化目前尚不完全清楚。由于研究分化中的细胞尚有技术困难，所以直到21世纪初，有关复制时序的研究基本都被局限于在已建立的转化细胞系中研究有限的基因位点的变化。因此，在细胞系中观察到的复制时序的变化有可能是由于细胞在长期培养过程中发生的遗传或表观遗传改变所导致的。直到2004年，科学家利用胚胎干细胞（ESC）定向分化培养体系获得足够匀质的细胞进行分子水平分析，并首次直接揭示了由于细胞分化诱导的常染色体上基因位点的复制时间改变[44,45]。然而仅对在神经分化过程中一个含约100个基因片段的复制时序分析并证明它的可调节性，这一观察很难推知在发育过程中复制时序改变究竟是常态还是特例。与细胞遗传学研究的结果一致，多个研究得出许多基因在所有的细胞类型中的复制时序是一致的结论[46,47]。因此这些报道也为分化诱导的复制时序变化提供了明确的证据。但限于样品数量，只有通过在基因组水平的分析来从统计学决定变化的范围。

第一个全基因组水平复制时序分析是在出芽酵母中完成的[48]。出乎意料的是，科学家未发现复制时序和转录之间有什么联系，此发现随即在裂殖酵母中也获得了验证[49]。此后不久，利用微阵列技术针对果蝇和哺乳动物细胞的一系列的研究发现，在比酵母更高等的真核细胞中，早期复制和转录活动之间有着很紧密的联系[7,50~58]，这提示两个过程的相关性可能仅限于多细胞生物[59]。由于分化过程中也发生基因表达的改变，因而存在如下一种可能性，即相当程度的复制时序变化是在发育过程中发生的。然而第一篇报道发现，在人类纤维母细胞及淋巴母细胞的22号染色体上，复制时序的差异只有1%[53]。此外，几篇类似的研究发现，复制时序与哺乳动物染色体的静态序列特征包括GC含量和基因密度等息息相关[7,52~56]。事实上，就在2008年，许多研究者认为复制时序的变化是非常少见的，因此它们在发育过程中的重要性遭到质疑[60~62]。然而，现有研究的低分辨率和有限的基因组覆盖率，以及利用细胞系进行比较使得数据缺乏完全的说服力，因而对这一基本问题目前仍然没有解答。

随着针对全基因组的高密度寡核苷酸探针芯片的出现，结合ESC分化体系的建立，为研究细胞命运转变过程中的复制时序变化提供了前所未有的便利。在2008年，我们完成了对小鼠胚胎干细胞到神经前体细胞（NPC）分化过程中的复制时序分析[7]，发现几种多态胚胎干细胞细胞系具有几乎完全一致的复制时序谱，并共同表现了协调复制的、由数百万碱基形成的染色体复制域的模式。从ESC到NPC的分化过程中，约20%的基因组发生复制时序变化[7]。结合后续研究，[8]我们建构了代表10种不同细胞类型的22种细胞系的复制谱，这10种细胞类型则模拟了早期小鼠发育阶段三胚层分化（图4-2）。

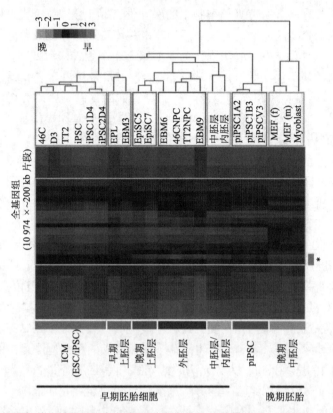

图 4-2 基于复制时序谱分析获得的各小鼠胚胎发生细胞培养模型之间的关系。利用基因芯片分析 22 个小鼠细胞系的复制时序谱分类分层图[8]。系统发生树揭示晚期外胚层来源干细胞（EpiSC）同来自于早期外胚层（EPL 和 EBM3）及 ICM［ESC（46C、D3、TT2）和完全重新编码的 iPSC］等不同细胞类型之间的表观遗传分离。晚期外胚层来源干细胞与早期胚胎中已定性的各胚层细胞［如外胚层（EBM6）、神经外胚层（46CNPC、TT2NPC 和 EBM9）、新生中胚层和内胚层］更为接近。3 个发生部分重编程的诱导多能干细胞系（piPSC）不仅有别于晚期胚胎的细胞类型（如 MEF 和成肌母细胞），亦有别于 ICM、外胚层及早期各胚层细胞类型，因此其自成一体。在右边的星号所代表的基因片段是在外胚层植入后阶段已完成了独立于细胞系的复制时序从早期向晚期转变（EtoL）的片段（大小约 155Mb）。这些片段的晚期复制特征在随后的细胞中得以稳定维持，并且在部分重编程的细胞中（piPSC）未能逆转，显示出对位于这些片段上的基因进行转录重编码的难度。此图采自 Hiratani 等文章并获得了 Genome Research 杂志的版权许可[8]。

实验方法：用 BrdU 脉冲标记细胞，先用流式细胞术将细胞分为 S 期早期和晚期两部分，然后用抗-BrdU 的抗体进行免疫沉淀将各部分细胞中 BrdU 标记的 DNA 分离出。将早期和晚期复制的 DNA 样本分别进行不同标记并与全基因组寡核苷酸芯片进行共杂交。计算早期和晚期复制片段中各探针丰度比［"复制时序率" = log2（早期复制片段丰度/晚期复制片段丰度）］被用来绘制全基因组的复制时序谱。芯片的探针密度为每 5.8kb 一个探针。之后整个基因组被分割为 10 974 个大约 200kb 的片段，计算每一片段平均的复制时序率，通过分层聚类的方法对不同细胞系进行比较。上面的热图展示了 10 974 个 200kb 片段复制时序率［= Log2（早期复制片段丰度/晚期复制片段丰度）］，其中的红色及绿色分别代表早期复制片段及晚期复制片段。蓝色框代表了在不同细胞类型间显著的差异的片段，这些片段占到了约 45% 的基因组。（另见图版）

这些结果揭示，由发育所引起的细胞类型特异的复制时序谱改变影响了几乎一半的基因组[8]。在较大的基本复制区域，复制时序变化仅发生在 400~800kb 的染色体片段，这解释了为什么我们无法通过细胞学检测来观察到这些复制时序变化。此外，尽管研究常表明 GC 含量或基因密度与早期复制之间存在正比例关系，且在所有细胞类型中，GC 含量和 AT 含量最丰富的基因片段分别出现在复制的早期和晚期，但实际上是 GC 含量及基因密度都居中的片段改变了复制时序，这些变化可以从本质上将整体水平转变为仅与复制时序相关的静止序列水平[8]。

一种在进化上相对保守的表观遗传学印记

现在已经清楚，哺乳动物发育早期广泛存在着复制时序差异。而且，通过对比来自于果蝇胚胎或成虫盘的两种培养细胞系发现了在复制时序上存在有大约 20% 的差异，这说明，在高等真核生物的发育过程中，复制时序亦普遍存在变化。但是，这些改变是对于动物的发育有意义的，还是说它们对于有机体的适应性来说只是不具重要意义的一种随机事件？我们目前对于复制调控机制的了解还不足以采用直接的手段来回答这一问题，取而代之的方法是评估在进化过程中特定细胞种类的复制时序是否赢得选择。为了达到这个目的，我们把研究扩展到了人的胚胎干细胞分化。与小鼠中获得的数据一样，我们发现多种人 ES 细胞系具有几乎完全相同的复制时序谱并在分化成 NPC 的过程中有大约 20% 的基因组发生时序改变（T. Ryba、I. H. 和 D. M. G，未发表）。与小鼠中相似，复制时序的改变通常都发生在染色体上 400~800kb 的区域内，这说明，可能存在着一种保守的复制时序改变的单元，而这一单元很可能同步调节着至少 2~3 个复制子。与上述相似点相对，人的胚胎干细胞与小鼠胚胎干细胞中进化保守的同源基因间却表现出明显不同的复制时序谱（T. Ryba、I. H. 和 D. M. G，未发表）。事实上，这些区域的复制时序谱与来源于胚胎植入后的小鼠外胚层的干细胞——EpiSC（外胚层源干细胞）[63,64]的谱系更为接近了。这一观察也在基因组分析水平上为将人胚胎干细胞定义来源于外胚层样阶段的假说提供了支持，而小鼠的胚胎干细胞则来源于处于外胚层样阶段上游的内细胞团（ICM）样阶段（T. Ryba、I. H. 和 D. M. G，未发表）。除此以外，人的细胞中早期复制与 GC 含量的相关性较低，同时人和小鼠基因组间的 GC 含量远不及复制时序保守（T. Ryba、I. H. 和 D. M. G，未发表）。

最近还未发表的数据对比了亲缘关系较远的裂殖酵母菌（N. Rhind，个人通讯）和出芽酵母菌（K. Lindstrom 和 B. Brewer，个人通讯），发现这两种酵母菌间虽然缺乏保守的复制起始位点，但却具有相当保守的复制时序程序。这与之前的许多观察结果发现复制时序与复制起始点无明确关系的结论相一致。例如，人的 β-球蛋白基因座通常从两个在空间上相邻的复制起始子中的一个开始复制，而小鼠的 β-球蛋白基因座复制则使用更分散的复制起始子，但人和小鼠的 β-球蛋白基因座的复制时序仍是保守的[65]。除此以外，在每个细胞周期的 G_1 期的特定时间点（TDP；定时决定点）染色体结构域的复制时序都会发生重建，复制起始位点的确定发生于 G_1 期晚期的 ODP

(原始决定位点),而 TDP 时间点发生则早于且独立于复制起始位点确定之前[66~69]。总之,这些发现表明,进化过程对于复制时序程序的正向选择较之对于基因组 GC 含量或复制起始位置的选择更严格。

复制时序的意义,以及它是否反映了与其他与进化选择压力有关的染色体性质之间的机制性关联还有待进一步探索。例如,早期复制与转录活性区、转录活化染色质(H3K4me3、H3K36me3)及非活化染色质(H3K9me2)(虽然不是与已知的非活化修饰染色质 H3K9me3、H3K27me3)的组蛋白修饰有明确关系[7,70]。然而,去除几个染色质修饰酶(如 Mll、Mbd3、Eed、Suv39h1/h2、G9a、Dnmt1/3a/3b、Dicer)却对复制时序只产生很微小的影响[1,70~72]。最近一项在裂殖酵母菌中的研究显示,异染色质蛋白 1(HP1)的同源分子——Swi6 通过加载复制起始因子 Sld3 来调节复制时序[73]。虽然在哺乳动物细胞中尚未有 Sld3 同源分子的报道,并且 HP1 分子似乎也不参与调节小鼠着丝点附近异染色质的晚期复制时序[71],在裂殖酵母中的研究结果提示了一种有意义的可能,即作用于结构域的染色质因子可以通过调控对复制起始因子结合来调控复制时序。例如,间接观察提示,组蛋白 H1 及 HMG-I/Y 蛋白在大染色体片段上的竞争可以调控分化过程中的复制时序[74]。

复制时序作为染色体三维结构的定量指标

在 X 染色体失活过程中,染色体复制时序的转变导致整条 X 染色体几乎同步的晚期复制程序。通过一个被称为"莱昂化"的过程,使 X 染色体发生压缩,形成了我们所熟知的巴氏小体并定位于细胞核的边缘区域(图 4-3A)[23]。染色体的复制时序及其在细胞核内部的定位之间的相关性并不仅限于 X 染色体,各个常染色体的复制结构域的核内定位也有类似关系。利用核苷酸类似物对染色体复制进行脉冲标记并应用这些类似物的特定抗体,可以将在 S 期不同时间进行复制的染色体结构域在细胞核的定位区域直观地显示出来(图 4-3B)。这些复制域呈现出点状的标记,被称为"复制焦点"[1]。在经过带荧光核苷酸标记的活细胞中也可观察到这些复制焦点,说明它们并不是由于细胞固定过程所产生的人工假象[75]。在几乎所有检测过的动物细胞中,在 S 期的前半程的细胞核内部都有数百个散在的复制位点,但在接近 S 期中期,这些位点会戏剧性地变位形成在细胞核边缘更为聚集性的复制焦点[75,76]。在接下来的细胞周期过程中,这些被标记焦点不会发生相互混合、分离,亦或形状、大小或强度的改变。这说明,同时进行复制的 DNA 片段形成了一个在结构和功能上稳定的细胞间期染色体结构单元[77]。在人类细胞中使用显微定量的方法初步估计,每个复制焦点大约含有将近 100 万碱基对(1Mb)的 DNA[78]。这一推测使得人们推想在小鼠及人类细胞分化过程中 DNA 复制改变时 40 万~80 万碱基对(400~800kb)大小的单元可能是复制焦点在分子水平的等价物,然而这一假设目前还很难用实验验证。

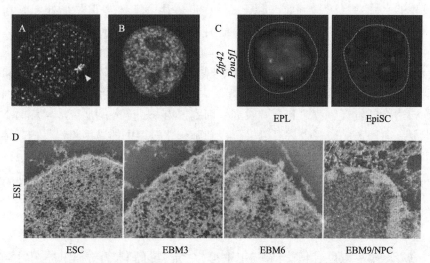

图 4-3 DNA 复制研究揭示的亚核水平基因组结构。A. 经过 10min 的 BrdU 脉冲标记显示，在 MEF 细胞 S 期中晚期复制、浓缩的失活 X 染色体（Xi，箭头处）聚集在核外周。图片改编自 Wu 等的文章并经 Journal of Cell Biology 许可[71]。B. 在 C127 小鼠成纤维细胞中通过"脉冲追踪脉冲"（"pulse-chase-pulse"）实验显示的在 S 期早期（绿色，在 S 期早期经 10min CldU 脉冲）及 S 期晚期（红色，在同一细胞周期中的 S 期晚期经 10min IdU 脉冲）的 DNA 复制位点。注意到在 S 期晚期复制的 DNA 主要位于核及核仁边缘位置，而在 S 期早期复制的 DNA 则发生在核内不包括上述两个亚核区域的部分。C. 2D DNA-FISH 方法定位比较在 EPL 细胞（小鼠胚胎早期外胚层）及 EpiSC 细胞（小鼠胚胎晚期外胚层）中 Zfp42（也称为 Rex2）及 Pou5f1（也称为 Oct4）基因的典型分布[8]。Zfp42 基因座是一个典型的非细胞系依赖的 EtoL 基因座，在由早期向晚期外胚层阶段过渡时完成 DNA 复制由早期到 S 期的中晚阶段的 EtoL 改变。相比之下，Pou5f1 基因座则一直属于早期复制。在 EpiSC 中，Zfp42 基因座（红色信号）由核中心区重新定位于核周，但在 EPL 细胞中无此现象。Pou5f1 基因座（绿色信号）在两种细胞类型中都保持在核中心位置。图片经 Genome Research 杂志许可，采用 Hiratani 等发表的数据[8]。D. 利用电子分光成像（ESI）对 ESC 分化过程中的胞核进行分析[8]。图像从左至右分别为：胚胎干细胞、EBM3（分化第 3 天的胚胎干细胞）、EBM6（分化第 6 天的胚胎干细胞）及 EBM9（分化第 9 天的神经前体细胞 NPC）。磷及氮的相对水平被用来区分染色体（黄色）及蛋白质和核糖核蛋白体（蓝色）[81]。胚胎干细胞的胞核是一个相对匀质的、由 10nm 染色质纤维构成的网状结构。在沿核膜及核浆内仅有低水平的染色质压缩。EBM3 与 ESC 的核结构非常相似。值得注意的是，进一步分化的 EBM6，出现紧密压缩的染色体明显积累于核周，在 EBM9（NPC）中这一边界变得更加清晰。基于对几个基因位点上基因表达、复制时序及亚核定位的分析得出，EBM3～EBM9 的转变大致与外胚层由早期向晚期的转变相一致[8]。经 Genome Research 允许，数据来自 Hiratani 及其他人员发表的文章[8]。（另见图版）

如果在常染色体中也存在类似于"莱昂化"的机制，那就可预测在常染色体复制时序改变时，将伴随染色体发生亚核水平的定位改变。对于神经分化过程中的 7 个基因位点的研究都证实了这一推测[7,8,79]。有趣的是，复制时序改变通常需要经历几个细胞周期才能达到稳定，而染色体在亚核水平的重定位通常是在复制时序从 S 期中期向晚期转变时突然发生的(图 4-3C)[8]。例如，对于复制时间发生显著推迟但是仍处于 S 期前半程的基因，其定位仍然会位于细胞核内部，然而一个即便很小但使得复制时间进入 S 期中晚阶段的变化，将会伴随基因定位向细胞核周边的移动[8]。除此以外，在 S 期前半程复制的所有基因具有同样的表达潜能；而在 S 期中晚阶段发生复制的基因，其转录与复制时间之间有最强有力的相关性[1]。所有研究结果都预测在 S 期后半程复制的基因定位于接近细胞核边缘的位置。考虑到现阶段进行细胞核内基因定位的研究非常耗时费力，对于基因组复制时序谱的分析也许可以推测在细胞分化过程中的基因组水平的基因细胞核定位。事实上，我们最近利用全基因组水平的 Hi-C（一种新的染色体构象获得分析方式，或称为3C）[80]，测量染色体之间在空间上的接近程度，发现染色体的相互空间关系与各自复制时序谱之间体现出细胞类型特定的高度相关性（T. Ryba、I. H. 和 D. M. G，未发表数据）。

复制时序反映表观遗传特征的改变：常染色体在外胚层阶段的莱昂化

那些揭示了小鼠发育早期细胞命运决定与复制时序改变之间的紧密关系的早期研究对于我们理解复制时序改变的生物学意义有着最重要的意义（图 4-2）[7,8]。复制时序改变与基因的转录行为改变相互协调，尤其是缺乏 CpG 岛的弱启动子的基因其表达与复制时序有着更强的关联性。小鼠发育过程中最早发生的事件有 DNA 复制从早期到晚期（EtoL）复制时序改变。这一变化过程在胚胎植入后外胚层阶段完成（图 4-2，用星号标出）。与此同时，复制从早期到晚期（EtoL）复制时序改变位点重新定位于细胞核周边（图 4-3C）及失活 X 染色体（Xi）转向晚期复制[8]。通过被称为电子分光成像（ESI）的电子显微镜分析技术[81]，发现在 EtoL 位点发生核内重定位的同时（图 4-3C），细胞核内出现一种显著的染色体构型改变，一些浓缩的染色质向核周聚集（图 4-3D）[8]。在晚期外胚层来源的干细胞（EpiSC）中已观察到核内染色体结构的明显改变，说明这一过程发生在 Oct4/Nanog/Sox2 基因表达下调及各胚层分化决定发生之前[8]，提示构成细胞多潜能性的核心细胞因子[82]在晚期外胚层来源干细胞（EpiSC）及 ICM/早期外胚层来源干细胞[ESC 或早期原始外胚层样（EPL）细胞][83]中发挥作用时所处的表观遗传学内环境有相当的差异。与早期到晚期（EtoL）改变不同，晚期至早期（LtoE）改变发生在各胚层分化的后期，并依所属细胞系形成细胞类型特异性的改变（图 4-2）。总之，在外胚层形成之前和之后的阶段都发生了广泛的改变，分别对应于发育过程中细胞系非依赖及细胞系依赖的改变。这些结果提示复制时序对应于特定的细胞分化阶段，并反映了细胞核内染色体的组织结构。因而复制时序的改变代表或反映了在关键的细胞命运转变过程中细胞的表观遗传学变化。通过对复制谱的比较研究，我们可以区分非常相近的两种细胞模型在表观遗传学上的差别，以及它们分别代表外胚层细胞早期及晚期发育阶段的特征。

ICM/早期外胚层与晚期外胚层细胞培养方式之间的表观遗传学上的显著差别与它们之间多种功能表型上的差异相一致[84]，虽然这两种细胞的基因表达只有微小的不同[85]。

第一，与着床前 ICM 细胞不同，着床后外胚层细胞尽管表达多种多能性标记基因却不能定植囊胚。第二，从 129 小鼠系的外胚层中成功分离胚胎干细胞（ESC）的效率在 E5.0 天和 E6.0 天的胚胎中呈现直线下降[84]。第三，对早期及晚期外胚层细胞的体外模型研究发现，晚期外胚层细胞似乎已经越过了某种目前还未明确的表观遗传学屏障，这种屏障很难通过核重编程来打破。因此，早期的外胚层细胞如 EPL 细胞，虽然已经失去了形成嵌合体小鼠的能力，但是可通过在含有 LIF 的 ESC 培养基中培养而恢复到胚胎干细胞阶段[63,64]，并且能形成嵌合体小鼠[83]。与此相对比的是，晚期着床后外胚层细胞 EpiSC，已经很难恢复到胚胎干细胞阶段[63,64]。事实上，利用 EpiSC 得到胚胎干细胞样 iPSC（诱导多潜能干细胞）的效率很低（0.1‰～1‰），就如同从体细胞类型产生 iPSC 一样[86]。通过这些观察到的现象，外胚层的发育似乎伴随细胞命运的关键改变。这一改变并未产生基因转录上大的变化，但却反映出染色体复制时序谱及基因组的空间结构的明显改变（图 4-2 和图 4-3）。

前面已经讨论过，在哺乳动物胚胎形成早期两个主要的细胞命运决定伴随失活 X 染色体（Xi）的复制时序改变。这一过程发生在滋养胚外层出现的 E3.5 天及原始内胚层出现的 E4.0 天。两个发育阶段都代表了胚外细胞系从胚体的分离。就这一点而言，着床后外胚层的分化代表了在这两个胚外细胞系产生后主要的细胞命运转变。在着床后胚体的外胚层细胞中的失活 X 染色体（Xi）向晚期复制的特征性改变首先在胚胎内观察到，我们推测，这是晚期外胚层细胞失去其可逆性转变到 ICM 阶段能力的决定性步骤。很显然，在胚胎形成早期，并非只有雌性胚胎经历了这一系列的细胞命运决定，雄性胚胎也同样经历了。因此，我们可推测，Xi 独特的复制行为变化是一个大范围"莱昂化"的结果，这一过程并不仅限于 X 染色体失活，还包括了常染色体，而且很可能还与核边缘附近出现的浓缩染色体相关（核边缘的巴氏小体类似物[23]）。从这个意义来讲，外胚层阶段发生的不依赖细胞系的常染色体 EtoL 改变（图 4-2，星号标出）主要发生在属于 X 染色体的特征性序列的缺乏 GC 而富于 LINE-1 的染色体片段[8]。因此，X 染色体的失活很可能仅仅是细胞命运改变时细胞预定模式下的行为改变，这种改变恰好以某种原因与缺乏 GC 而富于 LINE-1 的序列的染色体片段表现出相同的行为。如果这一推测成立，那么保留了活性的 X 染色体携带了特定的印记以逃脱这种调节作用。这一假设在 1984 年已由 Lyon 和 Rastan 提出[87]。他们指出，胚胎外细胞中印记 X 染色体失活的实验数据与这一假设的推论更加吻合，即活性的母源 X 染色体正是通过获得印记而保留其活性，而不是由于父源染色体获得失活印记从而导致失活。

复制时序及细胞重编程：常染色体莱昂化的进一步证明

从成体体细胞衍生出的诱导多能干细胞（iPSC）不仅对于 21 世纪的再生医学具有重要意义，而且也为了解细胞核重编程过程提供了可能。在小鼠中，与复制时序谱可反映细胞特征的结论保持一致，诱导多能干细胞的复制时序谱与胚胎干细胞几乎相同[7,8]。因此，仔细分解在体细胞重编程过程中染色体复制时序程序是如何变化的，可能从一个全新的角度认识重编程过程。特别是那些基于细胞形态学或报告基因表达获得的发生部分重编程的诱导多能干细胞细胞系（piPSC），为研究复制时序重编程中间阶段提供了可能。对这些细胞系的研究还为评价不同染色体区域重编程的效率提供了可能。发生部分重编程的诱导多能干细胞中许多多能性基因都不表达，并且不能形成嵌合体小鼠，这些都表明它们被抑制在一个重编程过程的中间阶段[88,89]。

研究由小鼠胚胎成纤维细胞（MEF）衍生出的三个独立的 piPSC 细胞系的复制时序谱，发现它们的复制时序与胚胎干细胞或多能性干细胞、早期外胚层样 EPL 细胞及三胚层来源的 EpiSC 或原代细胞明显不同（图 4-2）[8]。与此相对应，三个 piPSC 细胞系之间虽然具有各自独立的逆转录病毒整合位点，但它们之间的复制时序谱却非常相似，这表明它们的重编程过程都在同一表观遗传学阶段被阻抑的[88]。有趣的是，在外胚层阶段发生的不依赖于细胞系的 EtoL 复制改变的多数染色体片段在 piPSC 细胞中保持着晚期复制的状态（图 4-2，星号标出）[8]，这样的染色体片段还包括在雌性细胞中晚期复制的失活 X 染色体（Xi），突出了在外胚层中完成的 EtoL 转换的染色体复制时序的稳定性。相反，在随后的发育阶段中形成的细胞特异性的（如 MEF 特异的）复制时序模式则更易发生重编程。对于基因表达谱的对照研究发现，定位于细胞非依赖性 EtoL 片段的基因在 piPSC 与 ESC 之间表现出最小相似性[8]。同样，雌性 piPSC 细胞中失活的 X 染色体（Xi）也不能恢复转录活性[88,89]。总之，这些结果说明，很多在体细胞常染色体上发生的复制时序改变，并且尤其是在外胚层阶段的 EtoL 片段发生复制时序改变是一种稳定的表观遗传学改变。正如试图对 X 染色体失活重编程一样，很难对这些基因片段进行重编程，这也支持了"常染色体莱昂现象"的观点。

复制时序程序的维持和改变及其潜在作用

正如我们前面提到的那样，在 G_1 阶段早期发生复制时序重建[67]。有趣的是，这一过程与有丝分裂后核染色质结构域的重定位同时发生[67,90]。在不同细胞类型间，特定时间点（TDP）在 G_1 期的发生精确定时各不相同。在哺乳动物细胞中通常是在进入 G_1 期后的 1～3h[67]，而在出芽酵母菌中则发生在有丝分裂及 START 间的某个时间点。在一系列平行研究中发现，在 G_1 期的前 1～2h 中，染色体具有较高的可移动性，而此后分裂间期的剩余时间里，染色体将被限制于一个较局部的位置[92,93]。对特定位点的核纤层诱导定位研究发现，这一过程需要经历有丝分裂，并且染色体的重定位过程发生在细胞分裂终末期及 G_1 期早期之间[94,95]。由此看来，G_1 期早期似乎提供了一个临时的窗口，以实现每一细胞周期中染色体三维组织的重建或建立新的染色体三维组织模式。而染色体的三维组织可能决定了在细胞周期 S 期中染色体的复制时序。因此，对于细胞周期的调节可能成为对于发育过程中复制时序调节的一种方式。有可能对 G_1 期长度自身的调节可以影响在 DNA 复制起始前细胞核重组的程度，从而影响复制时间的程序。事实上，不同细胞类型 G_1 期的长度有很大差异。当小鼠胚胎干细胞发生分化时，G_1 期延长并同时观察到一个很大程度上的复制时序改变[8]。

至于复制时间程序的调控作用，需要强调的一点是，染色体在复制叉处被组装，因而 DNA 的复制为组装过程的调节提供了一个方便的窗口。将外源质粒分别注入到 S 期早期或晚期的哺乳动物细胞核中，质粒可被组装为过乙酰化（S期早期细胞）或乙酰化不足（S期晚期细胞）的染色体，为S期的不同阶段形成不同染色体结构提供了证据[96]。在一系列连续的细胞周期中，牛乳头瘤病毒质粒发生复制的时间不同[97]。利用这一特点，上述研究的作者最近在实验中发现，当病毒质粒在下一细胞周期的早期发生复制时，原来折叠紧密且晚期复制的染色体提前变为较松散折叠的状态[98]。这一研究报告采用了一个人工试验系统支持了染色体结构与复制时间之间的一个正反馈调节模式，即复制时间可以决定染色体结构状态，而染色体的结构又可调控随后细胞周期中的染色体复制时间[46]。虽然还缺乏足以证明在S期不同阶段复制叉上结合不同的染色体修饰成分的体内证据，但这一模式为特定染色体状态的表观稳定遗传提供了一个有吸引力的情景。

结论

复制时序程序的精确作用以及这一程序受发育过程的调节的意义还有待进一步的阐明。然而，从人类到出芽和裂殖酵母菌中，基因组复制的时空调节以及多个复制子结构域的存在均是保守的[49,99,100]。DNA的复制与很多受细胞周期及细胞发育过程调节的基础细胞生物学过程相联系，而且复制时序的缺陷也在不同疾病模型中被观察到[101]。最近的研究促使我们得出在果蝇（flies）[57]、小鼠[7]及人类中广泛存在着复制时序改变的结论（T. Ryba、I. H 和 D. M. G.，未发表）。此外，对人类及小鼠之间相似的细胞类型中保守的同源基因区域进行对比时，我们可以发现复制时序程序具有相当大的保守性（T. Ryba、I. H 和 D. M. G.，未发表）。正如我们上述讨论中发现的，小鼠发育早期最初发生的两个主要的细胞命运决定伴随了失活X染色体（Xi）复制时序的改变[35]。我们提出，着床后的外胚层可能利用"莱昂化"过程来经历接下来主要的细胞命运转变，这一过程包括了对于Xi及常染色体的复制时序程序的改变并将这种改变在后代细胞中稳定延续[8]。无论复制时序的具体作用是什么，无论它与细胞特性之间的关系是因果关系的还是反映性的，复制时序程序都具细胞类型特异性并且成为描述完整细胞特征的一部分。正因为如此，在发现DNA双螺旋结构并且在细胞核中观察到DNA复制过程50年后的今天，对于DNA复制的研究仍然不断为我们带来关于染色体的结构及其在分化中的变化的新见解。

致谢

感谢与 N. Rhind、K. Lindstrom 和 P. Tesar 有价值的讨论，感谢 J. Rathjen 对文稿的指正及富有意义的讨论。由于受限于这一章的题目及篇幅，对于非哺乳动物模型以及非同步复制只进行了十分有限的讨论，对此表示抱歉。Gilbert 实验室的研究得到了 NIH GM83337 及 GM085354 的资助。

（胡　晓　阮　峥　译）

参 考 文 献

1. Hiratani I, Takebayashi S, Lu J et al. Replication timing and transcriptional control: beyond cause and effect—part II. Curr Opin Genet Dev 2009; 19:142-9.
2. Stamatoyannopoulos JA, Adzhubei I, Thurman RE et al. Human mutation rate associated with DNA replication timing. Nat Genet 2009; 41:393-5.
3. Chang BH, Smith L, Huang J et al. Chromosomes with delayed replication timing lead to checkpoint activation, delayed recruitment of Aurora B and chromosome instability. Oncogene 2007; 26:1852-61.
4. Tabancay AP Jr, Forsburg SL. Eukaryotic DNA replication in a chromatin context. Curr Top Dev Biol 2006; 76:129-84.
5. MacAlpine DM, Bell SP. A genomic view of eukaryotic DNA replication. Chromosome Res 2005; 13:309-26.
6. Morishima A, Grumbach MM, Taylor JH. Asynchronous duplication of human chromosomes and the origin of sex chromatin. Proc Natl Acad Sci USA 1962; 48:756-63.
7. Hiratani I, Ryba T, Itoh M et al. Global reorganization of replication domains during embryonic stem cell differentiation. PLoS Biol 2008; 6:e245.
8. Hiratani I, Ryba T, Itoh M et al. Genome-Wide Dynamics of Replication Timing Revealed by In Vitro Models of Mouse Embryogenesis. Genome Res 2010; 20:155-69.
9. Holmquist GP. Role of replication time in the control of tissue-specific gene expression. Am J Hum Genet 1987; 40:151-73.
10. Pope BD, Hiratani I, Gilbert DM. Domain-Wide Regulation of DNA Replication Timing During Mammalian Development. Chromosome Research 2010; 18:127-36.
11. Taylor JH. Intracellular localization of labeled nucleic acid determined by autoradiographs. Science 1953; 118:555-7.
12. Meselson M, Stahl FW. The Replication of DNA in Escherichia Coli. Proc Natl Acad Sci USA 1958; 44:671-82.
13. Taylor JH, Woods PS, Hughes WL. The Organization and Duplication of Chromosomes as Reveals by Autoradiographic Studies Using Tritium-Labeled Thymidine. Proc Natl Acad Sci USA 1957; 43:122-8.
14. Taylor JH. The mode of chromosome duplication in Crepis capillaris. Exp Cell Res 1958; 15:350-7.
15. Taylor JH. Asynchronous duplication of chromosomes in cultured cells of Chinese hamster. J Biophys Biochem Cytol 1960; 7:455-64.
16. Latt SA. Microfluorometric detection of deoxyribonucleic acid replication in human metaphase chromosomes. Proc Natl Acad Sci USA 1973; 70:3395-9.
17. Latt SA. Fluorescent probes of chromosome structure and replication. Can J Genet Cytol 1977; 19:603-23.
18. Stubblefield E. Analysis of the replication pattern of Chinese hamster chromosomes using 5-bromodeoxyuridine suppression of 33258 Hoechst fluorescence. Chromosoma 1975; 53:209-21.
19. Latt SA. Fluorescence analysis of late DNA replication in human metaphase chromosomes. Somatic Cell Genet 1975; 1:293-321.
20. Lima De Faria A. Incorporation of tritiated thymidine into meiotic chromosomes. Science 1959; 130:503-4.
21. Epplen JT, Siebers JW, Vogel W. DNA replication patterns of human chromosomes from fibroblasts and amniotic fluid cells revealed by a Giemsa staining technique. Cytogenet Cell Genet 1975; 15:177-85.
22. Stambrook PJ, Flickinger RA. Changes in chromosomal DNA replication patterns in developing frog embryos. J Exp Zool 1970; 174:101-13.
23. Barr ML, Bertram EG. A morphological distinction between neurones of the male and female and the behaviour of the nucleolar satellite during accelerated nucleoprotein synthesis. Nature 1949; 163:676.
24. Lyon MF. Gene action in the X-chromosome of the mouse (Mus musculus L.). Nature 1961; 190:372-3.
25. Heard E, Disteche CM. Dosage compensation in mammals: fine-tuning the expression of the X chromosome. Genes Dev 2006; 20:1848-67.
26. Deakin JE, Chaumeil J, Hore TA et al. Unravelling the evolutionary origins of X chromosome inactivation in mammals: insights from marsupials and monotremes. Chromosome Res 2009; 17:671-85.
27. Ho KK, Deakin JE, Wright ML et al. Replication asynchrony and differential condensation of X chromosomes in female platypus (Ornithorhynchus anatinus). Reprod Fertil Dev 2009; 21:952-63.
28. Koina E, Chaumeil J, Greaves IK et al. Specific patterns of histone marks accompany X chromosome inactivation in a marsupial. Chromosome Res 2009; 17:115-26.
29. Takagi N, Sugawara O, Sasaki M. Regional and temporal changes in the pattern of X-chromosome replication during the early post-implantation development of the female mouse. Chromosoma 1982; 85:275-86.
30. Lock LF, Takagi N, Martin GR. Methylation of the Hprt gene on the inactive X occurs after chromosome inactivation. Cell 1987; 48:39-46.
31. Chow J, Heard E. X inactivation and the complexities of silencing a sex chromosome. Curr Opin Cell Biol 2009; 21:359-66.

32. Keohane AM, O'Neill LP, Belyaev ND, Lavender JS, Turner BM. X-Inactivation and histone H4 acetylation in embryonic stem cells. Dev Biol 1996; 180:618-30.
33. Chaumeil J, Okamoto I, Guggiari M et al. Integrated kinetics of X chromosome inactivation in differentiating embryonic stem cells. Cytogenet Genome Res 2002; 99:75-84.
34. Wutz A, Jaenisch R. A shift from reversible to irreversible X inactivation is triggered during ES cell differentiation. Mol Cell 2000; 5:695-705.
35. Sugawara O, Takagi N, Sasaki M. Correlation between X-chromosome inactivation and cell differentiation in female preimplantation mouse embryos. Cytogenet Cell Genet 1985; 39:210-9.
36. Snow MHL. Gastrulation in the mouse: growth and regionalization of the epiblast. J Embryol Exp Morphol 1977; 42:293-303.
37. Balazs I, Brown EH, Schildkraut CL. The temporal order of replication of some DNA cistrons. Cold Spring Harb Symp Quant Biol 1974; 38:239-45.
38. Burke W, Fangman WL. Temporal order in yeast chromosome replication. Cell 1975; 5:263-9.
39. Hatton KS, Dhar V, Brown EH et al. Replication program of active and inactive multigene families in mammalian cells. Mol Cell Biol 1988; 8:2149-58.
40. Goldman MA, Holmquist GP, Gray MC et al. Replication timing of genes and middle repetitive sequences. Science 1984; 224:686-92.
41. Epner E, Reik A, Cimbora D et al. The beta-globin LCR is not necessary for an open chromatin structure or developmentally regulated transcription of the native mouse beta-globin locus. Mol Cell 1998; 2:447-55.
42. Taljanidisz J, Popowski J, Sarkar N. Temporal order of gene replication in Chinese hamster ovary cells. Mol Cell Biol 1989; 9:2881-9.
43. Gilbert DM. Temporal order of replication of Xenopus laevis 5S ribosomal RNA genes in somatic cells. Proc Natl Acad Sci USA 1986; 83:2924-8.
44. Hiratani I, Leskovar A, Gilbert DM. Differentiation-induced replication-timing changes are restricted to AT-rich/long interspersed nuclear element (LINE)-rich isochores. Proc Natl Acad Sci USA 2004; 101:16861-6.
45. Perry P, Sauer S, Billon N et al. A dynamic switch in the replication timing of key regulator genes in embryonic stem cells upon neural induction. Cell Cycle 2004; 3:1645-50.
46. Gilbert DM. Replication timing and transcriptional control: beyond cause and effect. Curr Opin Cell Biol 2002; 14:377-83.
47. Azuara V, Brown KE, Williams RR et al. Heritable gene silencing in lymphocytes delays chromatid resolution without affecting the timing of DNA replication. Nat Cell Biol 2003; 5:668-74.
48. Raghuraman MK, Winzeler EA, Collingwood D et al. Replication dynamics of the yeast genome. Science 2001; 294:115-21.
49. Hayashi M, Katou Y, Itoh T et al. Genome-wide localization of preRC sites and identification of replication origins in fission yeast. EMBO J 2007; 26:1327-39.
50. Schubeler D, Scalzo D, Kooperberg C et al. Genome-wide DNA replication profile for Drosophila melanogaster: a link between transcription and replication timing. Nat Genet 2002; 32:438-42.
51. MacAlpine DM, Rodriguez HK, Bell SP. Coordination of replication and transcription along a Drosophila chromosome. Genes Dev 2004; 18:3094-105.
52. Woodfine K, Fiegler H, Beare DM et al. Replication timing of the human genome. Hum Mol Genet 2004; 13:191-202.
53. White EJ, Emanuelsson O, Scalzo D et al. DNA replication-timing analysis of human chromosome 22 at high resolution and different developmental states. Proc Natl Acad Sci USA 2004; 101:17771-6.
54. Jeon Y, Bekiranov S, Karnani N et al. Temporal profile of replication of human chromosomes. Proc Natl Acad Sci USA 2005; 102:6419-24.
55. Karnani N, Taylor C, Malhotra A et al. Pan-S replication patterns and chromosomal domains defined by genome-tiling arrays of ENCODE genomic areas. Genome Res 2007; 17:865-76.
56. Farkash-Amar S, Lipson D, Polten A et al. Global organization of replication time zones of the mouse genome. Genome Res 2008; 18:1562-70.
57. Schwaiger M, Stadler MB, Bell O et al. Chromatin state marks cell-type- and gender-specific replication of the Drosophila genome. Genes Dev 2009; 23:589-601.
58. Desprat R, Thierry-Mieg D, Lailler N et al. Predictable Dynamic Program of Timing of DNA Replication in Human Cells. Genome Res 2009; 19:2288-99.
59. Gilbert DM. Replication timing and metazoan evolution. Nat Genet 2002; 32:336-7.
60. Schmegner C, Hameister H, Vogel W et al. Isochores and replication time zones: a perfect match. Cytogenet Genome Res 2007; 116:167-72.
61. Grasser F, Neusser M, Fiegler H et al. Replication-timing-correlated spatial chromatin arrangements in cancer and in primate interphase nuclei. J Cell Sci 2008; 121:1876-86.
62. Costantini M, Bernardi G. Replication timing, chromosomal bands and isochores. Proc Natl Acad Sci USA 2008; 105:3433-7.

63. Brons IG, Smithers LE, Trotter MW et al. Derivation of pluripotent epiblast stem cells from mammalian embryos. Nature 2007; 448:191-5.
64. Tesar PJ, Chenoweth JG, Brook FA et al. New cell lines from mouse epiblast share defining features with human embryonic stem cells. Nature 2007; 448:196-9.
65. Aladjem MI. Replication in context: dynamic regulation of DNA replication patterns in metazoans. Nat Rev Genet 2007; 8:588-600.
66. Wu J-R, Gilbert DM. A distinct G1 step required to specify the chinese hamster DHFR replication origin. Science 1996; 271:1270-2.
67. Dimitrova DS, Gilbert DM. The spatial position and replication timing of chromosomal domains are both established in early G1-phase. Mol Cell 1999; 4:983-93.
68. Li F, Chen J, Solessio E et al. Spatial distribution and specification of mammalian replication origins during G1 phase. J Cell Biol 2003; 161:257-66.
69. Dimitrova DS, Prokhorova TA, Blow JJ et al. Mammalian nuclei become licensed for DNA replication during late telophase. J Cell Sci 2002; 115:51-9.
70. Yokochi T, Poduch K, Ryba T et al. G9a Selectively Represses a Class of Late-Replicating Genes at the Nuclear Periphery. Proc Natl Acad Sci USA 2009; 106:19363-8.
71. Wu R, Singh PB, Gilbert DM. Uncoupling global and fine-tuning replication timing determinants for mouse pericentric heterochromatin. J Cell Biol 2006; 174:185-94.
72. Jorgensen HF, Azuara V, Amoils S et al. The impact of chromatin modifiers on the timing of locus replication in mouse embryonic stem cells. Genome Biol 2007; 8:R169.
73. Hayashi MT, Takahashi TS, Nakagawa T et al. The heterochromatin protein Swi6/HP1 activates replication origins at the pericentromeric region and silent mating-type locus. Nat Cell Biol 2009; 11:357-62.
74. Flickinger R. Replication timing and cell differentiation. Differentiation 2001; 69:18-26.
75. Panning MM, Gilbert DM. Spatio-temporal organization of DNA replication in murine embryonic stem, primary and immortalized cells. J Cell Biochem 2005; 95:74-82.
76. Wu R, Terry AV, Singh PB et al. Differential subnuclear localization and replication timing of histone H3 lysine 9 methylation states. Mol Biol Cell 2005; 16:2872-81.
77. Gilbert DM, Gasser SM. Nuclear Structure and DNA Replication. In: DePamphilis ML, ed. DNA Replication and Human Disease. Cold Spring Harbor, New York: Cold Spring Harbor Laboratory Press, 2006.
78. Berezney R, Dubey DD, Huberman JA. Heterogeneity of eukaryotic replicons, replicon clusters and replication foci. Chromosoma 2000; 108:471-84.
79. Williams RR, Azuara V, Perry P et al. Neural induction promotes large-scale chromatin reorganisation of the Mash1 locus. J Cell Sci 2006; 119:132-40.
80. Lieberman-Aiden E, van Berkum NL, Williams L et al. Comprehensive Mapping of Long-Range Interactions Reveals Folding Principles of the Human Genome. Science 2009; 326:289-93.
81. Bazett-Jones DP, Li R, Fussner E et al. Elucidating chromatin and nuclear domain architecture with electron spectroscopic imaging. Chromosome Res 2008; 16:397-412.
82. Jaenisch R, Young R. Stem cells, the molecular circuitry of pluripotency and nuclear reprogramming. Cell 2008; 132:567-82.
83. Rathjen J, Lake JA, Bettess MD et al. Formation of a primitive ectoderm like cell population, EPL cells, from ES cells in response to biologically derived factors. J Cell Sci 1999; 112 (Pt 5):601-12.
84. Gardner RL, Brook FA. Reflections on the biology of embryonic stem (ES) cells. Int J Dev Biol 1997; 41:235-43.
85. Pfister S, Steiner KA, Tam PP. Gene expression pattern and progression of embryogenesis in the immediate post-implantation period of mouse development. Gene Expr Patterns 2007; 7:558-73.
86. Guo G, Yang J, Nichols J et al. Klf4 reverts developmentally programmed restriction of ground state pluripotency. Development 2009; 136:1063-9.
87. Lyon MF, Rastan S. Parental source of chromosome imprinting and its relevance for X chromosome inactivation. Differentiation 1984; 26:63-7.
88. Maherali N, Sridharan R, Xie W et al. Directly reprogrammed fibroblasts show global epigenetic remodeling and widespread tissue contribution. Cell Stem Cell 2007; 1:55-70.
89. Sridharan R, Tchieu J, Mason MJ et al. Role of the murine reprogramming factors in the induction of pluripotency. Cell 2009; 136:364-77.
90. Li F, Chen J, Izumi M et al. The replication timing program of the Chinese hamster beta-globin locus is established coincident with its repositioning near peripheral heterochromatin in early G1 phase. J Cell Biol 2001; 154:283-92.
91. Raghuraman M, Brewer B, Fangman W. Cell cycle-dependent establishment of a late replication program. Science 1997; 276:806-9.
92. Walter J, Schermelleh L, Cremer M et al. Chromosome order in HeLa cells changes during mitosis and early G1, but is stably maintained during subsequent interphase stages. J Cell Biol 2003; 160:685-97.

93. Thomson I, Gilchrist S, Bickmore WA et al. The radial positioning of chromatin is not inherited through mitosis but is established de novo in early G1. Curr Biol 2004; 14:166-72.
94. Kumaran RI, Spector DL. A genetic locus targeted to the nuclear periphery in living cells maintains its transcriptional competence. J Cell Biol 2008; 180:51-65.
95. Reddy KL, Zullo JM, Bertolino E et al. Transcriptional repression mediated by repositioning of genes to the nuclear lamina. Nature 2008; 452:243-7.
96. Zhang J, Xu F, Hashimshony T et al. Establishment of transcriptional competence in early and late S phase. Nature 2002; 420:198-202.
97. Gilbert DM, Cohen SN. Bovine papilloma virus plasmids replicate randomly in mouse fibroblasts throughout S-phase of the cell cycle. Cell 1987; 50:59-68.
98. Lande-Diner L, Zhang J, Cedar H. Shifts in replication timing actively affect histone acetylation during nucleosome reassembly. Mol Cell 2009; 34:767-74.
99. Versini G, Comet I, Wu M et al. The yeast Sgs1 helicase is differentially required for genomic and ribosomal DNA replication. EMBO J 2003; 22:1939-49.
100. McCune HJ, Danielson LS, Alvino GM et al. The Temporal Program of Chromosome Replication: Genomewide Replication in clb5{Delta} Saccharomyces cerevisiae. Genetics 2008; 180:1833-47.
101. Hiratani I, Gilbert DM. Replication timing as an epigenetic mark. Epigenetics 2009; 4:93-7.
102. Gilbert DM. Temporal order of DNA replication in eukaryotic cells: it's relationship to gene expression. Genetics. Stanford: Stanford University, 1989.

第5章 小鼠胚胎干细胞基因组完整性的保持

Peter J. Stambrook*, Elisia D. Tichy

摘要: 胚胎干细胞 (embryonic stem cell, ES 细胞) 和生殖细胞 (germ cell) 具有分化为完整机体的潜能, 而它们都必须具备一套健全的、能够保持基因组完整性的机制。由于体细胞具有非常高的突变率, 在体内大约高达 10^{-4}, 可达致病水平, 从而导致胎儿损伤和遗传性缺陷。选择性地检测内源性杂合位点的突变事件, 其 70%~80% 是由于有丝分裂重组而导致的杂合性丧失 (loss of heterozygosity, LOH)。此机制可影响报告基因和交叉重组位点之间的众多杂合位点。本章节阐述了小鼠胚胎干细胞保持自身基因组完整性的三种机制。第一种机制是, 与同基因小鼠胚胎成纤维细胞相比 (其与成人体细胞的突变率相仿), 胚胎干细胞可抑制同源染色体之间的突变及重组使之减少两个数量级。第二种机制是, 改变胚胎干细胞对外界环境危险的敏感性, 并通过自我更新的方式去除受损的细胞。小鼠 ES 细胞缺乏 G_1 期检测点, 因此受外界伤害 (如电离辐射) 的细胞不会停留在 G_1/S 期检测点, 而是进入 S 期, 受损的 DNA 进行复制, 由此损伤恶化, 驱使细胞凋亡。第三种机制与小鼠 ES 细胞对双链 DNA 断裂的修复机制有关。体细胞主要利用有错误修复倾向的非同源末端连接 (nonhomologous end joining, NHEJ) 修复机制, 从目的论的观点来看, 这种修复方式对于 ES 细胞是不利的, 因为这样会促进突变的累积。通过检测 ES 细胞修复双链 DNA 断裂的首选途径, 我们发现它们主要利用高保真的、同源介导的修复途径 (high fidelity homology-mediated repair), 因此尽可能地减少了修复过程中突变的发生。当小鼠 ES 细胞被诱导分化时, 主要的修复途径从同源介导的修复转向体细胞特征性的修复途径, 即非同源末端连接。

引言和历史观点

多年来, 单个细胞产生一个生物体多种不同类型细胞的能力一直是研究的热点领域。核等价 (nuclear equivalence) 的问题一直存在, 即是否多细胞生物的所有细胞具有相同数量、相同质量的基因组。核等价的检测最早是在 1902 年, Spemann 将蝾螈胚胎第一次分裂形成的两个卵裂细胞分离, 发现两个卵裂细胞都可以发育为一个完整的胚胎, 支持功能等价 (functional equivalence)[1]。而此发现早在 11 年前就被 Hans Driesch 预计到, Hans Driesch 的工作显示, 尽管从海胆胚胎分离出的卵裂细胞体型较小[2], 但

---

* Corresponding Author: Peter J. Stambrook—Department of Molecular Genetics, Biochemistry and Microbiology, University of Cincinnati College of Medicine, 231 Albert Sabin Way, Cincinnati, Ohio 45267-0524, USA. Email: stambrpj@ucmail.uc.edu

它们能够发育成正常的海胆。

半个世纪后陆续有了很多重要的、有深远意义的发现，下面主要介绍两个开创性的实验。Hämmerling 利用绿色藻类中的伞藻（*Acetabularium*）作为模型生物设计的经典、巧妙的实验显示，细胞核包含所有决定细胞形态的信息[3]。*Acetabularium* 是单细胞生物，分为基底（含有细胞核）、茎和帽子结构。Hämmerling 利用 *Acetabularium* 的两个种属，即具有光滑帽子的 *A. mediterrania* 和褶皱帽子的 *A. crenulata*，将 *A. mediterrania* 的茎和帽子移植到 *A. crenulata* 的基底部，植入的帽子转化成了褶皱的。反过来的移植实验也显示了同样的现象，结果表明决定帽子形态的遗传信息由基底内部的细胞核决定。大约同一时间，Briggs 和 Kingzao 在 Driesch 及 Spemann 发现的基础上成功地将未分化的蛙囊胚细胞核转入去核受精卵中。大约 1/3 的实验中，植入的细胞核能够指导受精卵发育为正常胚胎。然而，使用从晚期胚胎（如神经胚和尾芽）中获取的细胞核却不能使受体卵细胞正常发育，大部分都不能形成完整的原肠胚[4]，说明细胞成熟后分化能力受限。

1962 年，John Gurdon 的发现再次提出了是否多细胞生物已分化细胞的细胞核具有核等价及多能性的疑问。他报道了把非洲爪蟾（*Xenopus*）蝌蚪的肠细胞的核转入去核的爪蟾卵细胞，可以发育成一个完全成形的、可摄食的蝌蚪，证实肠细胞核包含了发育成为一个完整的多细胞生物体所需的遗传信息[5]。然而，值得注意的是，大量的核转移实验中仅有略大于 1% 的受体卵细胞可以成功发育为成熟的蝌蚪。那么具有多能性的细胞核究竟是来自于干细胞还是真正分化了的细胞呢？

克隆技术获得巨大进展是在 1981 年，剑桥的 Martin Evans 和旧金山的 Gail Martin 同时分别从小鼠的囊胚中成功培养了胚胎干细胞（embryonic stem cell，ES 细胞）[6,7]。胚胎干细胞的多能性，结合运用同源重组技术靶向特定基因序列至其同源位点（cognate site），使我们能够克隆特异性和选择性灭活或修饰相关基因的小鼠。Oliver Smithies 和他的同事们首次利用靶向同源重组的方法在小鼠的 ES 细胞内矫正了突变的 *Hprt* 基因[8]。另外，Capecchi 和同事们以类似的方法灭活了 ES 细胞内的 *Hprt* 基因[9]。3 年后，已有报道通过把靶向修饰过的 ES 细胞导入小鼠囊胚腔中，可以培育出转基因小鼠[10~12]。小鼠 ES 细胞可以发育形成完整的机体，并且可以导入胚系发育形成没有丝毫发育和生理障碍的转基因小鼠，提示 ES 细胞具备全基因组的效能（full genomic competency）。

随着模仿 Gurdon 实验方法的体细胞核转移（somatic cell nuclear transfer，SCNT）技术的发展，8 年后，人们证实了哺乳动物体细胞存在同样的核等价现象[5]。Ian Wilmut 和同事们[13]把绵羊乳腺细胞的核移植入激活的去核卵细胞，待其体外发育至囊胚阶段时，把它植入代孕母体的子宫内，生出了一只有生命力的羔羊，命名为"多莉"。此技术的成功率非常低，277 例中最后只有 3 例可以孕育生命，而且只有 1 只羔羊（即"多莉"）存活至成年。然而自首次报道后，SCNT 被成功运用到小鼠、马和雪貂等一系列哺乳动物中[14]。SCNT 也被用于制备 ES 细胞以矫正小鼠模型中的缺陷基因[15]，从而使 ES 细胞不必再从受精卵中获取，而且避免了使用从受精卵获取 ES 细胞带来的伦理问题。SCNT 的成功依然避免不了一些技术上的问题，移植后的卵细胞能够发育到囊胚期的比例较低,且在发育成熟的动物模型常常存在发育异常。"多莉"羊在正常绵羊一半寿命的时候就被施予安乐死,她罹患罕有的关节炎及进行性肺病,随后的尸检证实为肺腺癌。"多莉"羊体细胞的端粒很短,如同年老的动物,可能是其本身的端粒过早地缩短促进快速衰老,也可能是由于核转移使用的成体细胞核已经存在端粒侵蚀(eroded telomeres)。

最近,两个实验室分别报道先后在小鼠[16]和人类[17,18]的成体细胞中同时导入四个基因的编码 cDNA 后,可以被诱导去分化产生诱导多能干细胞(induced pluripotent stem cell,IPS cell),其在形态上和 ES 细胞类似,且可以再分化产生多种类型细胞。起初,人们认为 Oct4、Sox2、Klf4 和 c-Myc 四个基因即足以诱导 IPS,但是似乎其他某些基因,如 Nannog 和 Klf 家族其他成员对于诱导 IPS 细胞也很重要。在此,我们主要想说明的是,实验资料证实了体细胞包含可以调控细胞向各型细胞分化的所有的核信息。在去分化过程中,表观遗传学的调控依然是个难题,体细胞中核等价的程度、SCNT 技术所使用的核问题都仍然很受关注。

体细胞和小鼠 ES 细胞的基因组完整性在程度上有着本质的区别。也就是说,作为 SCNT 和 IPS 技术基础的体细胞,在核等价程度上依然不能达到从囊胚中获得的真正 ES 细胞的水平。SCNT 技术的低成功率可能与此有一定关系。而且,体细胞较生殖细胞和 ES 细胞更加耐受有害突变,体细胞突变积累引起老化,最终导致机体疾病和死亡[19~22]。从进化论和利己的角度来讲,衰老过程如同 Weil 和 Radman 所言:"我们显然是终其一生地服务于自私的进化目的:利用胚系的不死性和解决繁殖后的机体细胞来维持物种。虽然进化论不'关心'我们繁殖后的机体,我们当然是要关注的"[23]。与生殖细胞同样具有多能性特征的 ES 细胞,具有不同于或优于体细胞的众多机制来维持基因组的完整性。我们主要探讨在小鼠 ES 细胞中的情况并介绍其促进基因组完整性的机制。

体细胞的突变频率

哺乳动物体细胞在体内有高达 10^{-4} 的突变频率[21,22,24,25]。而且,杂合子位点发生的突变事件的 70%~80% 存在同源染色体重组(有丝分裂重组)导致的杂合性丧失(LOH),从而累及交叉重组位点附近的所有杂合位点[24~27]。点突变、小的插入或缺失等突变的频率可以通过转基因小鼠来判定,转基因模型含有原核报告基因如 *lacI*[26]、*lacZ*[28]、*gpt*[29],或 λ 噬菌体 *c11* 基因[30]。在啮齿类动物模型中,主要作为报告基因的内源性基因为选择性地编码促进嘌呤再利用的酶,即次黄嘌呤鸟嘌呤磷酸核糖转移酶(hypoxanthine guanine phosphoribosyltransferase,Hprt)[27]和腺嘌呤磷酸核糖转移酶(adenine phosphoribosyltransferase,Aprt)的基因[25,27]。在这些模型中,辨别小鼠或人中引起大部分杂合位点突变的有丝分裂重组最为有效的是 *Aprt* 模型。这个模型用靶向敲除 *Aprt* 位点的小鼠(如 129/sv 小鼠)与野生型 *Aprt* 的另一品系小鼠(如 C57B1/6 小鼠)杂交。F_1 代小鼠在 *Aprt* 位点及位点两侧品系特异性的多态性位点都是杂合性的。8~10 周龄时,处死小鼠,取皮肤成纤维细胞及脾 T 淋巴细胞进行培养,皮肤成纤维细胞通常取自耳部。将细胞立即培养在含有 5-氟腺嘌呤(5-fluoroadinine)或 2,6-二氨基嘌呤（2，6-diaminopurine，DAP）的培养基中。其中 5-氟腺嘌呤和 DAP 对于有功能性 Aprt 活性的细胞具有细胞毒性，而缺乏 Aprt 活性的细胞可以存活。缺乏 Aprt 活性的细胞在选择性培养基中可以形成克隆，挑克隆并用 PCR 的方法对 Aprt 位点首先进行 LOH 分析。如果确定 Aprt 位点存在 LOH，进一步分析其两侧的多态性位点，有必要的话将 *Aprt* 基因进行 PCR 扩增后测序。这样可以发现几乎所有导致 Aprt 功能缺陷的突变。70%～80% 的突变是由于 Aprt 位点及其两侧的多态性位点的 LOH，提示发生了有丝分裂重组。另外一些突变，如中间缺失、基因转换、点突变或功能性位点的表观遗传失活，大约占剩余突变事件的 20%～30%。

小鼠 ES 细胞基因组的保护

尽管胚系的基因突变对于遗传多样性和物种进化很重要，若突变频率或 LOH 发生率达到 10^{-4}，即达到体细胞内水平（图 5-1），则导致产生大量有缺陷的胚胎和具有先天性畸形的新生个体。

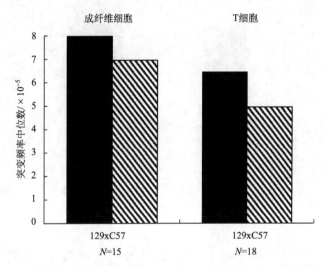

图 5-1　体内皮肤成纤维细胞和脾 T 细胞中突变率及 LOH 发生率。黑色柱表示总突变频率，斜条纹柱表示有丝分裂重组及 LOH 的发生率

那么对于发生的逐渐增多的突变，多能干细胞是否能够免受其害呢？在小鼠 ES 细胞中，至少存在三种相关保护机制：抑制突变和同源染色体重组；促进细胞凋亡；选择性利用高保真同源介导的机制来修复 DNA 双链损伤（DNA double strand break，DSB），而不是主要存在于体细胞中的、具有错误修复倾向的非同源末端连接（nonhomologous end-joining，NHEJ）。上述的保护机制是针对小鼠 ES 细胞的，在真正的生殖细胞，甚至人 ES 细胞中，也有着不完全相同的机制。它们让我们认识到多能干细胞保持基因组完整性的重要性。

小鼠 ES 细胞中突变受到抑制

如上所述，体细胞的高突变率，如果发生在生殖细胞或多能干细胞中，则导致生殖和发育功能损伤。运用 $Aprt$ 小鼠模型，用靶向敲除 $Aprt$ 位点的 129 小鼠和 $Aprt$ 野生型的 C3H 小鼠杂合获得 F_1 代，从其囊胚中获得 ES 细胞。从 14 天的胚胎中获取同基因小鼠胚胎成纤维细胞（mouse embryonic fibroblast，MEF）作为对照。ES 细胞和 MEF 细胞在 DAP 或 6-硫鸟嘌呤（6-thioguanine，6-GT）的作用下形成 $Aprt$ 或 $Hprt$ 功能缺陷细胞克隆的能力显示：ES 细胞中 $Aprt$ 位点的突变率（约 10^{-6}）比 MEF 细胞中（约 10^{-4}）大约低 100 倍，而 MEF 细胞中的突变率与成体细胞体内突变率近似（图 5-2）。

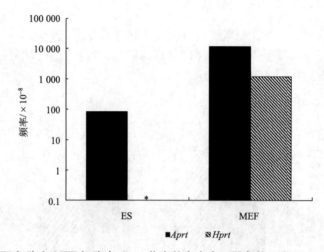

图 5-2　ES 细胞和 MEF 细胞中 $Aprt$ 位点的突变率（黑色柱）和 $Hprt$ 位点的突变率（斜条纹柱）。＊表示在接种的 10^8 个 ES 细胞中没有检测到 $Hprt$ 功能缺陷的克隆。纵坐标显示的为对数值

尽管 $Hprt$ 位点突变率较 $Aprt$ 低，但却显示了相同的趋势。突变率不同的原因和 $Aprt$ 位于常染色体可以进行有丝分裂重组（mitotic recombination）而 $Hprt$ 是 X 染色体连锁有部分关系。在男性，只有一条染色体具有该基因；在女性，由于 X 染色体灭活（X-inactivation），也相当于仅有一条染色体有该基因。大约 60% 的突变源于 LOH，但与 MEF 和成体细胞的 LOH 主要因有丝分裂重组导致不同，ES 细胞的 LOH 是由于姐妹染色单体的不分离而产生单亲源二体（uniparental disomy）[31]。$Aprt$ 基因位于小鼠 8 号染色体，可能在此染色体上发生染色单体不分离的概率很高，从而出现与其他染色体大相径庭的数据及相关诠释[32]。小鼠 ES 细胞中关于有丝分裂重组发生率亦有不一致的报道，将两个编码不同荧光蛋白的基因分别靶向插入 6 号染色体的两条同源染色单体上的 $ROSA26$ 基因位点构建小鼠模型，研究该小鼠模型中有丝分裂重组发生率[33]。当双色荧光变成单色荧光时意味着丢失了其中一个位点，其发生率大约为 10^{-4}。其中大约一半可能是由于有丝分裂重组，因为在 $ROSA26$ 基因附近的一个位点仍然保持了杂合

性。大部分其他的单色荧光的克隆在 ROSA26 位点和着丝粒区都丧失了杂合性，提示是由于姐妹染色单体不分离产生的单亲源二体所致。其报道的重组发生率大概比 Aprt 小鼠模型高两个数量级，可能的原因为 ES 细胞中不同的染色体的重组发生率有很大差别或 ROSA26 位点插入的荧光蛋白基因促进了有丝分裂重组的发生。值得注意的是，还有两项研究报道的数值却相当低，但仍然高于用 Aprt 作为报告基因的数值[34,35]。关于 ES 细胞的突变频率和 LOH，有人对相关的研究做了总结分析[36]。

原核生物中错配修复障碍促进了重组和物种间侧向基因转移（lateral genomic transfer）的发生[37,38]。在细菌中[37,38]，且很有可能在酵母菌中[39]，如果没有有效的错配修复，将导致超重组表型（hyper-recombination phenotype）。从进化的角度看，在 ES 细胞中，错配修复蛋白可能在控制有丝分裂重组方面发挥一定作用。ES 细胞裂解物的 Western Blot 结果显示，在小鼠 ES 细胞中，错配修复蛋白 MSH2 和 MSH6 的含量比 MEF 中大大增加；错配修复蛋白 MLH1 和 PMS2 也有所增加，但是程度远低于 MSH2 和 MSH6（图 5-3）。为了验证错配修复抑制有丝分裂重组的作用，在 Aprt 位点为杂合子的 ES 细胞转染了降解 MSH2 mRNA 的 siRNA，然后检测转染后的细胞突变频率和有丝分裂重组导致的 LOH 的发生率，同时对比由于 MSH2 基因 G674A 错义突变无错配修复活性的 ES 细胞系[40]。图 5-4 显示当小鼠 ES 细胞中错配修复功能受损或缺失时，突变频率和有丝分裂重组的发生率增加。在 MSH2 活性下降约 60% 的细胞克隆中，突变频率增高约 20 倍，有丝分裂重组所致的 LOH 也成比例地增高。在存在 MSH2 基因 G674A 错义突变，而错配修复功能缺失的细胞系中，突变频率和有丝分裂重组的发生率更为增加，几乎达到 MEF 细胞中的水平。总之，结果提示小鼠 ES 细胞保持基因组完整性的机制之一是通过抑制整体上的突变，使有丝分裂重组，尤其是其所致的 LOH，而后者由参与错配修复蛋白部分介导。

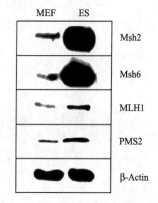

图 5-3　ES 细胞和 MEF 细胞中的错配修复蛋白的含量。ES 细胞和 MEF 细胞裂解液在凝胶电泳后使用 Msh2、Msh6、Mlh1 和 Pms2 的抗体做 Western Blot 检测。β-actin 为内参蛋白

ES 细胞群通过清除 DNA 受损的细胞来维持初始状态基因组

小鼠 ES 细胞对外源性刺激和 DNA 损伤高度敏感，导致细胞的凋亡[41~43]。这一点有利于从自我更新的干细胞群中清除损伤的细胞，从而保持细胞的初始基因组状态。小鼠 ES 细胞缺乏细胞周期 G_1 期检查点（G_1 checkpoint）[44,45]，且在体细胞中 DNA 双链损伤后激活 G_1 检查点的两条主要通路在 ES 细胞中缺如[44]。该通路可参见图 5-5 所示的简化示意图。简而言之，DNA 损伤被 MRN 复合体（MRE11、RAD50、NBS1）感知，可以使 ATM 磷酸化而活化。当双链 DNA 出现缺损时，ATM 活化此两条通路，使细胞周期阻滞于 G_1/S 检查点。一条通路是，ATM 磷酸化 p53 蛋白的第 15 位丝氨酸，从而使 p53 活化，活化的 p53 随后诱导 Cdk 抑制物 p21 的转录，导致 G_1/S 期阻滞。另一

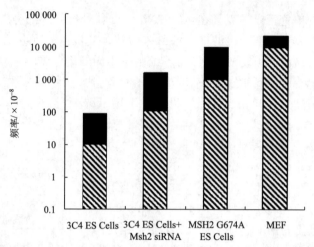

图 5-4 ES 细胞中 Msh2 活性的降低对于突变频率和有丝分裂重组的影响。在 *Aprt* 位点为杂合性的 ES 细胞中转染 Msh2 的 siRNA,然后培养在含有 DAP 的培养基中。定量其形成的克隆,将其突变频率与未经过选择性培养的对照进行比较。随后将 DNA 进行 PCR,分析 Aprt 活性缺失的原因

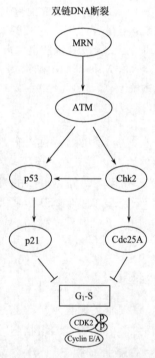

图 5-5 双链 DNA 损伤活化两条通路导致 G_1 期阻滞示意图

条通路是,ATM 磷酸化检查点激酶 Chk2 的第 68 位苏氨酸,从而促进其活化。活化的 Chk2 紧接着磷酸化 Cdc25A 的第 123 位丝氨酸,促使其通过蛋白酶体介导的途径降解。Cdc25A 双功能磷酸酶对 Cdk2 第 14 位磷酸化的苏氨酸、Cdk2/CyclinE 和 CyclinA 复合体的第 15 位磷酸化的酪氨酸进行去磷酸化,使细胞由 G_1 期进入 S 期。随着 Cdc25A 被蛋白水解酶降解而逐渐减少或缺失,Cdk2 无法被去磷酸化而使细胞又被阻滞于 G_1 期。上述的两种通路在小鼠 ES 细胞中缺如[44]。对于第一种通路,照射后小鼠 ES 细胞中虽然存在 p53 蛋白,但主要存在于细胞核外,而体细胞核外则无 p53 蛋白。图 5-6 很好地展示了当 ES 细胞克隆和 1 个 MEF 细胞经相同剂量辐射后,胞内 p53 蛋白分布位置的不同。MEF 细胞中检测到的 p53 蛋白大都位于核内,而在 ES 细胞克隆中主要位于胞质。与此现象一致,p21 蛋白在小鼠 ES 细胞中未检测到,而且即使是在给予离子辐射后,仍然未检测到[45]。

第二条通路,可能一定程度上被 Chk2 调控。在小鼠 ES 细胞中,Chk2 和 γ-微管蛋白共同位于中心体内,因此不太可能磷酸化它的底物,如 Cdc25A。ES 细胞中的 Cdc25A 在经过离子辐射后并不像在体细胞中一样被降解,而 Cdk2 仍主要处于低磷酸化状态,能够顺利进入 S 期。当 ES 细胞中转染编码 Chk2 的质粒而获得异位表达的 Chk2 时,辐射后的 Cdc25A

被蛋白酶酶解且 Cdk2 被磷酸化，预示着 G_1 期阻滞的发生[45]。与上述结论相符，结果显示异位表达 Chk2 的 ES 细胞在离子辐射后出现 G_1 期阻滞（图 5-7）。

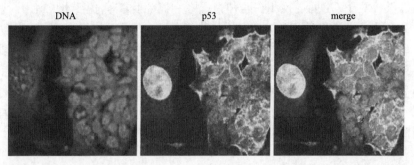

图 5-6　ES 细胞和 MEF 细胞在经过 10Gy×射线辐射后，在同一张盖玻片上通过免疫荧光显示其内的 p53 蛋白分布。左图显示 DRAQ5 染色后的细胞核，注意 MEF 细胞核和 ES 细胞克隆焦距相同。中图为辐射后 p53 蛋白染色。右图为前两个图片的叠加效果。（引自 Hong et al.，PNAS 2004；101：14443-14448）

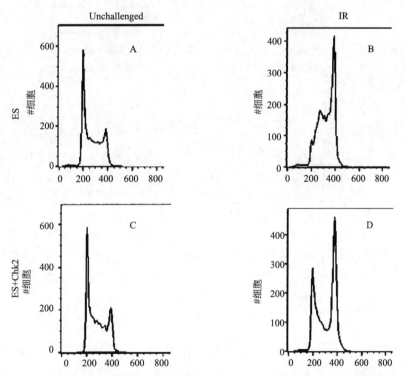

图 5-7　给予及未给予离子辐射的野生型 ES 细胞（A 和 B），以及转染了编码 Chk2 的质粒的 ES 细胞（C 和 D），经过固定、碘化丙啶染色及流式细胞仪检测进行细胞周期分析。在未给予辐射情况下（A 和 C），两种细胞的周期分布类似。离子辐射后，野生型 ES 细胞（B）G_1 期细胞缺乏，S 期细胞积累，且伴有明显 G_2/M 期阻滞。辐射后，异位表达 Chk2 的 ES 细胞也阻滞于 G_2/M 检查点，同时还存在 G_1/S 期阻滞（D）。（引自 Hong et al.，PNAS 2004；101：14443-14448）

假说认为小鼠 ES 细胞中缺乏 G_1 期检查点,使得受损的小鼠 ES 细胞进入 S 期,损伤的 DNA 经过复制后被扩大化而引起细胞死亡。异位表达 Chk2 的 ES 细胞在经受离子辐射后能够停滞于 G_1 期,预测在这种情况下细胞受保护而免于死亡。图 5-8 所示内容证实了此预测。野生型的 ES 细胞和异位表达 Chk2 的 ES 细胞在离子辐射后,用 PI 染色显示死细胞,用 AnnexinV 标记早期凋亡细胞。野生型的 ES 细胞和异位表达 Chk2 的 ES 细胞中的 AnnexinV 阳性细胞比例分别为 11% 和 16%,伴随少量的 PI 阳性死细胞。辐射后,40% 的野生型 ES 细胞 AnnexinV 阳性,且死亡细胞明显增多。相反,异位表达 Chk2 的 ES 细胞中的比例却没有变化(图 5-8)。该数据再次证实 G_1 检查点的缺失可以清除细胞群体中 DNA 受损的细胞,从而保持了整体基因组的完整性。

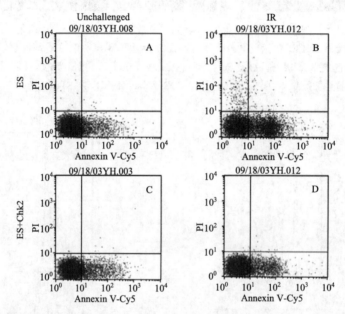

图 5-8 野生型 ES 细胞(A 和 B)和转染了编码 Chk2 的质粒的 ES 细胞(C 和 D),在给予离子辐射(B 和 D)及未给予辐射(A 和 C)情况下,用 Annexin V 和碘化丙啶(PI)染色后,进行二元流动分析(bivariate flow analysis)。水平线以上(PI 阳性)的细胞为死亡细胞,垂直线右侧的细胞(Annexin V 阳性)为凋亡细胞。(引自 Hong et al., PNAS 2004;101:14443-14448)

小鼠 ES 细胞首选高保真同源介导的修复而不是非同源末端连接来修复双链 DNA 损伤

当细胞在接触依托泊苷、拓扑异构酶Ⅱ抑制剂或使用放射类治疗手段时,出现双链 DNA 断裂等 DNA 损伤[46,47]。这种 DNA 损伤的两种主要修复机制是有错误修复倾向的 NHEJ,以及高保真的、同源介导的修复。虽然在体细胞中,上述两种修复机制都在发挥作用,但 NHEJ 起主要作用且在各个细胞周期都发挥功能[48]。考虑到保持生殖细胞和 ES 细胞基因组完整的重要性,任何有错误倾向的 DNA 修复机制自然都是有害的。为了验证 NHEJ 不是修复 ES 细胞中双链 DNA 断裂的主要途径，用 Western Blot 方法检测比较了 ES 细胞和同基因 MEF 细胞内参与两条途径的蛋白质的含量。显然，ES 细胞内同源介导的修复途径的蛋白质含量明显高于 MEF 细胞裂解物内的含量。而 ES 细胞内 NHEJ 途径的蛋白质水平相较于 MEF 细胞却是变化不一的（图 5-9）。有意思的是，ES 细胞中参与 NHEJ 修复途径的限速酶——DNA 连接酶 IV 显著低于 MEF 细胞。

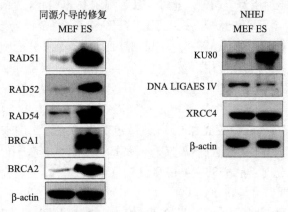

图 5-9　ES 细胞和 MEF 细胞裂解液通过 Western Blot 分析比较同源介导的修复途径（左图）和 NHEJ 修复途径中蛋白质的相对含量

虽然蛋白质的含量有一定提示意义，但并不一定能反映其功能。体细胞和 ES 细胞中同源介导及 NHEJ 两条修复途径的相对活性，通过转染一套包含 3 个报告基因的质粒来大体加以比较。评估同源介导修复途径活性的质粒是 pDR-GFP[49]，包含被嘌呤霉素耐药标志隔离开的两个串联无功能的 GFP 基因，其通过嘌呤霉素耐药标志来筛选稳定转染的细胞。5′端基因通过插入一个 I-Sce1 限制性内切核酸酶位点，产生移码和不成熟末端信号，其下游的 GFP 基因两端被破坏而失活。I-Sce1 限制性内切核酸酶位点被 I-Sce1 酶切割后导致双链断裂。GFP 基因两个无功能片段连接后，恢复为功能性片段，产生荧光信号可被量化。另外的两个报告质粒被用于评估 NHEJ 活性。质粒 pEGFP-PEM1-AD2[50] 同样是利用 GFP 活性恢复产生的荧光来衡量修复途径，但是它的荧光信号只有在 DNA 双链断裂被 NHEJ 途径修复后才能产生。此报告基因的构建体中，GFP 基因的两个部分被一个来自于 PEM 同源盒（homeobox）基因的含有剪接信号的内含子所分离。PEM 内含子内插有一个腺病毒 AD2 基因，它具有很强的剪接受体和剪接供体位点，干扰剪接活动并阻断 GFP 荧光的恢复。AD2 基因两侧同时存在成对的反向 Hind III 和 I-Sce1 剪接位点。用于瞬时转染分析时，在质粒转入细胞之前，AD2 序列在 I-Sce1 酶的消化作用下被移除，这同时也移除了 AD2 内含有的剪接位点，留下的非互补末端只有通过 NHEJ 修复。修复后，PEM 序列被剪除，使得 GFP 被其分开的两部分可以结合产生荧光信号从而可被识别和量化。第三个报告质粒是 pINV-CD4[51]，当双链断裂通过 NHEJ 途径修复时，可以产生细胞表面标志 CD4。母代质粒含有一个 CMV 启动子操控组织相容表面抗原（histocompatibility surface antigen）H-2Kd 的表达。H-2Kd 序列后为一个反向的 CD8 cDNA 和无启动子的 CD4 cDNA 序列。H-2Kd 和反向的

CD8 cDNA 两侧有反向的 I-Sce1 剪接位点，I-Sce1 酶可以剪除 H-2Kd 和 CD8 cDNA，从而产生非互补末端。非互补末端经 NHEJ 修复后，CD4 序列靠近 CMV 启动子而获得表达。利用 NHEJ 途径修复报告质粒的细胞可以通过 CD4 进行流式分选和定量。第一个和第三个报告质粒在稳定转染和瞬时转染中都有应用。

如前报道，体细胞主要通过 NHEJ 修复双链 DNA 损伤。通过使用上述的质粒转染证实，MEF 也是主要利用修复机制，而小鼠 ES 细胞并非如此，除非在被诱导分化时。结果如图 5-10 所示，MEF 和经全反式维甲酸（all-trans retinoic acid，ATRA）诱导分化的 ES 细胞不易被报告质粒稳定整合，因此通过瞬时转染来分析胞内的修复途径。两种细胞 DNA 在 I-Sce1 酶消化作用下被线性化，随后通过电转染 pDR-GFP 评估其胞内同源性介导的修复机制，通过电转染 pEGFP-PEM1-AD2 评估 NHEJ 介导的修复机制。这两种转染都是同时转染发射红色荧光的 pDs-Red 来评价其转染效率。不同机制的修复效率都是根据同时发射红色荧光和 GFP 的绿色荧光的细胞比例来衡量的。数据证实体细胞主要利用 NHEJ 修复双链 DNA 损伤，而 ES 细胞则倾向于利用重组介导的修复。当 ES 细胞被诱导分化时，它们的主要修复机制从重组介导修复转变成 NHEJ。在此过程中，伴随着大量蛋白质水平的下调，如参与重组介导修复过程的 Rad 51。同时，DNA 连接酶 IV 的水平也明显上调，而在 ES 细胞中过表达 DNA 连接酶 IV 和抑制其降解的各种途径都无效，且 DNA 连接酶 IV 的增多可导致 ES 细胞迅速死亡。因此，虽然不被 ES 细胞耐受，DNA 连接酶 IV 水平上的上调却伴随在 ES 细胞分化过程中，并且可能有助于调控体细胞内的 NHEJ 修复机制。

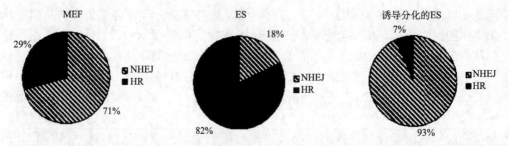

图 5-10 MEF 细胞、ES 细胞和 ATRA 诱导分化的 ES 细胞中修复双链 DNA 损伤的 NHEJ 途径（条纹扇形）和同源介导修复途径（黑色扇形）的相对活性。每种细胞 DNA 在 I-Sce1 酶消化作用下被线性化后，电转染 pDR-GFP 评估其胞内同源介导修复机制的相对水平，电转染 pEGFP-PEM1-AD2 评估 NHEJ 的相对水平

结论

体细胞的胞核转入无核的卵细胞后，可以形成一个完整的生物体，证实了体细胞的胞核保留了遗传信息的完整性和有效性。IPS 细胞技术的出现证实了体细胞的胞核可以进行去分化并恢复多能性。这类技术的成功率通常很低，反映了体细胞核，特别是来自老年个体的体细胞核的基因组，并不能和生殖细胞或 ES 细胞的基因组完全等同。IPS

技术的低成功率原因可能与体细胞核内突变累积以致其不能向多能性转化或端粒被降解不能维持有效长度有部分关系。由于这些问题仍有待解决，目前生殖细胞和 ES 细胞还是保持基因组完整性不可替代的细胞。

本章主要探讨的是小鼠 ES 细胞维持基因组完整性的三条重要的机制，并不是说这些机制是小鼠所特有的，也不是说它们是保持基因组完整性仅有的机制。例如，对于诸如氧化损伤引起的 DNA 链偶联、8-羟基脱氧鸟苷损伤和双链断裂等 DNA 损伤，人的 ES 细胞比体细胞有着更为有效的 DNA 修复机制[52]。小鼠 ES 细胞比成纤维细胞对氧化应激有着更为有效的应激反应[43]。人 ES 细胞的抗氧化基因 *SOD2* 和 *GPX2* 的表达水平要高于分化细胞，而活性氧自由基（reactive oxygen species，ROS）表达水平低于分化细胞。当人 ES 细胞受到内源性损伤时，其 DNA 受损较轻可能与此相关[53]。虽然体内细胞相对于体外培养的细胞不好进行实验操控和分析，仍有确证显示体内基因组亦存在保护机制。例如，雄性小鼠的生殖细胞有着特征性成批发生的大量凋亡[54,55]。在原始生殖细胞向性腺迁移过程中及出生后 10 天首次精子发生时，可能是为了清除受损的生殖细胞，小鼠的雄性生殖细胞特异性地大量凋亡[56]。同时值得注意的是，分裂中的精原细胞对辐射极其敏感，而起支持作用的 Sertoli 细胞却并非如此[57]。因此，Weill 和 Radman 重申了 Dawkins 的观点[58]，即"我们实际上如同行尸走肉般地活着，只是生殖系统基因组的奴隶"[23]。越来越多的证据表明多能干细胞，尤其是生殖细胞，有着一套维持自我更新、多能性，尤其是不死性的策略。

致谢

感谢 James Stringer、Alvaro Puga、El Mustapha Bahassi 和 Carolyn Price 博士仔细阅读了手稿并提出了许多有益的建议。这项工作由 NIEHS 和辛辛那提大学环境遗传学中心的基金支持。

（陈　芳　韩忠朝　译）

参 考 文 献

1. Ebert JD. Cell interactions: the roots of a century of research. Biol Bull 1985; 168:80-87.
2. Hämmerling J. "Nucleo-cytoplasmic relationships in the development of Acetabularia". J Intern Rev Cytol 1953; 2:475-498.
3. Briggs R, King TJ. Transplantation of living nuclei from blastula cells into enucleated frogs' eggs. Proc Natl Acad Sci USA 1952; 38:455-463.
4. Briggs R, King TJ. Nuclear transplantation studies on the early gastrula (Rana pipiens). Develop Biol 1960; 2:252-270.
5. Gurdon JB. The developmental capacity of nuclei taken from intestinal epithelium cells of feeding tadpoles. J Embryol Exp Morphol 1962; 10:622-640.
6. Evans M, Kaufman M. Establishment in culture of pluripotential cells from mouse embryos. Nature 1981; 292:154-156.
7. Martin G. Isolation of a pluripotent cell line from early mouse embryos cultured in medium conditioned by teratocarcinoma stem cells. Proc Natl Acad Sci USA 1981; 78:7634-7638.
8. Doetschman T, Gregg RG, Maeda N et al. Targeted correction of a mutant HPRT gene in mouse embryonic stem cells. Nature 1987; 330:576-578.

9. Thomas KR, Capecchi MR. Site-directed mutagenesis by gene targeting in mouse embryo-derived stem cells. Cell 1987; 51:503-512.
10. Koller BH, Hagemann LJ, Doetschman T et al. Germ-line transmission of a planned alteration made in a hypoxanthine phosphoribosyltransferase gene by homologous recombination in embryonic stem cells. Proc Natl Acad Sci USA 1989; 86:8927-8931.
11. Zijlstra M, Li E, Sajjadi F et al. Germ-line transmission of a disrupted beta 2-microglobulin gene produced by homologous recombination in embryonic stem cells. Nature 1989; 342:435-438.
12. Thomas KR, Capecchi MR. Targeted disruption of the murine int-1 protooncogene resulting in severe abnormalities in midbrain and cerebellar development. Nature 1990; 346:847-850.
13. Wilmut I, Schnieke AE, McWhir J et al. "Viable offspring derived from fetal and adult mammalian cells". Nature 1997; 385:810-813.
14. Campbell KHS, Fisher P, Chen WC et al. Somatic cell nuclear transfer: Past, present and future perspectives. Theriogenology 2007; 68(Suppl 1):S214-S231.
15. Rideout III WM, Hochedlinger K, Kyba M et al. Correction of a genetic defect by nuclear transplantation and combined cell and gene therapy. Cell 2002; 109:17-27.
16. Takahashi K, Yamanaka S. Induction of pluripotent stem cells from mouse embryonic and adult fibroblast cultures by defined factors. Cell 2006; 126:663-676.
17. Takahashi K, Tanabe K, Ohnuki M et al. Induction of pluripotent stem cells from adult human fibroblasts by defined factors. Cell 2007; 131(5):861-872.
18. Yu J, Vodyanik MA, Smuga-Otto K et al. Induced pluripotent stem cell lines derived from human somatic cells. Science 2007; 318:1917-1920.
19. Szilard L. On the nature of the aging process. Proc Natl Acad Sci USA 1959; 45:30-45.
20. Kirkwood TB, Proctor CJ. Somatic mutations and ageing in silico. Mech Ageing Dev 2003; 124:85-92.
21. Suh Y, Vijg J. Maintaining Genetic integrity in aging: a zero sum game. Antioxid Redox Signal 2006; 8:559-571.
22. Stambrook PJ. Do embryonic stem cells protect their genomes? Mech Ageing Dev 2007; 329:313-326.
23. Weil JC, Radman M. How good is our genome? Philos Trans R Soc London B 2004; 359:95-98.
24. Shao C, Deng L, Henegariu O et al. Mitotic recombination produces the majority of recessive fibroblast variants in heterozygous mice. Proc Natl Acad Sci USA 1999; 96:9230-9235.
24. Stambrook PJ, Shao C, Stockelman M et al. High frequency in vivo loss of heterozygosity is primarily a consequence of mitotic recombination. Cancer Res 1997; 157:1188-1193.
25. Stambrook PJ, Shao C, Stockelman M et al. Tischfield JA.APRT: a versatile in vivo resident reporter of local mutation and loss of heterozygosity. Environ Mol Mutagen 1996; 28:471-482.
26. Kohler SW, Provost GS, Fieck A et al. Analysis of spontaneous and induced mutations in transgenic mice using a lambda ZAP/lacI shuttle vector. Environ Mol Mutagen 19991; 18:316-321.
27. Van Sloun PP, Wijnhoven SW, Kool HJ et al. Determination of spontaneous loss of heterozygosity mutations in Aprt heterozygous mice. Nucleic Acids Res 1998; 26:4888-4894.
28. Gossen J, de Leeuw WJ, Verwest A et al. High somatic mutation frequencies in a LacZ transgene integrated on the mouse X-chromosome, Mutat Res 1991; 250:423-429.
29. Nohmi T, Katoh M, Suzuki H et al. A new transgenic mouse mutagenesis test system using spi- and 6-thioguanine selections, Environ Mol Mutagen 1996; 28:465-470.
30. Jakubczak JL, Merlino G, French JE et al. Analysis of genetic instability during mammary tumor progression using a novel selection-based assay for in vivo mutations in a bacteriophage lambda transgene target. Proc Natl Acad Sci USA 1996; 93:9073-9078.
31. Cervantes RB, Stringer JR, Shao C et al. Embryonic stem cells and somatic cells differ in mutation frequency and type. Proc Natl Acad Sci USA 2002; 99:3586-3590.
32. Liu X, Wu H, Loring J et al. Trisomy eight in ES cells is a common potential problem in gene targeting and interferes with germ line transmission. Dev Dyn 1997; 209:85-91.
33. Larson JS, Yin M, Fischer JM et al. Expression and loss of alleles in cultured mouse embryonic fibroblasts and stem cells carrying allelic fluorescent protein genes. BMC Mol Biol 2006; 7:36.
34. Yusa K, Horie K, Condo G et al. Genome-wide phenotype analysis in ES cells by regulated disruption of Bloom's syndrome gene. Nature 2004; 429:896-899.
35. Luo G, Santoro IM, McDaniel LD et al. Cancer predisposition caused by elevated mitotic recombination in Bloom mice. Nat Genet 2000; 26:424-429.
36. Fischer JM, Stringer JR. Visualizing loss of heterozygosity in living mouse cells and tissues. Mutat Res 2008; 645:1-8.
37. Vulic M, Lenski RE, Radman M. Mutation, recombination and incipient speciation of bacteria in the laboratory. Proc Natl Acad Sci USA 1999; 96:7348-7351.
38. Matic I, Taddei F, Radman M. No genetic barriers between Salmonella enterica serovar typhimurium and Escherichia coli in SOS-induced mismatch repair-deficient cells. J Bacteriol 2000; 182:5922-5924.
39. Manivasakam P, Rosenberg SM, Hastings PJ. Poorly repaired mismatches in heteroduplex DNA are hyperrecombinagenic in Saccharomyces cerevisiae. Genetics 1996; 142:407-416.

40. Lin DP, Wang Y, Scherer SJ et al. An Msh2 point mutation uncouples DNA mismatch repair and apoptosis. Cancer Res 2004; 64:517-522.
41. Van Sloun PPH, Jansen JG, Weeda G et al. The role of nucleotide excision repair in protecting embryonic stem cells from genotoxic effects of UV-induced DNA damage. Nucleic Acids Res 1999; 27:3276-3282.
42. Roos WP, Christmann M, Fraser ST et al. Mouse embryonic stem cells are hypersensitive to apoptosis triggered by the DNA damage O6-methylguanine due to high E2F1 regulated mismatch repair. Oncogene 2007; 26:186-197.
43. Saretzki G, Armstrong L, Leake A et al. Stress defense in murine embryonic stem cells is superior to that of various differentiated murine cells. Stem Cells 2004; 22:962-971.
44. Aladjem MI, Spike BT, Rodewald LW et al. ES cells do not activate p53-dependent stress responses and undergo p53-independent apoptosis in response to DNA damage. Curr Biol 1998; 8:145-155.
45. Hong Y, Stambrook PJ. Restoration of an absent G1 arrest and protection from apoptosis in embryonic stem cells after ionizing radiation. Proc Natl Acad Sci USA 2004; 101:14443-14448.
46. Ferguson LR, Baguley BC. Topoisomerase II enzymes and mutagenicity. Environ Molec Mutagenesis 1994; 24:245-261.
47. Mclendon AK, Osheroff N. Topoisomerase II, genotoxicity and cancer. Mutat Res 2007; 623:83-97.
48. Mao Z, Bozzella M, Seluanov A et al. Comparison of nonhomologous end joining and homologous recombination in human cells. DNA Repair 2008; 7:1765-1771.
49. Pierce AJ, Johnson RD, Thompson LH et al. XRCC3 promotes homology-directed repair of DNA damage in mammalian cells. Genes Dev 1999; 13:2633-2638.
50. Seluanov A, Mittelman D, Pereira-Smith OM et al. DNA end joining becomes less efficient and more error-prone during cellular senescence. Proc Natl Acad Sci USA 2004; 101:7624-7629.
51. Guirouilh-Barbat J, Huck S, Bertrand P et al. Impact of the KU80 Pathway on NHEJ-Induced Genome Rearrangements in Mammalian Cells. Molec Cell 2004; 14:611-623.
52. Maynard S, Swistowska AM, Lee JW et al. Human embryonic stem cells have enhanced repair of multiple forms of DNA damage. Stem Cells 2008; 26:2266-2274.
53. Saretzki G, Walter T, Atkinson S et al. Downregulation of multiple stress defense mechanisms during differentiation of human embryonic stem cells. Stem Cells 2008; 26:455-464.
54. Brinkworth MH, Weinbauer GF, Schlatt S et al. Identification of male germ cells undergoing apoptosis in adult rats. Journal of Reproduction and Fertility 1995; 105:25-33.
55. Richburg JH. The relevance of spontaneous- and chemically induced alterations in testicular germ cell apoptosis to toxicology. Toxicol Let 2000; 112-113:79-86.
56. Rodriguez I, Ody C, Araki K et al. An early and massive wave of germinal cell apoptosis is required for the development of functional spermatogenesis. EMBO J 1997; 16:2262-2270.
57. Hasegawa M, Zhang Y, Niibe H et al. Resistance of differentiating spermatogonia to radiation-induced apoptosis and loss in p53-deficient mice. Radiat Res 1998; 149:263-270.
58. Dawkins R. The Selfish Gene Oxford University Press 1990.

第6章 胚胎干细胞的转录调控

Jian-Chien Dominic Heng, Huck-Hui Ng*

摘要：转录调控是赋予细胞特性和调节细胞内生物学活动的一个关键步骤。在胚胎干细胞（embryonic stem cell，ESC）中，转录因子与它们在基因组模板上的靶点相互进行复杂、精细的作用，作为结构单元形成整个转录调控网络，从而操控细胞的自我更新和多能性。位居复杂网络中心的是三个转录因子，即Oct4、Sox2和Nanog，它们构成ESC转录调控的核心。通过各种不同调控机制，如自动调节环路和前馈环路，组成ESC的调控框架并且维持ESC的稳态。通过大规模的研究，如应用RNA干扰（RNAi）导致功能缺失的筛查和转录组学的分析，发现了更多维持细胞多能性的因子。通过在基因组范围对转录因子的定位研究，进一步揭示了ESC转录调控线路的内在关联性。转录因子同时也同表观遗传学调控因子作用，共同维持ESC的稳定状态。本章节阐述了ESC转录调控的意义，并且追溯ESC转录调控网络解析工作的最新进展。

引言

多细胞生物体的形成，表现为精细复杂的特征性细胞分化过程，即细胞逐渐丧失发育的可塑性并呈现独特功能。虽然同一个有机体的每个细胞都具有完全相同的一套遗传物质，但是不同细胞在其形态和功能方面却展示了完全不同的特性。细胞特性的不同，主要是因为蕴含在基因组DNA内的信息受到各不相同的动态调节。基因的可控性表达受一类叫做转录因子的蛋白调控。人类基因组中，转录因子归属于最大的一类蛋白质家族[1]，它们作为DNA结合蛋白，与增强子、沉默基因或启动子等调控元件结合。转录因子对基因的调控方式与环境相关，可以激活也可以抑制基因的表达。一旦转录因子与DNA基因调控区域结合，RNA聚合酶（RNA Pol）和其他有关因子随即被募集到基因启动子区域，启动转录过程，继而产生信使RNA（mesenger RNA，mRNA）[2]。

除了与DNA直接相互作用，转录因子还参与募集其他的转录调节蛋白进入基因组。转录调节蛋白能够改变基因启动子的可接近性，如表观修饰因子通过翻译后修饰组蛋白和重建核小体结构，从而改变染色质的结构和特性。转录调节蛋白还能够丰富转录调控方式。例如，虽然转录因子在基因组中有许多结合位点，但是它们只能与某些靶点选择性结合，而不能在染色质不易接近的位点结合。

---

* Corresponding Author: Huck-Hui Ng—Genome Institute of Singapore, 60 Biopolis Street, #02-01, Genome Building, Singapore 138672. Email: nghh@gis.a-star.edu.sg

一个细胞内合成的所有特征性 mRNA 分子，称为转录组。转录组的复杂性不仅体现在同一个机体内的不同细胞之间各有不相同，不同的机体之间亦大相径庭，并且转录调控的复杂度与机体的复杂度相关[3]。由于转录因子和辅调节因子操控的转录，表观为联合修饰因子精细调节的遗传学，此转录是决定机体细胞独特性的内在机制。

胚胎干细胞可作为研究转录调控的模型细胞

干细胞具有自我更新和分化成多种不同类型细胞的能力。因此，干细胞是研究如何操控发育和自我更新机制的转录调控理想模型。不同类型的干细胞具有不同的发育潜能，例如，造血干细胞具有多能性（multipotent），可以分化成不同类型的血细胞，而生殖细胞仅具有单能性。在可进行培养的众多干细胞中，胚胎干细胞（embryonic stem cell，ESC）是具有多能性（pluripotent）的一种干细胞。所谓多能性（pluripotency），是指细胞具有三个胚层，即内胚层、外胚层和中胚层，每个胚层都有分化来源细胞的能力。ESC 从着床前的囊胚中获得，它具有精细的转录调控网络，维持细胞的自我更新、多能性，并使其处于可分化状态。由于 ESC 在体外可以无限培养，大量扩增，且具有相对好的均一性，所以可用于分子生化实验研究其自我更新的机制。另外，ESC 具有分化成所有三个胚层来源的细胞的能力也使得它们成为研究细胞分化分子机制的良好模型。

转录因子决定胚胎干细胞的多能性

ESC 被分离获取以来，研究发现了几种维持 ESC 多功能性的重要转录因子。$Pou5f1$ 基因表达的 Oct4 是 ESC 中关键的转录因子，在细胞分化下游表达。Oct4 属于 POU（Pit/Oct/Unc）家族同源结构域蛋白，是八聚体转录因子，可结合 8 个碱基对的 DNA 序列。Oct4 除了在内细胞群表达（inner cell mass，ICM），还在外胚层和原始生殖细胞（primordial germ cell，PGC）表达[4]。在原肠胚形成后，Oct4 仅在生殖细胞中有表达。缺失 $Pou5f1$ 的胚胎产生异常 ICM 细胞，其不具备多能性，更倾向于表达滋养细胞标志，在着床期 ICM 细胞就会死亡[5]。同样，当 ESC 中 Oct4 表达被抑制或敲除时，细胞即丧失自我更新的能力，并且自发地分化为滋养外胚层细胞系[5,6]。Oct4 抑制系特异性基因如 $Cdx2$ 的表达（图 6-1），$Cdx2$ 是滋养外胚层细胞系发育所必需的基因[7]。ESC 维持一定水平的 Oct4 对于其保持多能性是至关重要的，Oct4 的表达水平比正常水平无论是增高还是降低 50% 以上，ESC 都将出现分化[6]。下调 Oct4 表达后，ESC 向滋养外胚层细胞系分化；而上调 Oct4 表达后，ESC 就会向包括原始内胚层细胞系的多种细胞类型分化[6]。

ESC 内的另一个重要转录因子是 Sox2，为性别决定域 Y（sex-determining region-Y，SRY）相关转录因子，其含有称为高动力组盒（high mobility group box，HMG 盒）的 DNA 结构域。Sox2 不仅在 ESC 中表达，同样也在其他细胞中表达，如神经祖细胞（neural progenitor cell，NPC）。Sox2 缺失型突变体不能形成正常的外胚层[8]，而

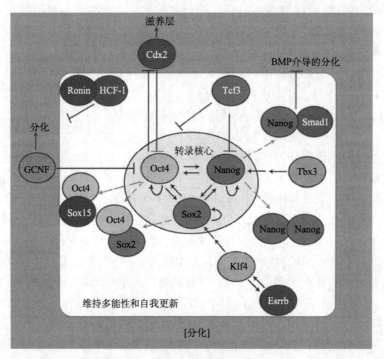

图 6-1　ESC 转录调控网络模型。ESC 转录调控网络核心包括 Oct4、Sox2 和 Nanog（椭圆形区域）。这些因子参与自我调控（弯箭头）和前馈环路等调控机制。转录因子之间可能存在相互调节，如转录核心区内因子之间、Esrrb 和 Klf4 之间（双向箭头）。转录因子可以与其他因子或与其自身相互作用（虚线）来调控另外基因的表达。例如，Oct4 可以与 Sox2 或 Sox15 形成异二聚体，而 Nanog 则与自身形成同二聚体。核心调控蛋白亦参与抑制细胞分化。例如，Nanog 可以与 Smad1 结合抑制 BMP 介导的分化，Oct4 能够抑制 Cdx2 的表达。此外，Ronin 在其他因子的参与下，可以与 HCF-1 相互作用抑制分化。反过来，分化相关因子，如 GCNF 和 Cdx2，可以抑制 Oct4 的表达。位于白色区域内的因子或基因与维持多能性和自我更新有关，而位于绿色区域内的因子或基因则与分化有关。此图的彩色版本详见 www.landesbioscience.com/curie

ESC 在敲除 Sox2 表达后开始向滋养外胚层细胞分化[9]，这与敲除 Oct4 表达后的结果相似[5,6]。而在 Sox2 缺失的细胞中导入 Oct4 后，可以恢复细胞的自我更新能力[9]。因此，Sox2 被认为是通过维持一定水平的 Oct4 来保持 ESC 的稳定性[9]。另一方面，Sox2 表达水平增高可以诱导小鼠 ESC 细胞向神经细胞系分化[10]。除了受 Sox2 调控，Oct4 还与 Sox2 形成异二聚体（图 6-1）相互作用并协同结合成 Oct4-Sox2 结合元件，调控包括其自身在内的[18,19]一些 ESC 特异性基因的转录[11~17]。研究显示 Sox2 和 Oct1 晶体结构如何结合 DNA 的实验数据[20,21]，使得我们能够更好地理解 Sox2 和 Oct4 如何结合 DNA。例如，通过 Sox2 的 HMG 结构域和 Oct1 的 POU 结构域相互结合，形成纤维细胞生长因子（fibroblast growth factor 4，Fgf4）的增强子，建立三元复合物晶体结构的同源模拟的 Oct4-Sox2-DNA 复合物的结构模型[20]。尽管 Sox2 在 ESC 基因表达调控中起关键性作用，令人惊讶的是，在 Sox2 缺失的细胞中，Oct-Sox 增强子依然是激活的，这暗示着其他的 Sox 蛋白也可能是参与活化 Oct-Sox 的调控元件[9]。与此现象一致，人们发现其他的 Sox 蛋白，如 Sox4、Sox11 和 Sox15 也可以结合 Oct-Sox 元件[9]。与 Sox2 类似，Sox15 也与 Oct4 形成异二聚体，尽管结合力要弱些，但也能与几种 Oct-Sox 调控元件结合[22]。然而，Sox15 与 Sox2 功能上有所不同，缺失 Sox15 的 ESC 表现正常[22]，而缺失 Sox2 的 ESC 倾向于向滋养外胚层细胞分化[9]。

Nanog 是 ESC 中的另一个关键转录因子，在 ICM 和 PGC 细胞内特异性表达。Nanog 是个同源结构域转录因子，以二聚体的形式发挥作用（图 6-1），且二聚体化的过程对于其维持 ESC 的自我更新和多能性很重要[23,24]。而且，Nanog 二聚体化对于其自身与其他的多能性相关蛋白的相互作用至关重要[24]。Nanog 是在白血病抑制因子（leukemia inhibitory factor，LIF）缺失的情况下，筛查维持 ESC 多能性新因子时发现的[25]。另外一个独立研究小组通过对 ESC 和着床前胚胎中特异性表达基因的硅片分析，发现 Nanog 对于维持 LIF-Stat3 通路、无活化 ESC 细胞的多能性很重要[26]。Nanog 缺失的胚胎不能形成外胚叶，而 Nanog 缺失的 ICM 细胞可以分化成颅顶部内胚层样细胞[26]。与 Oct4 相似，Nanog 在 ESC 分化过程中表达下调，而 Nanog 缺失的 ESC 向胚外内胚层细胞系分化[27]。因此，Nanog 和 Oct4 两种转录因子对于 ESC 多能性的保持都起着至关重要的作用[2,7,28]。然而，与以前研究结果相反，有研究发现 Nanog 缺失的 ESC 可以保持 Oct4 和 Sox2 的表达水平并维持细胞的自我更新[29]。有报道，小鼠 ESC 内 Nanog 的表达存在异质性[29,32]。Chambers 等发现 ESC 中有一个亚群并不表达 Nanog，此亚群细胞表达正常水平的 Oct4 和 Sox2，并且具有自我更新功能[29]。但是，这些不表达 Nanog 的细胞，大大减弱了自我更新能力，更倾向于向原始内胚层分化[29]。有意思的是，Nanog 表达下调的 ESC 能够再次表达 Nanog。所以 Nanog 联合其他的多能性相关因子在 ESC 中发挥着非常重要的作用，并且被认为是介导产生 ESC 的多能性基态[33]。也就是说，Nanog 对于 ESC 获得多能性是必不可少的，而对于 ESC 细胞自我更新和维持多能性却是非必需的。

除了 Oct4、Sox2 和 Nanog，很多其他的转录因子也参与维持 ESC 的多能性与自我更新。例如，Krüppel 样因子（Krüppel-like factor，Klf），是锌指结构转录因子，在维持 ESC 自我更新中有着明确的作用[34]。过度表达 Klf2 和 Klf4 可以促进 ESC 的自我更新[35]。Klf5 同样也参与维持小鼠 ESC 的细胞状态[36]。与 Sox2 相似，家族蛋白成员 Klf，同样也存在转录调控功能重叠，如 Klf5，与 Klf2 和 Klf4 不仅结合位点存在重叠，而且这些 Klf 转录本的敲除实验证实它们在 ESC 中的功能也存在重叠[34]。

最近发现，含有锌指结合的模序 Ronin 蛋白，在小鼠 ESC 中起着维持其多能性的作用[37]。Ronin 缺失的胚胎出现着床期损伤，证明 Ronin 蛋白在胚胎发生的过程中也很重要。Ronin 敲除的 ESC 细胞无法存活，而 Ronin 的过度表达则抑制 ESC 细胞分化。Dejosez 等发现 Ronin 的异位表达可以恢复 Pou5f1 敲除引起的 ESC 表型的改变[37]。他们进一步推测 Ronin 通过抑制分化相关的基因来维持细胞的多能性。Ronin 可以与一种关键的转录调节蛋白——宿主细胞因子（host cell factor-1，HCF-1）（图 6-1）及其他蛋白质相互作用，形成一个多聚体复合物从而抑制基因的表达。

大规模 RNA 干扰（RNAi）功能缺失筛查实验发现了更多的参与维持 ESC 细胞多能性和自我更新的转录因子。通过 RNAi 筛查，Ivanova 等发现另外两个转录因子，即 Tbx3 和 Esrrb，在维持 ESC 自我更新中起着重要作用[27]。然而，这些调控自我更新的蛋白质被干扰而引起分化，可以通过过度表达 Nanog 得到逆转，提示 Nanog 的作用不受这些因子缺失的影响[27]。在另外一项独立研究中，对 Oct4 和 Nanog 的几个潜在靶基因的干扰实验同样提示 Esrrb 对于 ESC 多能性发挥着重要作用[28]。

除了转录因子，大规模的 RNAi 筛查还发现了维持 ESC 稳定状态的其他重要因子。Ivanova 及其同事发现了 Akt 信号通路的一个辅因子 Tcl1，对于 ESC 的自我更新也很重要[27]。通过进行基因组范围的 RNAi 筛查，Loh 等发现端粒蛋白 Rif1 和前面提及的 Esrrb，在维持 ESC 多能性方面也发挥一定作用[28]。一个研究小组报道了 Trim28 和 Cnot3 这两个转录共调节蛋白，对于细胞自我更新必不可少[38]。最近，研究发现基因组范围的 RNAi 和 Paf1C 的组分，即与 RNA 聚合酶Ⅱ（RNA polyⅡ）复合体相关的，是 Oct4 的调控蛋白[39]。另一项 RNAi 筛查研究发现染色质重构因子 Chd1，对于保持 ESC 细胞中染色质的开放状态很重要[40]。

随着更多的维持细胞多能性调控蛋白的发现，人们开始关注这些转录因子之间的相互关系，以及它们是如何与基因组模板相互作用来调控 ESC 特异性基因表达的。

转录调控网络

转录因子并不一定独立地作用于它的靶基因，而是与其他转录因子协同进行基因调控。Oct4 和 Sox2 转录调控的联合作用就是一个例子。转录因子之间精细的相互作用和协同，连同它们所结合的众多基因靶点，组成了复杂的调控网络，这些网络形成后，维持细胞的特性。

关键转录因子与蛋白质相互作用，形成网络调控的分析，可以让人们更深入地认识转录调控网络是如何调节 ESC 的。例如，通过亲和纯化和质谱分析，可以找到与 Nanog 物理上相互作用的蛋白质[41]。发现与 Nanog 相互作用的蛋白质包括：Sall4、Rif1、Oct4、Dax1、Hdac2、Nac1 和 Zfp281。另外一个研究小组也证明了 Sall4 与 Nanog 相互作用[42]。除了与因子相互作用维持细胞的多能性和自我更新，Nanog 还与某些蛋白质相互作用抑制细胞分化。例如，Nanog 不仅与 Smad1 相互作用阻断 BMP 介导的 ESC 分化[43]（图 6-1），还与 Stat3 联合作用[44]结合并抑制能够促进分化的核因子 kappaB（nuclear factor kappaB，NF-κB）。

转录调控网络的分析构建任务艰巨，需要利用各种基因组学的手段来阐明 ESC 转录组和转录因子，在基因组范围的结合位点情况与其相关的技术将在后一章节给予介绍。

转录调控网络分析技术

多样的技术平台使得我们可以深入地研究 ESC 的转录组。DNA 微阵列分析

(DNA microarray，DNA 芯片）是分析细胞转录组最常见并且广泛使用的技术。DNA 微阵列通过使用探针与细胞表达的转录本结合，高通量地分析基因的表达。这个技术使我们不但可以识别 ESC 内高表达基因，还可以研究 ESC 分化时或 ESC 中特定基因敲除后基因表达的变化情况。微阵列分析联合其他实验技术可以更好地识别 ESC 内的转录节点。例如，在 Ivanova 的研究中，联合采用了在分化 ESC 中快速下调基因微阵列分析和 RNA 干扰致基因功能缺失的综合性分析。诸如此类的综合分析方法是分析转录调控网络强有力的手段。

虽然微阵列分析是研究基因表达的强大手段，却不能分辨基因的不同剪接异构体，不能帮助我们发现新的基因。应用其他的技术系统地分析 ESC 转录组，可以发现 ESC 内存在很多罕见的新基因转录本[45,46]。微阵列分析的其他不足还包括不同微阵列之间数据的比较方法复杂，且杂交技术不可避免地会出现假阴性和假阳性结果。因此，有必要考虑应用其他基因组水平的通量更高且覆盖面更广的检测技术。例如，最近开展起来的 RNA 测序（RNA sequencing，RNA-seq）就应用深度测序技术来更好地分析转录组[47,48]。RNA-seq 采用新一代测序平台进行 cDNA 高通量测序，覆盖面更广且可以分辨基因的不同异构体，并对转录本进行更精确的定量。RNA-seq 技术的飞快发展使得测序工作可以不必合成 cDNA，而直接应用 mRNA 进行[49]。RNA-seq 的这些优点很大程度上克服了微阵列分析的不足，这类优化分析平台的应用让我们能够更好地分析 ESC 转录组。

当转录因子结合到特定的基因组序列时，即发生蛋白质-DNA 相互作用。染色质免疫共沉淀（chromatin immunoprecipitation，ChIP）技术是研究转录因子与其所结合的基因组 DNA 在体内相互作用的好方法。因此，ChIP 技术被广泛用于研究转录因子与基因调控区域，如启动子和增强子等的结合。简而言之，ChIP 首先通过使用甲醛令蛋白质与 DNA 发生交联，其后将染色质片段化并使用抗体沉淀特异性蛋白质，最后交联解除并获得免疫共沉淀富集的染色质（图 6-2）。科技的飞速发展使得通过各种高通量 ChIP 在基因组范围内进行转录因子的结合位点分析成为可能（图 6-2）。我们随后将对 ChIP 技术进行详细介绍。

转录调控网络核心：Oct4、Sox2 和 Nanog

转录因子及其所结合的靶点之间复杂精细的相互作用构成了 ESC 内的转录调控网络。Oct4、Sox2 和 Nanog 三者共同形成了转录网络的核心，维持 ESC 细胞的自我更新和多能性（图 6-1）。

Loh 等采用 ChIP-PET 技术在小鼠 ESC 中分析了 Oct4 和 Nanog 基因组范围内的结合位点，进一步构建 ESC 内的转录调控网络[28]。ChIP-PET 技术需要将 ChIP 富集的 DNA 两末端，也就是配对末端双标签（paired-end diTag，PET）克隆进入载体内，然后对载体进行测序。对 Oct4、Sox2 和 Nanog 基因组范围的结合位点分析显示，它们所结合的靶基因存在很大程度的重叠[28,50]。Boyer 等应用 ChIP-chip 技术研究了 OCT4、SOX2 和 NANOG 在人 ESC 内与启动子结合的情况[50]。ChIP-chip 技术对 ChIP 富集的

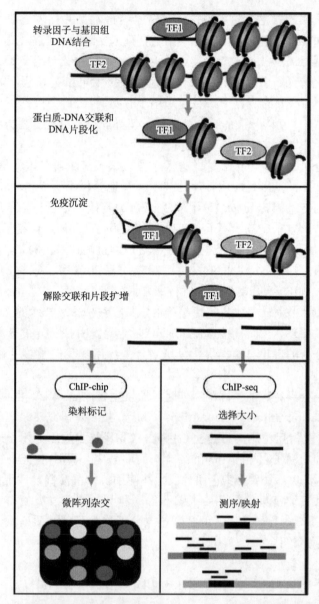

图6-2 ChIP为基础基因组范围分析转录因子结合位点技术。首先使用甲醛令蛋白质与DNA发生交联,再通过超声降解或核酶处理使DNA片段化。其后,使用特异性抗体免疫沉淀目标蛋白质。最后,将蛋白质与DNA之间的交联解除,分离出来的DNA片段进行扩增。在ChIP基础上,ChIP-chip和ChIP-seq等技术可以用于研究转录因子基因组范围的结合位点。ChIP-chip分析中,ChIP与微阵列技术同时应用,即将扩增后的片段进行染料标记后再与微阵列进行杂交。对于ChIP-seq,则对扩增后的DNA片段进行测序和定位

DNA进行扩增后再将其与微阵列进行杂交(图6-2)。上述两项研究一致发现这几种因子可同时结合许多靶基因。而且,研究发现Oct4、Nanog和Sox2可以结合它们自身的启动子形成自调控,通过前馈或反馈环来调节自身的表达[18,19,28,50]。自调控性质上可以

是正调控，也可以是负调控，是因子调节自身表达简单有效的途径。此类调控体系稳定 ESC 细胞状态，同时也让它们处于可分化的状态。核心转录因子 Oct4、Sox2 和 Nanog 参与的调控环路继而影响其下游的、与其有着相似稳态调节功能的基因。

总之，核心转录因子 Oct4、Sox2 和 Nanog 通过前馈和自调控环路等机制调控并维持 ESC 的多能性（图 6-1）。

扩大化的转录调控网络

Kim 等研究了 9 个转录因子（Oct4、Sox2、Klf4、c-Myc、Nanog、Dax1、Rex1、Zfp281 和 Nac1）在启动子上的结合位点来进一步阐明 ESC 的转录网络[51]。研究中采用了改良的 ChIP-chip 技术，叫做体内生物素化 ChIP-chip（biotinylation ChIP-chip，bio-ChIP-chip）（图 6-2）。有趣的是，用这种研究方法发现，许多相同的协作基因可以被不同转录因子同时调控。在小鼠 ESC 中，多于 4 个转录因子结合的基因通常处于转录活跃状态，而很少量转录因子结合的基因则转录受抑或不活跃。并且，此研究进一步发现 Oct4、Sox2、Klf4、Nanog、Dax1 和 Zfp281 拥有众多共同的靶基因。

众多靶基因被归于一个簇，c-Myc 和 Rex1 则形成另一个与之不同的簇[51]。另外还有一种基因组水平的 ChIP 技术，在 ChIP 之后进行大量的平行 DNA 测序工作，被称为 ChIP-测序（ChIP-sequencing，ChIP-seq）（图 6-2）[52~54]。与 ChIP-chip 不同的是，它不必受到微阵列上提前设计好的数量有限探针的限制，使得研究者在基因组范围内可以客观地寻找转录因子结合位点。并且，由于在测序之前不需要对 ChIP-DNA 进行特别修饰和繁琐的克隆工作，ChIP-seq 也优于 ChIP-PET。ChIP-seq 精细测序分析，构成了高精度的结合位点数据库。Chen 等采用了此技术在小鼠 ESC 细胞中分析定位了 13 个转录因子（Oct4、Sox2、Klf4、c-Myc、Nanog、STAT3、Smad1、Zfx、n-Myc、Esrrb、Tcfcp2l1、E2f1 和 CTCF）和 2 个转录共调节蛋白（p300 和 Suz12）结合位点。该研究同时报道了很多基因组区域存在多种转录因子密集的结合[52]。这些基因组"热点"即多转录因子结合位点（multiple transcription-factor-binding loci，MTL）。研究发现主要有两类转录因子簇与 MTL 结合。一类由转录因子 Nanog-Oct4-Sox2、Stat3 及 Smad1 组成，另一类包括 c-Myc、n-Myc、Zfx 和 E2F1（图 6-3）。c-Myc 与 Nanog、Oct4 和 Sox2 归属于不同的簇，这一点与 Kim 等的研究发现一致，并且两个研究都提示多种因子的转录活化对于基因的转录表达有累加效应[51,52]。转录因子的联合结合也可能对转录起到协同调控作用。

ChIP 技术为基础的高通量测序分析，其最大的优越性在于，研究者们可以在基因组水平和定位转录因子-DNA 相互作用方面进行研究。然而，需要强调的是，诸如 ChIP-PET、ChIP-chip 和 ChIP-seq 这类基因组范围的 ChIP 技术只能识别转录因子结合的靶基因，却不能证明靶基因的功能是否受到它们相应结合因子的调控。因此，ChIP 技术和基因表达谱的联合分析是识别转录因子结合和调控基因的理想手段[55~57]。例如，ESC 细胞基因微阵列分析显示 Oct4 表达发生变化，结合已有的基因组范围的 ChIP 数据可以识别 Oct4 的下游调控基因[57]。此研究发现了 *Tcl1* 基因，在 Ivanova 的 RNA 干

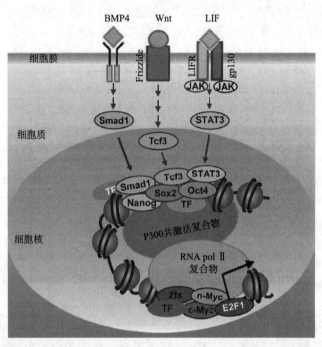

图 6-3 ESC 中调控基因表达的增强体复合物。主要由 Oct4、Sox2、Nanog、Smad1 和 Stat3 组成的转录因子簇倾向于共同结合在基因的增强子区域。另一个主要由 c-Myc、n-Myc、Zfx 和 E2f1 组成的簇主要结合在基因的启动子区域。TF 指可以和簇相互作用的其他转录因子。共活化蛋白 p300 可能被募集到 Oct4-Sox2-Nanog 簇,通过与 RNA Pol II 作用来促进增强子复合物和启动子复合物的相互作用。这种相互作用诱导 DNA 链环化,继而促发转录活化,维持 ESC 状态的重要外部信号,如 LIF 和 BMP,从细胞表面受体分别传递到转录调节蛋白 Stat3 和 Smad1,然后再将这些信号传递到核转录机器。Wnt 等其他信号传递到如 Tcf3 的下游因子

扰筛查实验中同样也发现了此基因[27]。Tcl1 基因是 Oct4 的直接靶基因,与 ESC 的增殖调节有关[57]。另一个研究小组分析了 ChIP-chip 数据,分化过程 ESC 随时间变化的微阵列数据库和 Pou5f1 及 Sox2 敲除 ESC 细胞的基因表达谱来推断决定未分化 ESC 命运的转录网络[56]。同样通过这类综合性分析手段,Zhou 等发现在 ESC 细胞基因调控网络中,Esrrb、Sall4、Lrh-1、Tcf7 和 Stat3 这些转录因子可能作为共调节蛋白与核心转录因子 Oct4、Sox2 和 Nanog 协同作用[55]。

总而言之,基因组范围内的转录因子结合位点分析结合表达谱分析给我们提供了丰富的信息,并且让我们更好地认识 ESC 内复杂的转录调控网络。

增强体:转录因子复合体

基因调控区域并不一定只与一个转录因子结合,可以同时与多种转录因子结合。这

种可以与基因的短增强子区域结合的蛋白复合体称为增强体[58]。由 NF-κB、干扰素调节因子 1 (interferon-regulatory factor 1, IRF1) 和活化转录因子 2 (activating transcription factor 2, ATF-2) /c-JUN 所组成的增强子复合体是一个典型的增强体,在病毒感染时,它们结合到人干扰素 β (interferon-beta, IFN β) 基因使其表达上调。另一个因子——HMG-I 能够稳定这个转录增强子复合体和促进这些转录因子的协同功能[59]。ESC 内被认为存在同样的增强体调控其转录过程。例如,Chen 等报道的 MTL 处,具有多种转录因子的密集结合,提示增强体的存在(图 6-3)。研究还发现转录共活化蛋白 p300 也主要与 Nanog-Oct4-Sox2 簇在增强子区域一起结合[52]。人们注意到 Nanog、Oct4 和 Sox2 可能起着募集 p300 的作用,因为三者中敲除任一因子都能减少 p300 的结合[52]。由于 Nanog-Oct4-Sox2 簇主要与基因的增强子区域结合,c-Myc、n-Myc 簇主要与基因的启动子结合,而 p300 也曾被报道在启动子区域与 c-Myc 相互作用[60](图 6-3),人们推测这两个主要转录因子簇可能在 p300 的协助下相互作用。这种相互作用可能反过来促进 DNA 环化使得增强子区域与 RNA Poly II 复合体结合的启动子区域相互靠近(图 6-3)。增强体的形成可能是细胞始动转录的调控策略,而对于增强体模型的深入了解尚需进一步的研究阐明。

信号通路参与转录网络

转录调控网络并不是孤立存在的,内部与细胞信号通路关联。维持 ESC 细胞状态的外部信号被细胞表面受体感知,通过下游的信号传递给转录因子(图 6-3)。Chen 等不仅构建了他们检测到的 ESC 内转录因子相互关联网络,还将 LIF 和 BMP 信号通路也整合于其中,与 Oct4-Sox2-Nanog 簇拥有共同的靶蛋白 Stat3 和 Smad1。Stat3 和 Smad1 分别是这两个主要信号通路的下游效应蛋白(图 6-3)。LIF-Stat 和 BMP 通路对 ESC 细胞的自我更新都起着至关重要的作用。然而,当分化通路 FGF-MEK 被抑制时,即便没有 Stat3 的活化,ESC 细胞状态仍可维持[61]。最近,研究显示 LIF 的下游 JAK-Stat 通路活化 Klf4,Klf4 随后再激活 Sox2(图 6-1)[30]。另外,磷脂酰肌醇 3-激酶-Akt (phosphatidylinositol-3-OH kinase-Akt, PI3K-Akt) 和丝裂原活化蛋白激酶 (mitogen-activated protein kinase, MAPK) 通路通过 LIF 激活 Tbx3,Tbx3 继而激活 Nanog (图 6-1)[30]。Sox2 和 Nanog 再活化 Oct4,后者是保持 ESC 细胞多能性的至关重要的因子。有趣的是,当 LIF 缺失时,异位表达 Klf4 和 Tbx3 即足以维持 ESC 的多能性[30]。因此,在小鼠 ESC 中,Klf4 和 Tbx3 通过传递 LIF 下游信号到核心调控网络,从而维持细胞多能性。此外,最近还发现 LIF-Stat 引起的 Klf4 活化与 Oct4 引起的 Klf2 活化效应偶联增强,共同维持 ESC 的自我更新能力[30]。

与小鼠 ESC 不同,LIF-JAK-Stat 不是人类 ESC 自我更新所必需的通路,而 activin/Nodal/TGF β 和 FGF 却是必需的。值得注意的是,LIF 信号通路的组分在人 ESC 细胞中并不表达,而 FGF 通路的大部分蛋白在小鼠 ESC 细胞中不表达。人和小鼠 ESC 转录组的内在区别可能缘于种基因表达的属特异性。这种基因表达的差异还有一个可能的原因就是人和小鼠 ESC 反映的是早期胚胎发育的不同多能状态。例如,相较于小鼠 ESC，人 ESC 与小鼠胚胎着床后晚期外胚层来源于外胚层干细胞（epiblast stem cell，EpiSC）较为相似[62,63]。与人 ESC 类似，EpiSC 通过 activin/Nodal 和 FGF 信号通路维持细胞状态，并且与人 ESC 具有类似的基因表达模式，如维持小鼠 ESC 细胞状态的转录因子 Dax1 的表达量比较低[63,41]。相较于小鼠 ESC，人 ESC 与 EpiSC 内 Oct4 结合的靶基因存在更大程度的重叠[63]。人与小鼠 ESC 转录组的差异说明需要在分子水平进一步地研究细胞特性。

虽然存在差异，小鼠及人的 ESC 细胞的多能性维持都需要 Wnt 信号通路[64]。转录因子 T 细胞因子-3（T-cell factor-3，Tcf3）是 Wnt 的下游效应蛋白（图 6-3）。尽管 Tcf3 与 Oct4 和 Nanog 共同结合于某些基因位点[65]，但它同时也抑制诸如 Oct4 和 Nanog 等一些维持多能性和自我更新重要基因的表达（图 6-1）[65~67]。而且，Tcf3 的缺失可以抑制 ESC 细胞分化[65~67]。所以，Tcf3 对 ESC 的自我更新和分化对于 ESC 细胞起着精细的调控作用。

总之，转录因子与信号通路的密切联系使得 ESC 内的转录调控网络更加广阔与复杂。

转录调控与表观遗传学调控的相互作用

遗传学和表观遗传学的调控，在维持 ESC 的状态中有着同等重要的作用。与此相符，Efroni 等报道了转录因子和染色质重塑因子在 ESC 内表达水平较高，而在细胞分化时出现沉默基因[68]。这种现象反映了 ESC 细胞的高转录状态，甚至在非编码区和沉默基因处也出现了转录现象[68]。与高转录状态一致，ESC 中的染色质蛋白，如连接组蛋白和异染色质蛋白 1（heterochromatin protein 1，HP1）也处于高动力结合状态[69]。修饰过的染色质蛋白使染色质发生某些改变，从而使变形后转录因子更易与它们的靶基因结合。当 ESC 细胞分化时，染色质相关蛋白的高动力结合与高转录状态同样减弱[69]。综合以上，高转录和染色质蛋白的高动力结合，说明了基因调控和表观遗传调控是相互交联的，因此，转录因子和表观遗传学修饰蛋白的相互作用，能够维持 ESC 细胞处于自我更新和多能性状态。

在 Wang 等构建的蛋白质相互作用网络中，物理学方面，Oct4 和 Nanog 与 SWI/SNF 染色质重塑复合体、多梳组（polycomb group，PcG）蛋白和组蛋白脱乙酰基酶复合体 NuRD 等表观遗传学调节蛋白相互作用[41]。PcG 蛋白组成的 polycomb 抑制复合体（polycomb repressive complex，PRC）如 PRC2 是表观遗传学修饰蛋白，对早期发育相当重要，能够诱导组蛋白 3 赖氨酸 27（histone 3 lysine 27，H3K27）甲基化，而 H3K27 是沉默基因的表观遗传学标志[70]。在人 ESC 中，OCT4、SOX2 和 NANOG 同时结合到数个 PRC2 的靶基因。Oct4 还与 PRC1 的一个成员 Ring1b 组成复合体，其中 Ring1b 与靶基因的结合依赖于 Oct4 的存在[71]。PRC2 促使产生抑制性表观遗传学标志，认为 PRC1 与这些抑制性标志结合后诱导染色质构象发生改变。Oct4 和 Nanog 还与其他的转录抑制复合物如 Sin3A 和 Pml 等存在相互作用[72]。

近来，人们发现 Oct4 可以募集 Eset 至数种特异性表达在滋养外胚层细胞系的基因，如 $Cdx2$ 和 $Tcfap2a$，抑制其表达[73~75]。Eset 属于组蛋白修饰因子，它在 H3K9 的靶基因处，如 $Cdx2$ 和 $Tcfap2a$，催化 H3K9 发生甲基化，进而抑制这些靶基因在 ESC 细胞中的表达[73~75]。与此一致，在缺失 $Eset$ 的桑葚胚细胞中，出现 $Cdx2$ 和 $Tcfap2a$ 的高表达[74]。将 $Eset$ 缺失的 ESC 显微注射至胚胎内，其分化形成滋养外胚层[74]。Loh 的一项研究显示，Oct4 正向调控 Jmjd1a 和 Jmjd2c 的表达，Jmjd1a 和 Jmjd2c 是 H3K9 去甲基化酶，此酶使 H3K9 发生去甲基化，进而调控 $Nanog$ 和 $Tcl1$ 等基因的表达[76]。总而言之，遗传学和表观遗传学调控的交叉作用是形成 ESC 复杂精细调控网络的基础。

结论

转录调控是细胞内基本的生物学活动。自 ESC 分离获取以来，人们已经获知操控 ESC 自我更新和多能性的转录调控的大量信息。科技的飞速发展使我们能够进行基因组范围的研究，并更加系统地构建转录调控网络。但是，由于大部分的实验都是用小鼠的 ESC 研究的，势必要进一步分析人 ESC 的转录调控网络。对于人 ESC 转录调控的深入研究，有助于将来将其应用于干细胞再生医学方面。

致谢

我们感谢 Yun-Shen Chan、Jia-Hui Ng 和 Lin Yang 对稿件的审阅。

（陈 芳 韩忠朝 译）

参考文献

1. Lander ES, Linton LM, Birren B et al. Initial sequencing and analysis of the human genome. Nature 2001; 409:860-921.
2. Johnson KM, Mitsouras K, Carey M. Eukaryotic transcription: the core of eukaryotic gene activation. Curr Biol 2001; 11:R510-3.
3. Levine M, Tjian R. Transcription regulation and animal diversity. Nature 2003; 424:147-51.
4. Pesce M, Scholer HR. Oct-4: gatekeeper in the beginnings of mammalian development. Stem Cells 2001; 19:271-8.
5. Nichols J, Zevnik B, Anastassiadis K et al. Formation of pluripotent stem cells in the mammalian embryo depends on the POU transcription factor Oct4. Cell 1998; 95:379-91.
6. Niwa H, Miyazaki J, Smith AG. Quantitative expression of Oct-3/4 defines differentiation, dedifferentiation or self-renewal of ES cells. Nat Genet 2000; 24:372-6.
7. Niwa H, Toyooka Y, Shimosato D et al. Interaction between Oct3/4 and Cdx2 determines trophectoderm differentiation. Cell 2005; 123:917-29.
8. Avilion AA, Nicolis SK, Pevny LH et al. Multipotent cell lineages in early mouse development depend on SOX2 function. Genes Dev 2003; 17:126-40.
9. Masui S, Nakatake Y, Toyooka Y et al. Pluripotency governed by Sox2 via regulation of Oct3/4 expression in mouse embryonic stem cells. Nat Cell Biol 2007; 9:625-35.
10. Kopp JL, Ormsbee BD, Desler M et al. Small increases in the level of Sox2 trigger the differentiation of mouse embryonic stem cells. Stem Cells 2008; 26:903-911.
11. Rodda DJ, Chew JL, Lim LH et al. Transcriptional regulation of nanog by OCT4 and SOX2. J Biol Chem 2005; 280:24731-7.
12. Kuroda T, Tada M, Kubota H et al. Octamer and Sox elements are required for transcriptional cis regulation of Nanog gene expression. Mol Cell Biol 2005; 25:2475-85.
13. Ambrosetti DC, Scholer HR, Dailey L et al. Modulation of the activity of multiple transcriptional activation domains by the DNA binding domains mediates the synergistic action of Sox2 and Oct-3 on the fibroblast growth factor-4 enhancer. J Biol Chem 2000; 275:23387-97.

14. Ambrosetti DC, Basilico C, Dailey L. Synergistic activation of the fibroblast growth factor 4 enhancer by Sox2 and Oct-3 depends on protein-protein interactions facilitated by a specific spatial arrangement of factor binding sites. Mol Cell Biol 1997; 17:6321-9.
15. Botquin V, Hess H, Fuhrmann G et al. New POU dimer configuration mediates antagonistic control of an osteopontin preimplantation enhancer by Oct-4 and Sox-2. Genes Dev 1998; 12:2073-90.
16. Nishimoto M, Fukushima A, Okuda A et al. The gene for the embryonic stem cell coactivator UTF1 carries a regulatory element which selectively interacts with a complex composed of Oct-3/4 and Sox-2. Mol Cell Biol 1999; 19:5453-65.
17. Yuan H, Corbi N, Basilico C et al. Developmental-specific activity of the FGF-4 enhancer requires the synergistic action of Sox2 and Oct-3. Genes Dev 1995; 9:2635-45.
18. Chew JL, Loh YH, Zhang W et al. Reciprocal transcriptional regulation of Pou5f1 and Sox2 via the Oct4/Sox2 complex in embryonic stem cells. Mol Cell Biol 2005; 25:6031-46.
19. Okumura-Nakanishi S, Saito M, Niwa H et al. Oct-3/4 and Sox2 Regulate Oct-3/4 Gene in Embryonic Stem Cells. J Biol Chem 2005; 280:5307-5317.
20. Remenyi A, Lins K, Nissen LJ et al. Crystal structure of a POU/HMG/DNA ternary complex suggests differential assembly of Oct4 and Sox2 on two enhancers. Genes Dev 2003; 17:2048-59.
21. Williams DC, Jr., Cai M, Clore GM. Molecular basis for synergistic transcriptional activation by Oct1 and Sox2 revealed from the solution structure of the 42-kDa Oct1.Sox2.Hoxb1-DNA ternary transcription factor complex. J Biol Chem 2004; 279:1449-57.
22. Maruyama M, Ichisaka T, Nakagawa M et al. Differential roles for Sox15 and Sox2 in transcriptional control in mouse embryonic stem cells. J Biol Chem 2005; 280:24371-24379.
23. Mullin NP, Yates A, Rowe AJ et al. The pluripotency rheostat Nanog functions as a dimer. Biochem J 2008; 227-231.
24. Wang W, Lin C, Lu D et al. Chromosomal transposition of PiggyBac in mouse embryonic stem cells. Proc Natl Acad Sci USA 2008; 105:9290-5.
25. Chambers I, Colby D, Robertson M et al. Functional expression cloning of Nanog, a pluripotency sustaining factor in embryonic stem cells. Cell 2003; 113:643-55.
26. Mitsui K, Tokuzawa Y, Itoh H et al. The homeoprotein Nanog is required for maintenance of pluripotency in mouse epiblast and ES cells. Cell 2003; 113:631-42.
27. Ivanova N, Dobrin R, Lu R et al. Dissecting self-renewal in stem cells with RNA interference. Nature 2006; 442:533-8.
28. Loh YH, Wu Q, Chew JL et al. The Oct4 and Nanog transcription network regulates pluripotency in mouse embryonic stem cells. Nat Genet 2006; 38:431-40.
29. Chambers I, Silva J, Colby D et al. Nanog safeguards pluripotency and mediates germline development. Nature 2007; 450:1230-4.
30. Niwa H, Ogawa K, Shimosato D et al. A parallel circuit of LIF signalling pathways maintains pluripotency of mouse ES cells. Nature 2009; 460:118-22.
31. Kalmar T, Lim C, Hayward P et al. Regulated fluctuations in nanog expression mediate cell fate decisions in embryonic stem cells. PLoS Biol 2009; 7:e1000149.
32. Carter MG, Stagg CA, Falco G et al. An in situ hybridization-based screen for heterogeneously expressed genes in mouse ES cells. Gene Expr Patterns 2008; 8:181-98.
33. Silva J, Nichols J, Theunissen TW et al. Nanog is the gateway to the pluripotent ground state. Cell 2009; 138:722-37.
34. Jiang J, Chan YS, Loh YH et al. A core Klf circuitry regulates self-renewal of embryonic stem cells. Nat Cell Biol 2008; 10:353-60.
35. Hall J, Guo G, Wray J et al. Oct4 and LIF/Stat3 additively induce Kruppel factors to sustain embryonic stem cell self-renewal. Cell Stem Cell 2009; 5:597-609.
36. Ema M, Mori D, Niwa H et al. Kruppel-like factor 5 is essential for blastocyst development and the normal self-renewal of mouse ESCs. Cell Stem Cell 2008; 3:555-67.
37. Dejosez M, Krumenacker JS, Zitur LJ et al. Ronin is essential for embryogenesis and the pluripotency of mouse embryonic stem cells. Cell 2008; 133:1162-74.
38. Hu G, Kim J, Xu Q et al. A genome-wide RNAi screen identifies a new transcriptional module required for self-renewal. Genes Dev 2009; 23:837-48.
39. Ding L, Paszkowski-Rogacz M, Nitzsche A et al. A genome-scale RNAi screen for Oct4 modulators defines a role of the Paf1 complex for embryonic stem cell identity. Cell Stem Cell 2009; 4:403-15.
40. Gaspar-Maia A, Alajem A, Polesso F et al. Chd1 regulates open chromatin and pluripotency of embryonic stem cells. Nature 2009; 460:863-8.
41. Wang J, Rao S, Chu J et al. A protein interaction network for pluripotency of embryonic stem cells. Nature 2006; 444:364-8.
42. Wu Q, Chen X, Zhang J et al. Sall4 interacts with Nanog and co-occupies Nanog genomic sites in embryonic stem cells. J Biol Chem 2006; 281:24090-4.
43. Suzuki A, Raya A, Kawakami Y et al. Nanog binds to Smad1 and blocks bone morphogenetic protein-induced differentiation of embryonic stem cells. Proc Natl Acad Sci USA 2006; 103:10294-9.
44. Torres J, Watt FM. Nanog maintains pluripotency of mouse embryonic stem cells by inhibiting NFkappaB and cooperating with Stat3. Nat Cell Biol 2008; 10:194-201.

45. Araki R, Fukumura R, Sasaki N et al. More than 40,000 transcripts, including novel and noncoding transcripts, in mouse embryonic stem cells. Stem Cells 2006; 24:2522-8.
46. Sharov AA, Piao Y, Matoba R et al. Transcriptome analysis of mouse stem cells and early embryos. PLoS Biol 2003; 1:E74.
47. Nagalakshmi U, Wang Z, Waern K et al. The transcriptional landscape of the yeast genome defined by RNA sequencing. Science 2008; 320:1344-9.
48. Wilhelm BT, Marguerat S, Watt S et al. Dynamic repertoire of a eukaryotic transcriptome surveyed at single-nucleotide resolution. Nature 2008; 453:1239-43.
49. Ozsolak F, Platt AR, Jones DR et al. Direct RNA sequencing. Nature 2009; 461:814-8.
50. Boyer LA, Lee TI, Cole MF et al. Core transcriptional regulatory circuitry in human embryonic stem cells. Cell 2005; 122:947-56.
51. Kim J, Chu J, Shen X et al. An extended transcriptional network for pluripotency of embryonic stem cells. Cell 2008; 132:1049-1061.
52. Chen X, Xu H, Yuan P et al. Integration of external signaling pathways with the core transcriptional network in embryonic stem cells. Cell 2008; 133:1106-17.
53. Johnson DS, Mortazavi A, Myers RM et al. Genome-wide mapping of in vivo protein-DNA interactions. Science 2007; 316:1497-502.
54. Jothi R, Cuddapah S, Barski A et al. Genome-wide identification of in vivo protein-DNA binding sites from ChIP-Seq data. Nucleic Acids Res 2008; 36:5221-31.
55. Zhou Q, Chipperfield H, Melton DA et al. A gene regulatory network in mouse embryonic stem cells. Proc Natl Acad Sci USA 2007; 104:16438-43.
56. Walker E, Ohishi M, Davey RE et al. Prediction and testing of novel transcriptional networks regulating embryonic stem cell self-renewal and commitment. Cell Stem Cell 2007; 1:71-86.
57. Matoba R, Niwa H, Masui S et al. Dissecting Oct3/4-regulated gene networks in embryonic stem cells by expression profiling. PLoS ONE 2006; 1:e26.
58. Thanos D, Maniatis T. Virus induction of human IFN beta gene expression requires the assembly of an enhanceosome. Cell 1995; 83:1091-100.
59. Du W, Thanos D, Maniatis T. Mechanisms of transcriptional synergism between distinct virus-inducible enhancer elements. Cell 1993; 74:887-98.
60. Faiola F, Liu X, Lo S et al. Dual regulation of c-Myc by p300 via acetylation-dependent control of Myc protein turnover and coactivation of Myc-induced transcription. Mol Cell Biol 2005; 25:10220-34.
61. Ying QL, Wray J, Nichols J et al. The ground state of embryonic stem cell self-renewal. Nature 2008; 453:519-23.
62. Brons IG, Smithers LE, Trotter MW et al. Derivation of pluripotent epiblast stem cells from mammalian embryos. Nature 2007; 448:191-5.
63. Tesar PJ, Chenoweth JG, Brook FA et al. New cell lines from mouse epiblast share defining features with human embryonic stem cells. Nature 2007; 448:196-9.
64. Sato N, Meijer L, Skaltsounis L et al. Maintenance of pluripotency in human and mouse embryonic stem cells through activation of Wnt signaling by a pharmacological GSK-3-specific inhibitor. Nat Med 2004; 10:55-63.
65. Cole MF, Johnstone SE, Newman JJ et al. Tcf3 is an integral component of the core regulatory circuitry of embryonic stem cells. Genes Dev 2008; 22:746-55.
66. Pereira L, Yi F, Merrill BJ. Repression of Nanog gene transcription by Tcf3 limits embryonic stem cell self-renewal. Mol Cell Biol 2006; 26:7479-91.
67. Tam WL, Lim CY, Han J et al. T-cell factor 3 regulates embryonic stem cell pluripotency and self-renewal by the transcriptional control of multiple lineage pathways. Stem Cells 2008; 26:2019-31.
68. Efroni S, Duttagupta R, Cheng J et al. Global transcription in pluripotent embryonic stem cells. Cell Stem Cell 2008; 2:437-47.
69. Meshorer E, Yellajoshula D, George E et al. Hyperdynamic plasticity of chromatin proteins in pluripotent embryonic stem cells. Dev Cell 2006; 10:105-16.
70. Cao R, Wang L, Wang H et al. Role of histone H3 lysine 27 methylation in Polycomb-group silencing. Science 2002; 298:1039-43.
71. Endoh M, Endo TA, Endoh T et al. Polycomb group proteins Ring1A/B are functionally linked to the core transcriptional regulatory circuitry to maintain ES cell identity. Development 2008; 135:1513-24.
72. Liang J, Wan M, Zhang Y et al. Nanog and Oct4 associate with unique transcriptional repression complexes in embryonic stem cells. Nat Cell Biol 2008; 10:731-9.
73. Yeap LS, Hayashi K, Surani MA. ERG-associated protein with SET domain (ESET)-Oct4 interaction regulates pluripotency and represses the trophectoderm lineage. Epigenetics Chromatin 2009; 2:12.
74. Yuan P, Han J, Guo G et al. Eset partners with Oct4 to restrict extraembryonic trophoblast lineage potential in embryonic stem cells. Genes Dev 2009; 23:2507-20.
75. Bilodeau S, Kagey MH, Frampton GM et al. SetDB1 contributes to repression of genes encoding developmental regulators and maintenance of ES cell state. Genes Dev 2009; 23:2484-9.
76. Loh YH, Zhang W, Chen X et al. Jmjd1a and Jmjd2c histone H3 Lys 9 demethylases regulate self-renewal in embryonic stem cells. Genes Dev 2007; 21:2545-57.

第7章 干细胞自我更新和分化中的选择性剪接

David A. Nelles，Gene W. Yeo*

摘要：本章将对信使RNA前体的选择性剪接在干细胞生物学中重要的前沿进展作一综述。大多数的转录前体mRNA经过剪接，去除内含子，外显子拼接形成成熟的mRNA序列。这种可调节的、选择性的剪接使得部分或全部外显子移除，为RNA丰度的控制及蛋白质多样性提供了可能性。本章我们重点讨论干细胞生物学中关键的选择性剪接事件的几个例子，并对近期发展的能在系统发生及基因组水平上评估选择性剪接的芯片技术和测序技术在干细胞分化中的应用进行概述。

引言

干细胞是研究细胞多能性、自我更新及定向分化的独一无二的工具。胚胎干细胞在体外培养过程中能够长时间保持未分化状态并随时可被诱导向三胚层分化，并且在体外能够分化为全部或绝大多数组成正常机体的组织和器官的细胞。因此，在很多方面，胚胎干细胞为健康和疾病状态的研究提供了很好的平台。通过富集定向分化得到特异的细胞群，并且对分化前后的细胞特征进行详细分析，将为细胞鉴定提供分子生物学基础。

基因表达的研究表明干细胞在向不同谱系分化时存在总体转录水平的差异[1~4]，但是目前的研究并未对相同基因位点不同的选择性剪接体进行区分。在本章中，我们将介绍几例在干细胞分化过程中重要的选择性剪接，并对现有的、可对选择性剪接进行识别和定量分析的技术进行综述，同时也将对影响选择性剪接因素的最新成果进行讨论。

选择性剪接的概述

在转录过程中，转录的前体mRNA由间隔相邻的、长的非编码内含子序列和短的150bp的外显子序列组成。内含子移除、外显子连接及剪接位点的选择是被严格调控的，因为错误剪接可以生成异常的蛋白质或抑制mRNA的翻译。这一过程由一种称为剪接体的蛋白质-RNA复合物所介导，复合物的构成体现了其严格的功能要求；剪接体包括5种RNA和几百种蛋白质[5,6]。前体mRNA编码的顺式作用元件与反式作用因子组合起来被称为剪接因子。

---

* Corresponding Author: Gene W. Yeo—Department of Cellular and Molecular Medicine, Stem Cell Program, University of California, San Diego, 9500 Gilman Drive, La Jolla, California 92037, USA. Email: geneyeo@ucsd.edu

第 7 章 干细胞自我更新和分化中的选择性剪接

顺式作用元件、剪接因子和剪接体的相互作用共同形成一套被称为"剪接代码"的规则。这一代码中的成分决定了选择性剪接过程中从同一 mRNA 前体的不同剪接位点剪接形成不同的成熟 mRNA（图 7-1）。剪接代码仍然是个谜，但其关键成分正被迅速确认。剪接因子包括剪接体元件、丝氨酸-精氨酸（SR）蛋白家族成员、核不均一核糖核蛋白（hnRNP）及辅助因子构成的经典 RNA 结合蛋白（RBP）[7,8]。近期研究表明大部分的人类基因都被选择性剪接，大大地增加了基因的蛋白质编码潜力[9,10]，并且已开始进行在基因组水平对干细胞选择性剪接的研究[11]。选择性剪接引起的蛋白质多样性在很大程度上还未被研究，这为更好地认识干细胞生物特性提供了机会。

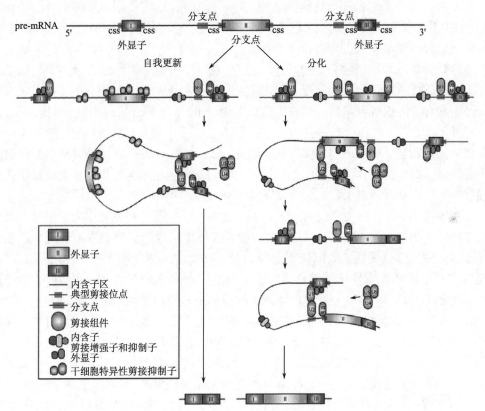

图 7-1　图中概括了在多能干细胞和已分化的细胞中 RNA 剪接过程中剪接因子的功能。图中左侧，干细胞特异性的剪接抑制子阻止了靠近外显子Ⅱ的剪接体的装配，导致外显子Ⅱ和它侧面的内含子的切除。图中右侧缺少剪接抑制子，正在分化的干细胞产生了包括所有 3 个外显子的 mRNA。剪接抑制子通过抑制剪接体的组装或是其他机制阻止剪接。在上述过程中，核心剪接体小核 RNA 蛋白被描绘出来（U1、U2、U4、U5、U6），它们的组装导致了剪接体从前体 mRNA 区域酶促移除。5′外显子末端由 U1 标记，而 3′外显子末端由 U2 辅助因子修饰（U2AF）。当剪切体装配时，SF1 标记的分支点招募 U2 并与 U1 形成一个环。然后，U4、U5 和 U6 被招募过来，结合在 5′外显子的内含子被切除，它们结合于分支点并使得外显子相连。产生的"套索"（未显示）和相连的外显子被释放出来，随后剪切体解离

干细胞干性维持及分化中涉及的选择性剪接基因

越来越多的证据证明了选择性剪接在干细胞中的影响。近期研究表明一些干细胞富集基因（干性基因）的剪接变异体表达水平与特定的分化阶段有关，一些类似的研究表明不相干和具有再生基因的表型与干性基因的剪接变体的过度表达相关。这些结果提示了单一基因位点的剪接变体对表型的不同影响及认识选择性剪接对转录后基因表达调节的重要性。这里，我们将介绍几例对干细胞生物学比较重要的选择性剪接基因。

POU5F1 是一种受选择性剪接调节的关键干性基因，这一基因编码 POU 结构域转录因子 OCT4，其是一种干细胞多能性的关键调节因子。OCT4 高表达于干细胞，它的表达对重编程成熟细胞为诱导多能干细胞状态是必需的[12~14]。尽管对 POU5F1 基因产物的精细分辨提示为不同的受控剪接体，但在一些体细胞中检测出了 OCT4[15~16] 之后，它作为干细胞标志物的有效性就受到了质疑。一种被称为 OCT4A 的剪接体仅存在于胚胎干细胞和胚胎肿瘤细胞中，并能在 OCT4 启动子的作用下起始表达[17]。相反，OCT4 启动子并不能起始 OCT4B 的表达[18,19]。更复杂的是，另一个叫做 OCT4B1 的异构体也在胚胎干细胞和胚胎肿瘤细胞中高表达，但是 OCT4B 在许多已分化的细胞类型中表达水平却很低。这些异构体的功能明显不同，但我们对此仍然知之甚少，这也提示了选择性剪接在另一个水平上对 OCT4 功能的调节。

虽然不像同一家族的 OCT4 那样已经被了解清楚，OCT2 在发育中的中枢神经系统和成年鼠的大脑中表达水平很高[20]。它的剪接突变体似乎也能影响干细胞的分化：在鼠胚胎干细胞中人为过表达 OCT2.2 异构体，能有效地诱导出神经样表型，而 OCT2.4 甚至能在其他已知可诱导神经样分化的诱导因素存在的情况下抑制对神经样表型的诱导[21]。因此，对 OCT2 剪接突变体的详细了解能为定向分化神经细胞提供线索。

在干细胞和分化细胞中，DNA 甲基转移酶组成了另一群具有完全不同的选择性剪接模式的基因。通过使 DNA 甲基化，甲基转移酶能表观上影响基因的转录并提供了表达调控的可遗传方式。最初的研究揭示了 DNA 甲基转移酶 3B（DNMT3B）被高度选择性剪接；已经鉴定出了将近 40 种异构体[22,23]。Gopalakrishnan 等最近发现一种 DNMT3B 剪接异构体，DNMT3B3Δ5 的剪接体在 NH_2-端调节结构域处缺失了外显子 5，相比 DNMT3B3，它在胚胎干细胞和脑组织中高表达，但在分化过程中表达降低，在重编程的成纤维细胞中表达上调，并像 DNMT3B3 一样缺少有关甲基化的催化片段[24]。重要的是，DNMT3B3Δ5 相比 DNMT3B3 提高了 DNA 的亲和力，提示它能作为 DNMT3B 激活形式的阻断剂以阻止 DNA 的超甲基化。已经知道 DNMT3B 能与许多其他的 DNA 甲基转移酶相互作用[25]，类似 DNMT3BΔ5 的其他剪接异构体可能给这些相互作用开辟了一个新的空间。

PKCδ 基因的表达受选择性剪接高度影响，同时其也是一个基因表达的重要调节子。很早已经知道 PKCδ 在凋亡级联的激活中起作用并正向调节许多凋亡蛋白的转录[26]。PKCδ 也与稳态和抗凋亡通路有关[27~29]。就 PKCδ 的剪接变异体来说，这种双重性能被更好地鉴别。PKCδⅠ由 Caspase 3 剪切，产生了已知能诱导凋亡的催化片段[30]。

Sakurai 等发现另一种异构体 PKCδⅡ能平衡 PKCδⅠ的活性，因为它对 Caspase 3 的剪切不敏感[31]。与神经分化相关的选择性剪接是 PKCδⅠ异构体，它能支持凋亡介导的发育中神经系统的重塑[32]。因此，PKCδⅠ和 PKCδⅡ的内含子碱基对的不同足以显著调节分化诱导前后畸胎瘤细胞的凋亡。

选择性剪接也介导了在自我更新中功能相反的剪接体的产生。已知成纤维细胞生长因子（FGF）对人胚胎干细胞的自我更新信号通路具有正调节功能[33,34]。Mayshar 等发现一种被称为 FGF4 的剪接变异体在人胚胎干细胞（hESC）分化之后表达下调，而另一种被称为 FGF4si 的剪接变异体在多能性的和分化的 hESC 中都表达[35]。FGF4 的自我更新潜能依赖于它对 ERK1/2 的磷酸化和激活 MEK/ERK 信号通路的能力，而可溶性的 FGF4si 似乎可以显著降低 ERK1/2 的磷酸化。这种在不同剪接体之间的相反作用证明了对多能性相关的的信号通路的强力调节并揭示了相同基因不同剪接体之间的调控网络。

成体干细胞的多能性和自我更新也受选择性剪接的高度影响。胰岛素样生长因子1（IGF-1）的多种剪接体能产生促进增殖并阻止肌肉干细胞分化（IGF-IEc）或者通过合成代谢途径诱导肌肉细胞生长（IGF-IEa）的作用。这些剪接变异体以一种顺序的方式表达以响应能促使肌肉生长和修复的机械应力[36]。低水平的 IGF-IEc 与受伤患者肌肉组织中的肌肉萎缩有关[37]，这提示了针对这种剪接体的临床治疗潜能。

血管内皮生长因子 A（VEGFA）强烈影响间充质干细胞（MSC）并受选择性剪接的调节。MSC 的治疗效用依赖于它们独特的旁分泌信号通路[38,39]，鼠的同源 VEGF 剪接变异体影响 MSC 中的旁分泌和其他信号通路。特别之处在于，Lin 等证明了 VEGF120 和 VEGF188 能诱导生长因子和免疫抑制因子的表达，而 VEGF164 影响并重塑与内皮细胞分化有关的基因的表达。VEGF188 也能诱导 MSC 的成骨表型[40]。未来的组织治疗依靠 VEGF 以提高 MSC 的再生潜能，选择合适的 VEGF 异构体的重要性最近才变得明显。

除了顺式选择性剪接，蛋白质的多样性也因反式剪接而增加。反式剪接是指来自于至少一个基因的前体 mRNA 片段集合而产生一个新的 mRNA 转录产物。很少几个与干细胞干性相关的反式剪接产物被发现，但是来自 RNA 结合基序蛋白 14（RBM14）和 RBM4 的反式剪接 mRNA 能间接影响重要基因产物的剪接。RBM14 单独能产生增强和抑制许多基因的共转录剪接体 CoAA 和 CoAM。在胚胎干细胞向胚体分化过程中，剪接产物转向形成异构体 CoAM，从而抑制 CoAA 的活性，并诱导分化标志物 SOX6 的表达[41]。当 RBM4 和 RBM14 同时反式剪接时，剪接体和剪接调节子 CoAZ、ncCoAZ 产生了一个复杂网络，能影响 tau 前体 mRNA 的外显子 10 的共转录剪接[42]。尽管这些剪接变异体的功能尚不清楚，它们在分化前后不同的表达谱暗示了它们对干细胞生物学的重要性。

这些初期的研究揭示了选择性剪接能够调节与干细胞状态相关所有方面的基因（表7-1）。DNA 甲基化受高度剪接的 DNA 甲基转移酶的影响。对干性发挥主要和次要作用的转录因子活性也受选择性剪接机制的调控，甚至剪接机制本身也受反式剪接产物 RBM4 和 RBM14 的调控。很多这些基因剪接模式与干细胞的不同时期相关，并且有时与干细胞的终末分化相关。大多数调控干细胞特性的剪接变异体尚未发现。目前，基于基因组水平的选择剪接体检测的新方法能够显著增加检测效率、减少费用。一旦确定功能，这些剪接变异体及剪接代码将揭示前所未有的转录后基因调控机制。

表 7-1　在干细胞分化前后具有不同基因表达谱的选择性剪接基因产物和/或介导重要干细胞过程的基因产物

基因	异构体	功能	参考文献
IGF-1	IGF-1Ec, IGF-1-Ea	IGF-1Ec促进肌肉前体细胞的增殖，抑制分化；IGF-1-Ea激活合成代谢通路	37
POU5F1（OCT4）	OCT4A OCT4B OCT4C	OCT4A 和 OCT4B1 表达于干细胞，OCT4B 表达于分化细胞	17
RNA 结合基序蛋白 4 基因（RBM4）和 RBM14（CoAA）	CoAZ	CoAZ 和 pc CoAZ 影响共转录剪接	42
DNA 甲基转移酶3B基因（DNMT3B）	DNMT3B3 和 DNMT3B3Δ5	DNMT3B3Δ5 表达于干细胞，在功能上与 DNMT3B3 不同	24
VEGFA	VEGFA120 VEGFA164 VEGFA188	都促进 MSC 的分化，有些放大旁分泌信号通路促进成骨细胞或内皮细胞分化	40
FGF4	FGF4 FGF4si	FGF4 对于干细胞干性维持十分重要；FGF4si 拮抗 FGF4 的功能	35
蛋白酶 Cδ 基因（PKCδ）	PKCδI PKCδII	PKCδI 和 PKCδII 可分别被 Caspase 剪切或不剪切	32
PU2F2 基因（OCT2）	OCT2.2 OCT2.4	OCT2.2 能有效诱导小鼠胚胎干细胞的神经分化；OCT2.4 能有效阻断神经分化	21
RNA 结合基序蛋白基因（RBM14, CoAA）	CoAA, CoAN	CoAA 在 CoAM 的影响下在胚胎发育早期表达下调	41

注：每个前体 mRNA 反式剪接为单一 mRNA。

基因组学方法鉴别、检测选择性剪接

目前对选择性剪接的检测主要采用对每一个 mRNA 片段逐一进行逆转录 PCR，高通量全基因组检测和剪接高敏感性寡核苷酸芯片加快了转录组学的系统检测[43~45]。通过设计已知及预测的外显子序列和外显子与内含子间接头处核苷酸探针，使得高通量基因组水平的选择性剪接检测成为可能。计算机算法被用来在芯片数据中鉴别不同的剪接外显子。我们的研究团队利用这个平台研究人胚胎干细胞的神经分化[46]。研究结果显示选择性剪接位点在丝/苏氨酸激酶和解旋酶类的基因群中广泛存在。比较基因组学分析在外显子选择性剪接位点附近的内含子区域发现了公认的顺式作用元件，可调控神经元分化过程中选择性剪接。这种方法为比较不同的祖细胞向不同的细胞类型分化过程中的不同选择性剪接提供了研究模式。例如,最近的研究比较了未分化的人胚胎干细胞和其来源的祖细胞,揭示了在心肌和神经祖细胞存在共有的和各自特有的剪接方式[47],未来的研究需要探讨相反的剪接模式和剪接变异体功能之间的联系对细胞表型的影响。

不幸的是,基因芯片技术存在明显的缺点,探针密度的物理性限制、探针的非特异性杂交、对稀少转录产物的不敏感性使得某些选择性剪接检测的成本变得昂贵或者不可能。除此之外,芯片不能检测未标注的基因。基于 Tag 标签的表达谱方法如大规模平行测序技术(MPSS)、基因表达系列分析(SAGE)、基因表达的 CAP 分析(cap-analysis gene expression,CAGE)及基因表达多元聚类分析(PMAGE)等具有高灵敏性,并且可以发现新转录变异体,但是费时、费力、低效。下一代的测序技术回避了这些问题,产生数以百万计的可读取 RNA 序列,并通过一种成本低、质量高的方式显示转录产物(图 7-2)。一项对小鼠胚胎干细胞和类胚体的比较研究发现了新的选择性剪接事件,显示了"鸟枪法"在转录谱研究中的重要作用[48]。

RNA 结合蛋白对选择性剪接的调控

一系列 RNA 结合蛋白参与多功能干细胞及分化细胞中选择性剪接的调控。虽然目前对维持干细胞特性的剪接因子知之甚少,但是有研究显示参与了神经的分化。过去认为,大部分的剪接事件通过直接与前体 mRNA 某些区域作用的剪接因子来调控。找到剪接因子作用的 RNA 靶点及作用机制,能更好地理解并破译干细胞分化过程中选择性剪接的规则。

在人胚胎干细胞神经分化研究中,我们发现'GCAUG'保守序列常出现在不同选择性剪接位点的附近[46]。其与 FOX1/2 剪接因子结合位点密切相关,提示 FOX 剪接因子在人胚胎干细胞中剪接位点的选择中有重要作用。

为了进一步检测 FOX2 RBP-RNA 的相互作用在选择性剪接中的功能,我们采用了紫外交联免疫共沉淀(CLIP)方法。这项技术通过采用紫外照射的方法加强了体内RNA-蛋白质间的稳定结合[49]。我们对此方法进行了改良,提取与蛋白质结合的 RNA 进行高通量测序分析(CLIP-seq,图 7-3)。应用 CLIP-seq 提取人胚胎干细胞中与 FOX2 结合的 RNA,能够发现 3000 多种与 FOX2 结合的 RNA 序列。FOX2 结合的 RNA 序列图谱验证了一些 FOX2 参与调控的选择性剪接位点[50]。FOX2 在哺乳动物中高度保守并在人胚胎干细胞中高表达。沉默 FOX2 会使人胚胎干细胞迅速死亡但不影响神经干细胞和其他细胞系。这些结果提示 FOX2 是人胚胎干细胞选择性剪接位点的重要选择器,进一步阐明其特性将揭示其与其他剪接体组分的相互作用。

多聚嘧啶串结合蛋白(PTB)是广泛表达于早期胚胎中的剪接因子,PTB 调节许多基因外显子的选择性剪接,并且提示参与 mRNA 多个方面的调控[51]。几项研究表明 PTB 蛋白的敲除在非神经细胞中足以引起神经特异性选择性剪接[52~54]。为了探测 *PTB* 在胚胎干细胞中的作用,Shibayama 等构建了 *PTB* 基因敲除的纯合鼠胚胎干细胞,这些胚胎干细胞的生长状态良好但是不能正常增殖[55]。

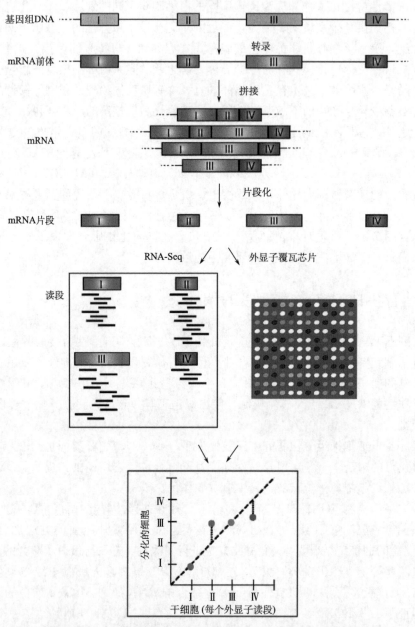

图 7-2 该图描述了两种选择性剪接的检测方法：RNA 测序和剪接敏感性芯片。在细胞中，基因组 DNA 转录为 RNA 并形成多种剪接变异体。这些剪接变异体被酶切为短片段并被测序或与基因芯片微杂交。在 RNA 测序中，读出的短序列与人类基因组比对，经电脑计算预测出表达的剪接变异体。基因芯片依赖于荧光标记的片段，它们在芯片上的相对密度显示了外显子在 mRNA 中的比例。通过比较已分化细胞和干细胞中剪接变异体的表达，就可以识别在干细胞富集或者被清除的变异体

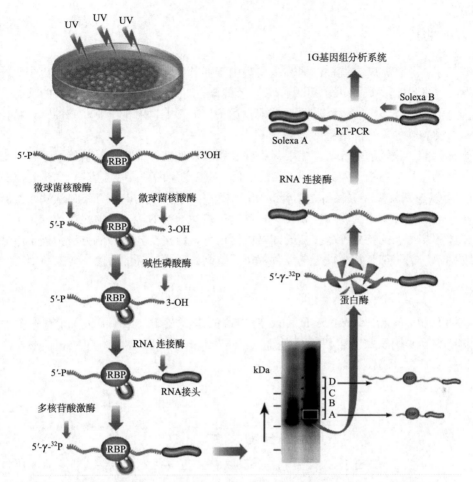

图 7-3 该图描绘了应用于干细胞的 CLIP-seq 技术。来源于紫外线照射后干细胞的 RNA 与 RNA 结合蛋白（RBP）形成的复合物与 RBP 抗体孵育。复合物中的 RNA 被两种不同浓度的 MNase 酶所修饰，之后进行上述放射自显影。与 A 和 B 带相应的蛋白质-RNA 共价复合物在 SDS-PAGE 胶上重现。最终，结合的 RNA 被纯化和测序。引用于文献 50 并有所修改

SAM68 是一种表达于细胞核的、参与剪接的 RNA 结合蛋白[56]，广泛表达于多种细胞，其过表达可抑制神经干细胞的增殖[57]。RNAi 技术和可变性剪接的芯片分析显示，小鼠成神经母细胞瘤中外显子受到该基因的调控[58]。Chawla 等也发现在维甲酸存在时，SAM68 基因沉默可以抑制小鼠胚胎肿瘤细胞的分化，SAM68 在神经分化过程表达上调并影响和/或调节对神经表型重要基因的剪接。这些预示着 SAM68 是神经分化过程中重要的剪接调节子。

随着基于基因组水平的检测技术的发展，大量 RNA 结合蛋白及基序（motif）被发现，RNA 剪接代码规则将逐渐清晰，这将有助于预测 RNA 剪接模式和更好地控制干细胞命运。

结论和展望

近年来,干细胞转录调控的研究主要集中于哺乳动物基因组的大量转录因子。这些努力揭示了干性相关转录因子可以作为有效的干性标记和重编程细胞到多能状态的工具。但是,鉴于很多细胞也受转录后基因表达调控的影响,对干细胞状态的认识将依赖于对选择性剪接的规则和结果的理解。

选择性剪接领域的研究目前主要集中在鉴别重要的剪接异构体和剪接因子及其对剪接体数量的影响。鸟枪法可以快速检测更多干细胞富集的剪接变异体,而其他技术如CLIP,可以识别剪接因子和与之结合的功能RNA要素。最初的研究结果表明干细胞分化与剪接成不同的剪接变异体密切相关,同时过表达某些剪接变异体足以改变干细胞表型。随着更多谱系的特异性剪接变异体被检测出来,剪接变异体探测方法为确定干细胞状态和揭示新的干细胞特异性分化事件提供了更高效和更准确的方法。

致谢

感谢 Scolnick 和 Stephanie Huelga 对初稿的审阅及 Jason Nathanson 对作图的帮助。该研究受美国国家卫生研究所基金(HG004659,GM084317)和加利福尼亚再生医学研究所基金(RB1-01413)的支持。

(王丽娜 李宗金 译)

参 考 文 献

1. Brandenberger R, Wei H, Zhang S et al. Transcriptome characterization elucidates signaling networks that control human ES cell growth and differentiation. Nat Biotechnol 2004; 22(6):707-716.
2. Brandenberger R, Khrebtukova I, Thies RS et al. MPSS profiling of human embryonic stem cells. BMC Dev Biol 2004; 4:10.
3. Miura T, Luo Y, Khrebtukova I et al. Monitoring early differentiation events in human embryonic stem cells by massively parallel signature sequencing and expressed sequence tag scan. Stem Cells Dev 2004; 13(6):694-715.
4. Bhattacharya B, Cai J, Luo Y et al. Comparison of the gene expression profile of undifferentiated human embryonic stem cell lines and differentiating embryoid bodies. BMC Dev Biol 2005; 5:22.
5. Jurica MS, Moore MJ. Pre-mRNA splicing: awash in a sea of proteins. Mol Cell 2003; 12(1):5-14.
6. Nilsen TW. The spliceosome: the most complex macromolecular machine in the cell? Bioessays 2003; 25(12):1147-1149.
7. Matlin AJ, Clark F, Smith CW. Understanding alternative splicing: towards a cellular code. Nat Rev Mol Cell Biol 2005; 6(5):386-398.
8. Martinez-Contreras R, Cloutier P, Shkreta L et al. hnRNP proteins and splicing control. Adv Exp Med Biol 2007; 623:123-147.
9. Blencowe BJ. Alternative splicing: new insights from global analyses. Cell 2006; 126(1):37-47.
10. Wang ET, Sandberg R, Luo S et al. Alternative isoform regulation in human tissue transcriptomes. Nature 2008; 456(7221):470-476.
11. Pritsker M, Doniger TT, Kramer LC et al. Diversification of stem cell molecular repertoire by alternative splicing. Proc Natl Acad Sci USA 2005; 102(40):14290-14295.
12. Takahashi K, Tanabe K, Ohnuki M et al. Induction of pluripotent stem cells from adult human fibroblasts by defined factors. Cell 2007; 131(5):861-872.
13. Yu J, Vodyanik MA, Smuga-Otto K et al. Induced pluripotent stem cell lines derived from human somatic cells. Science 2007; 318(5858):1917-1920.
14. Lowry WE, Richter L, Yachechko R et al. Generation of human induced pluripotent stem cells from dermal fibroblasts. Proc Natl Acad Sci USA 2008; 105(8):2883-2888.

15. Tai MH, Chang CC, Kiupel M et al. Oct4 expression in adult human stem cells: evidence in support of the stem cell theory of carcinogenesis. Carcinogenesis 2005; 26(2):495-502.
16. Zangrossi S, Marabese M, Broggini M et al. Oct-4 expression in adult human differentiated cells challenges its role as a pure stem cell marker. Stem Cells 2007; 25(7):1675-1680.
17. Atlasi Y, Mowla SJ, Ziaee SA et al. OCT4 spliced variants are differentially expressed in human pluripotent and nonpluripotent cells. Stem Cells 2008; 26(12):3068-3074.
18. Cauffman G, Liebaers I, Van Steirteghem A et al. POU5F1 isoforms show different expression patterns in human embryonic stem cells and preimplantation embryos. Stem Cells 2006; 24(12):2685-2691.
19. Lee J, Kim HK, Rho JY et al. The human OCT-4 isoforms differ in their ability to confer self-renewal. J Biol Chem 2006; 281(44):33554-33565.
20. He X, Treacy MN, Simmons DM et al. Expression of a large family of POU-domain regulatory genes in mammalian brain development. Nature 1989; 340(6228):35-41.
21. Theodorou E, Dalembert G, Heffelfinger C et al. A high throughput embryonic stem cell screen identifies Oct-2 as a bifunctional regulator of neuronal differentiation. Genes Dev 2009; 23(5):575-588.
22. Wang L, Wang J, Sun S et al. A novel DNMT3B subfamily, DeltaDNMT3B, is the predominant form of DNMT3B in nonsmall cell lung cancer. Int J Oncol 2006; 29(1):201-207.
23. Ostler KR, Davis EM, Payne SL et al. Cancer cells express aberrant DNMT3B transcripts encoding truncated proteins. Oncogene 2007; 26(38):5553-5563.
24. Gopalakrishnan S, Van Emburgh BO, Shan J et al. A novel DNMT3B splice variant expressed in tumor and pluripotent cells modulates genomic DNA methylation patterns and displays altered DNA binding. Mol Cancer Res 2009; 7(10):1622-1634.
25. Margot JB, Ehrenhofer-Murray AE, Leonhardt H. Interactions within the mammalian DNA methyltransferase family. BMC Mol Biol 2003; 4:7.
26. Brodie C, Blumberg PM. Regulation of cell apoptosis by protein kinase c delta. Apoptosis 2003; 8(1):19-27.
27. Peluso JJ, Pappalardo A, Fernandez G. Basic fibroblast growth factor maintains calcium homeostasis and granulosa cell viability by stimulating calcium efflux via a PKC delta-dependent pathway. Endocrinology 2001; 142(10):4203-4211.
28. Zrachia A, Dobroslav M, Blass M et al. Infection of glioma cells with Sindbis virus induces selective activation and tyrosine phosphorylation of protein kinase C delta. Implications for Sindbis virus-induced apoptosis. J Biol Chem 2002; 277(26):23693-23701.
29. Kilpatrick LE, Lee JY, Haines KM et al. A role for PKC-delta and PI 3-kinase in TNF-alpha-mediated antiapoptotic signaling in the human neutrophil. Am J Physiol Cell Physiol 2002; 283(1):C48-57.
30. Sitailo LA, Tibudan SS, Denning MF. Bax activation and induction of apoptosis in human keratinocytes by the protein kinase C delta catalytic domain. J Invest Dermatol 2004; 123(3):434-443.
31. Sakurai Y, Onishi Y, Tanimoto Y, Kizaki H. Novel protein kinase C delta isoform insensitive to caspase-3. Biol Pharm Bull. Sep 2001;24(9):973-977.
32. Patel NA, Song SS, Cooper DR. PKCdelta alternatively spliced isoforms modulate cellular apoptosis in retinoic acid-induced differentiation of human NT2 cells and mouse embryonic stem cells. Gene Expr 2006; 13(2):73-84.
33. Dvorak P, Dvorakova D, Koskova S et al. Expression and potential role of fibroblast growth factor 2 and its receptors in human embryonic stem cells. Stem Cells 2005; 23(8):1200-1211.
34. Li J, Wang G, Wang C et al. MEK/ERK signaling contributes to the maintenance of human embryonic stem cell self-renewal. Differentiation 2007; 75(4):299-307.
35. Mayshar Y, Rom E, Chumakov I et al. Fibroblast growth factor 4 and its novel splice isoform have opposing effects on the maintenance of human embryonic stem cell self-renewal. Stem Cells 2008; 26(3):767-774.
36. Hill M, Goldspink G. Expression and splicing of the insulin-like growth factor gene in rodent muscle is associated with muscle satellite (stem) cell activation following local tissue damage. J Physiol 2003; 549(Pt 2):409-418.
37. Ates K, Yang SY, Orrell RW et al. The IGF-I splice variant MGF increases progenitor cells in ALS, dystrophic and normal muscle. FEBS Lett 2007; 581(14):2727-2732.
38. Gnecchi M, He H, Liang OD et al. Paracrine action accounts for marked protection of ischemic heart by Akt-modified mesenchymal stem cells. Nat Med 2005; 11(4):367-368.
39. Tang YL, Zhao Q, Qin X et al. Paracrine action enhances the effects of autologous mesenchymal stem cell transplantation on vascular regeneration in rat model of myocardial infarction. Ann Thorac Surg 2005; 80(1):229-236; discussion 236-227.
40. Lin H, Shabbir A, Molnar M et al. Adenoviral expression of vascular endothelial growth factor splice variants differentially regulate bone marrow-derived mesenchymal stem cells. J Cell Physiol 2008; 216(2):458-468.
41. Yang Z, Sui Y, Xiong S et al. Switched alternative splicing of oncogene CoAA during embryonal carcinoma stem cell differentiation. Nucleic Acids Res 2007; 35(6):1919-1932.
42. Brooks YS, Wang G, Yang Z et al. Functional pre-mRNA trans-splicing of coactivator CoAA and corepressor RBM4 during stem/progenitor cell differentiation. J Biol Chem 2009; 284(27):18033-18046.
43. Yeakley JM, Fan JB, Doucet D et al. Profiling alternative splicing on fiber-optic arrays. Nat Biotechnol 2002; 20(4):353-358.

44. Johnson JM, Castle J, Garrett-Engele P et al. Genome-wide survey of human alternative premRNA splicing with exon junction microarrays. Science 2003; 302(5653):2141-2144.
45. Pan Q, Shai O, Misquitta C et al. Revealing global regulatory features of mammalian alternative splicing using a quantitative microarray platform. Mol Cell 2004; 16(6):929-941.
46. Yeo GW, Xu X, Liang TY et al. Alternative splicing events identified in human embryonic stem cells and neural progenitors. PLoS Comput Biol 2007; 3(10):1951-1967.
47. Salomonis N, Nelson B, Vranizan K et al. Alternative splicing in the differentiation of human embryonic stem cells into cardiac precursors. PLoS Comput Biol 2009; 5(11):e1000553.
48. Cloonan N, Forrest AR, Kolle G et al. Stem cell transcriptome profiling via massive-scale mRNA sequencing. Nat Methods 2008; 5(7):613-619.
49. Ule J, Jensen KB, Ruggiu M et al. CLIP identifies Nova-regulated RNA networks in the brain. Science 2003; 302(5648):1212-1215.
50. Yeo GW, Coufal NG, Liang TY et al. An RNA code for the FOX2 splicing regulator revealed by mapping RNA-protein interactions in stem cells. Nat Struct Mol Biol 2009; 16(2):130-137.
51. Ashiya M, Grabowski PJ. A neuron-specific splicing switch mediated by an array of premRNA repressor sites: evidence of a regulatory role for the polypyrimidine tract binding protein and a brain-specific PTB counterpart. RNA 1997; 3(9):996-1015.
52. Boutz PL, Stoilov P, Li Q et al. A posttranscriptional regulatory switch in polypyrimidine tract-binding proteins reprograms alternative splicing in developing neurons. Genes Dev 2007; 21(13):1636-1652.
53. Makeyev EV, Zhang J, Carrasco MA et al. The MicroRNA miR-124 promotes neuronal differentiation by triggering brain-specific alternative premRNA splicing. Mol Cell 2007; 27(3):435-448.
54. Spellman R, Llorian M, Smith CW. Crossregulation and functional redundancy between the splicing regulator PTB and its paralogs nPTB and ROD1. Mol Cell 2007; 27(3):420-434.
55. Shibayama M, Ohno S, Osaka T et al. Polypyrimidine tract-binding protein is essential for early mouse development and embryonic stem cell proliferation. FEBS J 2009; 276(22):6658-6668.
56. Matter N, Herrlich P, Konig H. Signal-dependent regulation of splicing via phosphorylation of Sam68. Nature 2002; 420(6916):691-695.
57. Moritz S, Lehmann S, Faissner A et al. An induction gene trap screen in neural stem cells reveals an instructive function of the niche and identifies the splicing regulator sam68 as a tenascin-C-regulated target gene. Stem Cells 2008; 26(9):2321-2331.
58. Chawla G, Lin CH, Han A et al. Sam68 regulates a set of alternatively spliced exons during neurogenesis. Mol Cell Biol 2009; 29(1):201-213.

第8章 微小RNA调节胚胎干细胞自我更新和分化

Collin Melton，Robert Blelloch*

摘要：干细胞从自我更新细胞到分化细胞的转变是需要一系列事件复杂协作的过程。干细胞可分为全能干细胞（能分化为所有的胚胎细胞）、多能干细胞（能分化为多种细胞）和单能干细胞（只能分化为单个细胞）。不管它们的潜能如何，所有干细胞分化时，它们的自我更新程序必须是沉默的。干细胞自我更新程序可以定义为使细胞增殖的同时维持其潜能的内外信号的整合。两个标志性的自我更新程序是自我增强的转录网络和特异的细胞周期模式。在本章中，我们讨论各种微小RNA对增强或抑制干细胞自我更新程序的影响并探讨这种整合的调节信号如何为细胞命运的决定提供支持。我们将侧重于胚胎干细胞（embryonic stem cell，ESC）相关微小RNA在自我更新、分化和逆分化中的功能。我们将把胚胎干细胞中微小RNA的功能与系特异的成体干细胞及肿瘤中微小RNA的功能相比较。

引言：自我更新程序

在胚胎和成体干细胞群中，干细胞自我更新程序在干细胞进行连续几轮复制的过程中维持其潜能。不同干细胞群的潜能及增殖的程度差别很大，取决于其进化压力和生物功能。ESC来源于发育的囊胚的内细胞层，与发育的上胚层细胞类似。上胚层细胞能发育成胚胎的内胚层、中胚层、外胚层及生殖系统，因此是全能的[1]。上胚层细胞具有短的细胞周期，但是，它们随着细胞周期的延伸而最终分化。与上胚层细胞类似，ESC具有短的细胞周期，也是全能的。但是，与上胚层细胞不同，ESC在培养中能无限的自我更新。

在胚胎发育过程中，上胚层细胞分化成具有有限潜能的、特定的胚胎干细胞群，包括神经干细胞和造血干细胞等，这些胚胎干细胞增殖率高，但是潜能有限[2,3]。最终，这些胚胎干细胞被成体系特异的胎干细胞取代，包括胚胎神经干细胞和造血干细胞的相应成体细胞。成体干细胞的潜能有限，但是和相应的胎细胞不同，其增殖率低。实际上，成体干细胞群大部分是静态的，尽管它们生产暂时的祖细胞群。祖细胞与其相应的胚胎干细胞类似，通常具有高的增殖率。静态成体干细胞能在进化过程中减少有害突变的概率，如一些导致肿瘤的突变[4]。

---

* Corresponding Author：Robert Blelloch—The Eli and Edythe Broad Center of Regeneration Medicine and Stem Cell Research，Center for Reproductive Sciences，Biomedical Science Graduate Program and Department of Urology，University of California San Francisco，San Francisco，California 94143，USA. Email：blellochr@stemcell. ucsf. edu

胚胎干细胞

研究人员已经对 ESC 自我更新程序进行了分子水平的深入研究。自我更新程序由许多因子的相互作用决定，这些因子的核心是不同的转录网络[5]。在 ESC，核心转录网络包括转录因子 Oct4、Sox2、Nanog、Tcf3 及 Myc 家族蛋白（cMyc 和 nMyc）。这些转录因子协同作用直接和间接地决定表观遗传状态，能在分化时使激活或者抑制三胚层各细胞系基因的转录保持平衡[5]。ESC 的转录网络以这种方式维持它的全能性。而且，ESC 转录网络驱动因子的表达，从而通过直接和间接地维持短的细胞周期的方式使细胞实现高增殖率。

随着对 ESC 进行诱导分化，细胞自我更新程序的许多成分必须被关闭，而新的分化程序必须被激活。因此，细胞命运的转变受多种因素的调节，这些因素既沉默自我更新程序，又诱导系特异的分化程序。这些因素可以被大概地划分为在染色质状态、转录、转录稳定、蛋白质翻译、蛋白质稳定或蛋白质功能水平影响基因表达。

在本章中，我们将侧重于促进自我更新和分化功能的微小 RNA。

微小 RNA 的生成和功能

miRNA 是小的非编码 RNA，能在转录后通过翻译抑制或 mRNA 降解沉默基因表达。miRNA 是通过 RNA 转录形成的（图 8-1）。微小 RNA 首先被转录为长链的 RNA 多聚酶 II 转录本，命名为原始的 miRNA（pri-miRNA）[6,7]。这些 pri-miRNA 可能是编码的，也可能是非编码的。如果是后者，miRNA 通常定位于编码基因的内含子区[8]。在细胞核，miRNA 被微处理器复合物识别及切割。微处理器复合物包括 RNA 结合蛋白 DGCR8 和 RNA 酶 III DROSHA[9~13]。此复合物识别长度大约 33 个碱基对的茎环结构，而且具有酶活性，能切割茎环 11 个碱基对形成 3′端 2 个核苷酸的黏性末端[14]。处理过的 RNA 命名为 miRNA 前体，被 Exportin V 从细胞核转运至细胞质。在细胞质，miRNA 前体被第二个包含 RNA 酶 III DICER 的复合物识别[15~18]。这个复合物识别 miRNA 前体的发夹结构，并在发夹环的基底部切割 miRNA 前体形成 3′端 2 个核苷酸黏性末端，同时生成大约 22 个核苷酸长度的成熟 miRNA 的双倍体[18]。这个成熟的 miRNA 双倍体一直保持双链直至被掺入到 RNA 诱导的沉默复合体（RISC）。仅小 RNA 双倍体的一条链掺入到 RISC，这条链通常含有较少稳定的 5′末端[19]。

被加载至 RISC 复合体的 miRNA 通过与开放阅读框和 3′非翻译区的碱基配对直接与靶 mRNA 作用。这种作用依赖于 miRNA 的 6~8 个核苷酸的种子序列（5′端的第 2~8 个核苷酸）与靶 mRNA 的碱基配对[20]。结合到靶 mRNA 的 RISC 复合体通过多种机制破坏蛋白质生成，包括：与 5′帽子作用破坏核糖体起始、阻止核糖体延伸、通过缩短 polyA 尾巴促进 RNA 降解[21]。

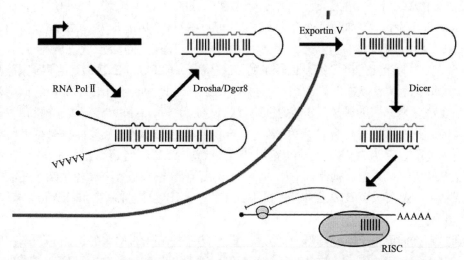

图 8-1 miRNA 生成。miRNA 首先被转录为长链的 RNA 多聚酶 Ⅱ 转录本。这些转录本的发夹结构被包含 Drosha 和 Dgcr8 的微处理器复合物识别并切割形成更小的 miRNA 前体。miRNA 前体被从细胞核转运至细胞质，并被 Dicer 切割形成成熟的 miRNA 双倍体。双倍体的一个链被加载至 RISC 复合体。加载了 miRNA 的复合物能降解和抑制靶 mRNA 的翻译

促进自我更新的 ESCC miRNA

许多 miRNA 共表达于同一个转录物，例如，miR-290 簇在 ESC 高表达，包括 7 个 miRNA。miR-290 簇的一个亚类共享一个共同的种子序列，并且都调节 ESC 细胞周期，因此被称为 ESCC miRNA 家族（ESC cell cycle promoting miRNA）[22]。与 ESCC miRNA 相关的家族包括 miR-302 家族及 miR-17/20/106 家族，后者 miRNA 的种子序列稍微不同。miR-290 簇在人类的同源性与 miR-370 簇稍微不同，但是 miR-302 簇在小鼠和人类非常类似[23,24]。

在小鼠和人类的多能干细胞中共同表达相似的 miRNA，提示这些 miRNA 在 ESC 中具有重要的功能。实际上，miRNA 功能的证据首次被揭示是通过敲除 ESC Dicer 或 *Dgcr8* 基因的模型[25~27]。这些 ESC 具有低的增值率和 G_1 期延长的细胞周期模式[27]。这些发现特别有趣，基于野生型小鼠 ESC 的特征是非典型的细胞周期，与体细胞相比较，ESC 的细胞周期的 G_1 期较短[28]。也就是说，这些最初的发现表明 ESC 表达的 miRNA 抑制体细胞周期结构。

ESC 短的细胞周期能促使它们快速增殖，至少部分是继发于 G_1/S 限制点的缓和[28]。在非典型的体细胞中，G_1/S 限制点阻止 S 期的起始及 DNA 复制。G_1/S 限制点包括一系列复杂的信号，这些信号在过渡到 S 期之前必须达到一个阈值。这个反应的关键信号成分包括但是不仅限于细胞周期素、细胞周期素依赖的蛋白激酶（CDK）、CDK 抑制剂（CDI）、Rb 家族蛋白及 E2F 家族蛋白[29]。D 和 E 型细胞周期素与 CDK 形成复合物驱动

Rb 家族蛋白磷酸化[30]。在鼠 ESC，细胞周期素 E 的高表达不依赖于细胞周期，但是细胞周期素 D 不表达[31]。细胞周期素 E 与 CDK2 形成复合物启动及灭活 Rb 家族蛋白磷酸化 (pRb、P107 和 P130)。当 Rb 家族蛋白处于低磷酸化的激活状态时能灭活 E2F (E2F1-3)，同时激活抑制 S 期基因转录的 E2F 蛋白 (E2F4 和 E2F 5)。当 Rb 家族蛋白处于高磷酸化的失活状态时，它们不再激活抑制的 E2F。同时，激活 E2F 的抑制被解除从而驱动 S 期基因的转录。S 期的进展可被包括 CIP 和 INK 家族成员在内的 CDK 抑制剂阻断。这些抑制剂阻断 CDK/Cyclin 复合物的活性[32]。INK 家族抑制剂在小鼠 ESC 是没有功能的，因为它们是通过细胞周期素 D 起作用的，但是细胞周期素 D 在小鼠 ESC 表达很低。CIP 家族抑制剂对 CDK/Cyclin 复合物的抑制效应比较复杂，还能够结合并灭活 CDK2/CyclinE 复合物[32]。在小鼠 ESC，CIP 家族抑制剂和 Rb 蛋白都表达较低[28,31]。

通过筛选能促进 Dgcr8 敲除 (-/-) ESC 增殖的 miRNA，ESCC miRNA 在细胞周期控制中的作用被揭示。这些 miRNA 不仅促进 Dgcr8 (-/-) ESC 增殖，而且降低处于 G_1 期的细胞数量。对 G_1 期的这种效应有一部分是通过直接作用于 CIP 家族的 CDK 抑制剂 P21、LATS2 及一些 Rb 家族蛋白如 pRb 和 P130。通过抑制这些以及其他预测的、与 G_1 期调节相关的 miRNA 的靶基因，ESCC miRNA 促进了 ESC 细胞周期 (图 8-2)[22]。

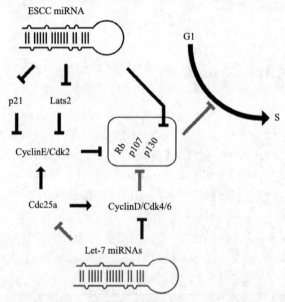

图 8-2 Let-7 和 ESCC miRNA 对 G_1-S 转变具有相反的作用。本图是 ESCC 和 Let-7 miRNA 对 ESC 与 G_1-S 转变相关因子的直接抑制效应的模式图。随着 ESC 从自我更新转变为分化状态，ESCC miRNA 下调而 Let-7 miRNA 上调。这些改变对细胞周期有直接的作用。黑色/加粗的箭头、线及文本表示在 ESC 中上调的 miRNA 和蛋白质的相互作用。灰色的箭头、线及文本表明在 ESC 中下调的 miRNA 和蛋白质的相互作用。请注意，Let-7 miRNA 对细胞周期的作用是在体细胞测试的，而不是在 ESC

最近，ESCC miRNA 对 ESC 转录组的影响被深入地分析。据发现，ESCC miRNA 不直接激活 cMyc 表达[33]。Myc 既是促进增殖，又是 ESC 自我更新必需的转录因子[34,35]。而且，在 ESC，抑制 cMyc 蛋白促进自我更新的丧失，但是，增加 cMyc 表达阻止缺少 LIF 时自我更新的丧失[35]。Lin 等最近确定了 Myc 蛋白促进 ESC 自我更新的机制。尤其是他们发现 cMyc 驱动多种促自我更新 miRNA 的转录，包括 miR-141、miR-200 和 miR-429。这些 miRNA 促进缺少 LIF 时自我更新的维持，但是这些效应的生物机制未知[36]。而且，cMyc 调节 ESCC miRNA 表达形成一个正反馈环，如下所述。

据证实，许多其他的因子如 DNA 甲基转移酶（DNMT3a 和 b）被 ESCC miRNA 间接上调[37,38]。这些 DNA 甲基转移酶表达增加是亚端粒区维持正确的 DNA 甲基化所必需的，DNA 甲基化又是阻止异常的端粒延长所必需的[37]。调节 DNMT3a 和 DNMT3b 是通过 ESCC miRNA 作用于 P130-DNMT3a 和 DNMT3b 转录的负性调节因子[37,38]。除了 DNA 甲基转移酶，许多的其他多能相关转录因子被 ESCC miRNA 间接上调，这些转录因子包括 Lin28、Trim71 和 Sall4[33]。这些大量的、被 ESCC miRNA 诱导的分子改变对促进细胞周期具有深远的影响，而且保持对端粒的维持，从而确保 ESC 在进行适当的自我更新的同时维持其多能性。

ESC 分化过程中诱导的 miRNA 阻止自我更新程序

miRNA 适于稳定自我更新状态，同样也适于促进自我更新到分化的转变。沉默自我更新的 miRNA 可以根据它们的靶标和表达模式分类。少量 miRNA 被发现直接作用于 ESC 核心转录网络的成分[39~41]。这些相同的 miRNA 在 ESC 分化为下游特异的谱系过程中被快速诱导。另一类 miRNA 在 ESC 分化为下游广泛的谱系过程中被诱导，并且广泛地抑制 ESC 相关基因但不是 ESC 核心转录因子[33]，它们也促进体细胞周期[42~44]。这两类促分化的 miRNA 可能在分化过程中的作用不同。第一类 miRNA 直接抑制 ESC 自我更新状态，但是第二类 miRNA 主要稳定分化状态——就像 ESC miRNA 稳定 ESC 状态。在小鼠 ESC，miR-134、miR-296 和 miR-470 被发现能直接抑制 Nanog、Pou5f1（即 Oct4）和 Sox2[40,41]。这些 miRNA 和靶标的相互作用主要通过在开放阅读框的相互作用。这些 miRNA 在维甲酸诱导分化时高度上调，维甲酸主要诱导向神经分化，提示这些 miRNA 可能与系特异的沉默 ESC 自我更新有关。在人的 ESC，miR-145 被发现通过作用于 Oct4、Sox2 和 Klf4[39] 直接抑制 ESC 自我更新。了解各种抑制自我更新的 miRNA 的生物功能和体内的作用将是未来研究的重要领域。

与直接抑制 ESC 自我更新的 miRNA 相反，let-7 家族的 miRNA 是细胞分化命运的稳定因素[33]。Let-7 变异最先在线虫中筛选阻止干细胞终末分化的变异基因时被发现[45]。自从在线虫中发现 let-7，let-7 的同源体在所有的多细胞动物中被发现[46]。在小鼠和人类，有 9 种不同的 let-7 家族成员，它们具有组织特异的表达模式[47-50]。在 ESC 中存在使 let-7 转录后沉默的精确机制。RNA 结合蛋白 lin28 和末端 uridyl-转移酶 TUT4 形成复合物结合并诱导前 let-7 转录本降解[51~56]。Lin28 表达在 ESC 分化过程中快速丢失[57,58]，导致 let-7 表达快速增加[55]。

最近,研究发现 let-7 家族成员能在 miRNA 缺陷的 Dgcr8-/-ESC 而不是野生型的 ESC 诱导自我更新沉默[33]。这些结果提示 ESC 表达的 miRNA 通常阻止 let-7 沉默 ESC 自我更新。实际上,ESCC miRNA 能够阻止 let-7miRNA 诱导的自我更新的丧失。Let-7 优先作用于在 ESC 中丰富的转录本,包括被多能转录因子 Oct4、Sox2、Nanog 和 Tcf3 调节的转录本。除此之外,许多 let-7 的直接靶基因被 ESCC miRNA 间接调节,这样可以解释 ESCC 如何拮抗 let-7。Let-7 和 ESCC 具有相反调节作用的靶基因包括 Myc、Sall4、Lin28 和 Trim7 [33]。

ESCC miRNA 和 let-7 之间的拮抗作用及两者对靶基因的调节方式相反,提示 ESCC miRNA 和 let-7 miRNA 之间存在一个相互排斥的表达和功能的网络(图 8-3)。在 ESC,ESCC miRNA 导致 lin28 上调,从而直接抑制 let-7 成熟。而且,ESCC 间接上调 cMyc 和其他促进 ESC 自我更新的 let-7 靶基因。通过上述机制,ESCC miRNA 抵抗 let-7 的作用。Oct4、Sox2 和 Nanog 促进 ESCC miRNA 表达[59]。随着 ESC 分化,Oct4、Sox2 和 Nanog 表达下降导致 ESCC 表达相应降低。在 ESCC 不存在的条件下,lin28 水平也下降。在分化状态下,let-7 不再抑制而反馈作用于 lin28,从而增强自身的表达。而且,let-7 使分化状态保持稳定,通过限制 ESC 命运决定必需因子的表达,这些因子包括被多能转录因子 Nanog、Oct4 和 Sox2 激活的转录本。

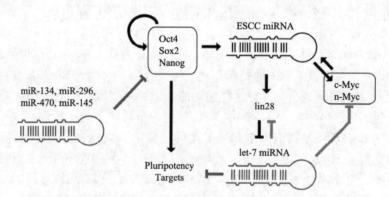

图 8-3 ESC 自我更新网络中 miRNA 的相互作用。本图是 ESCC、let-7、miR-134、miR-296、miR-470 和 miR-145 直接抑制和间接激活效应的模式图。黑色/加粗箭头、线和文本代表在 ESC 上调的 miRNA 与蛋白质的相互作用。灰色线和文本代表在 ESC 下调的 miRNA 与蛋白质的相互作用。随着 ESC 分化,miR-134、miR-296、miR-470 和 miR-145 动摇 Oct4/Sox2/Nanog 转录因子网络从而促进分化,但是 let-7 miRNA 抑制 Myc 和 Oct4/Sox2/Nanog 网络下游的靶基因从而稳定分化状态

let-7 miRNA 除了阻止 ESC 转录程序外,还促进体细胞周期(图 8-2)。Let-7 miRNA 直接和间接地作用于多种 G_1-S 期转变的激活子,包括 cdc25a、cdk6、cyclinD1 和 cyclinD2[42,44]。这些相互作用有助于 let-7 miRNA 增加 G_1 期细胞数的总体效应[42~44]。细胞周期是否及如何直接影响 ESC 自我更新还是未知。据推测,处于 G_1 期的细胞最易受促分化信号包括 MAPK 信号的影响。更详细地了解细胞周期和 ESC 转录网络间的相互

作用及了解 miRNA 对这些相互作用的影响非常重要。

控制 miRNA 表达的调节网络

在 ESC，miR-290 簇来源的 ESCC miRNA 表达受多能转录因子 Nanog、Oct4、Sox2、Tcf3，以及 Myc 转录因子 nMyc 和 cMyc 的控制[59,61]。ESCC miRNA 间接上调 cMyc 形成一个能增强自身表达的正反馈环。当 ESC 分化时，多能转录因子下调，ESCC miRNA 也是如此[59]。

直接抑制 ESC 自我更新的 miRNA 表达的转录控制仍然是一个未知的研究领域。但是，ESC 中染色质免疫沉淀因子高通量测序的结果使我们能了解一些它们的调节机制。在 ESC，miR-296 启动子与 Oct4、Sox2、Nanog 和 Tcf3 结合。但是，miR-296 启动子的特点是 H3K27 甲基化抑制及结合多梳蛋白 Suz12[59]。这些结果提示 miR-296 在 ESC 中存在被激活的机制。ESC 一旦分化，抑制的 H3K27 组氨酸标记快速丢失，早于 Oct4、Sox2 和 Nanog 的丢失，这些转录因子能暂时驱动 miR-296 转录。这种调节方式将形成一个负反馈环，导致 ESC 自我更新的更快丧失。在 ESC，miR-296 启动子区 H3K27 甲基化如何维持，以及随分化如何丢失是未知的。对 miR-134 和 miR-470 启动子的调节了解得也非常少。

同样，let-7 表达的转录控制仍然相对不清楚。在不同的组织，let-7 的转录表达不同，因此可能多种转录因子诱导 let-7 表达[47]。在 ESC，Oct4、Sox2、Nanog 驱动 let-7g 原始转录本[59]。原始转录本被加工为 miRNA 前体并被 lin28/Tut4 复合物所降解[51~56]。随着 ESC 分化，lin28/Tut4 的抑制作用丢失，接着成熟的 let-7 产生[55,57,58]。其他的 miRNA 在 ESC 中也以这种方式调节[53]。

最近，反现了一类新的调节性 RNA 结合蛋白——Trim-NHL 蛋白。在神经干细胞，Schwamborn 等报道 Trim32 表达增强了 let-7 抑制靶标的能力并与神经干细胞分化有关[62]。在 ESC 中，ESCC miRNA 促进 Trim71（也称为 Mlin-41）表达。Trim71 是 let-7 的靶基因，是小鼠分化所必需的[63]。Rybak 等证明 Trim71 作为 ESC 表达的 E3 泛素连接酶具有降解 Ago2 蛋白的功能，Ago2 蛋白是 RISC 复合物的一个成分[64]。Trim32 和 Trim71 是一个更大的 Trim-NHL 蛋白家族的成员，这个蛋白家族还包括果蝇蛋白 Brat 和 Mei-P26。这些果蝇蛋白具有通过与 Ago1 相互作用调节 miRNA 途径的功能[65]。了解 Trim71 是否仅通过降解 Ago2 简单地调节整个 miRNA 途径的活性，还是像 Trim32 一样增加特异的 miRNA 亚型的活性是非常重要的。

miRNA 能促进或抑制 IPS 细胞分化

ESCC miRNA 促进 ESC 自我更新，但是 let-7 miRNA 促进 ESC 自我更新的沉默。通过核转移或者引进外源的转录因子诱导体细胞重编程为诱导性的多能干细胞（iPS）[66]。与 ESCC miRNA 促进 ESC 自我更新的作用一致，加入这些 miRNA 到重编程的体系能增加重编程的效率[61]。同样，抑制 let-7 miRNA 增加重编程[33]。直接抑制 ESC

自我更新的 miRNA 对重编程的影响仍未知。这些发现一起证明控制 ESC 自我更新和分化的共同机制也控制脱分化过程。

此外，有及无 Myc 的转录因子的组合（Sox2、Oct4、Klf4、cMyc 或者 Sox2、Oct4、Klf4，无 cMyc）诱导重编程的能力使 miRNA 和 miRNA 抑制剂发挥功能的途径是否与这些因子相同成为疑问。例如，ESCC miRNA 在没有 cMyc 时而不是有 cMyc 时被证明促进重编程[61]。这些发现提示 ESCC 和 cMyc 具有多重作用。实际上，目前已知 ESCC 诱导 cMyc 间接上调，以及 cMyc 和 nMyc 都促进 ESCC miRNA 转录[33,61]。

另外，据发现，Myc 不存在时抑制 let-7 促进重编程的效果大于 Myc 存在时[33]。这一发现提示在体细胞中 let-7 部分通过 Myc 抑制 ESC 自我更新。实际上，cMyc 和 nMyc 都是 let-7 的直接靶基因[33,67]。是否存在一些 miRNA，它们的信号途径与多能的转录因子信号途径一致，以及是否这些 miRNA 能替代转录因子在 iPS 细胞重编程中的作用，了解上述问题将非常重要及有趣。

成体干细胞中的 miRNA

miRNA 在成体干细胞中的功能仍然研究得很少。实际上，除了 ESC 来源的神经祖细胞（NPC），没有研究详细的分析成体干细胞群的 miRNA 谱。在 NPC，let-7 miRNA 是主要的种类[59]。有意思的是，最近的研究证明 let-7 miRNA 是胚胎小鼠脑中神经干细胞分化所必需的，但不是神经干细胞扩增所必需的[62]。在这个模型中，神经干细胞不对称分裂使 RNA 结合蛋白 Trim32 分离进入定向分化的子细胞。在子细胞中，Trim32 除了其他功能，还能增加 let-7 的活性而促进分化。

肿瘤细胞中的 miRNA

ESCC miRNA 和 miR-17/20/106 家族具有相似的种子序列，miR-17/20/106 家族被证明在肿瘤中有重要的功能。例如，在多种肿瘤中，miR-93 和 miR-106 miRNA 作用于 p21 使 G_1/S 监测点放开并促进细胞快速增殖[68,69]。而且，体内研究证明这些 miR-NA 在肿瘤形成中具有重要作用。尤其是，miR-17-19b 多顺反子的过表达增加肿瘤形成并降低 Eμ-Myc B 细胞淋巴瘤小鼠中肿瘤细胞的凋亡[70]。这个模型中凋亡的降低可能至少部分是由于 miR-17 家族 miRNA 作用于促凋亡蛋白 Bim[71]。miR-17/92 簇也通过增加肿瘤血管新生而有助于肿瘤形成[72]。人 miRNA miR-372 和 miR-373 共享 ESCC 种子序列。这些 miRNA 与原癌基因 Ras 协作促进人纤维母细胞肿瘤形成，而且在生殖细胞肿瘤中高表达[73]。总体来说，这些结果证明，与 ESCC miRNA 共享种子序列的 miRNA 具有和原癌基因类似的功能，它们的信号途径与在 ESC 中的 miRNA 类似。

与 ESCC 相关 miRNA 相反，let-7 miRNA 是肿瘤抑制子。在乳腺癌模型，一个肿瘤细胞的亚群，肿瘤起始细胞（TIC）能再生肿瘤。当 TIC 分化，其不再具有形成整个肿瘤的能力。let-7 miRNA 足够诱导 TIC 分化。在这种背景下，let-7 通过抑制 Ras 来抑制增殖及 HMGA2 的表达，从而促进肿瘤细胞分化[74]。同样，在 K-Ras 诱导的小鼠肺肿瘤模型及肿瘤细胞系植入模型，加入外源的 let-7 miRNA 移植肿瘤形成，同时，抑制 let-7 活性促进肿瘤形成[75~77]。而且，最近的证据证实 lin28 通过抑制 let-7 活性促进肿瘤形成[78~81]。Let-7 被证明作用于多种原癌基因，包括 K-Ras、N-Ras、Hmga2、cMyc、nMyc 和其他降低细胞增殖的因子[82]。综上所述，这些数据强烈支持 let-7 作为肿瘤抑制因子的作用。

结论

本章中总结的结果支持多种 miRNA 在稳定干细胞自我更新状态或促进它们的分化方面发挥重要的作用。这些 miRNA 与其他基因表达的通用调节子类似，因为这些 miRNA 的不同亚型能促进或抑制干细胞自我更新。这些 miRNA 对自我更新的影响是通过调节细胞周期及干细胞转录程序。随着我们对影响干细胞自我更新的 miRNA 的了解越来越多，我们越来越清楚这些 miRNA 被复杂的分子网络紧密调节。调节可以发生在各种水平，包括转录及转录后。而且，不同种类的 miRNA 相互抑制或激活表达。了解发育过程中这些网络的成分及功能将大大增加我们对发育和疾病状态的认识。

（马凤霞　译）

参 考 文 献

1. Surani MA, Hayashi K, Hajkova P. Genetic and epigenetic regulators of pluripotency. Cell 2007; 128(4):747-762.
2. Kriegstein A, Alvarez-Buylla A. The glial nature of embryonic and adult neural stem cells. Annu Rev Neurosci 2009; 32(1):149-184.
3. Mikkola HKA, Orkin SH. The journey of developing hematopoietic stem cells. Development 2006; 133(19):3733-3744.
4. Arai F, Suda T. Quiescent stem cells in the niche. In: StemBook, ed. The Stem Cell Research Community, StemBook, doi/10.3824/stembook.1.6.1, http://www.stembook.org. 2008.
5. Jaenisch R, Young R. Stem cells, the molecular circuitry of pluripotency and nuclear reprogramming. Cell 2008; 132(4):567-582.
6. Cai X, Hagedorn CH, Cullen BR. Human microRNAs are processed from capped, polyadenylated transcripts that can also function as mRNAs. RNA 2004; 10(12):1957-1966.
7. Lee Y, Kim M, Han J et al. MicroRNA genes are transcribed by RNA polymerase II. EMBO J 2004; 23(20):4051-4060.
8. Rodriguez A. Identification of mammalian microRNA host genes and transcription units. Genome Res 2004; 14(10a):1902-1910.
9. Basyuk E, Suavet F, Doglio A et al. Human let-7 stem-loop precursors harbor features of RNase III cleavage products. Nucleic Acids Res 2003; 31(22):6593-6597.
10. Gregory RI, Yan K, Amuthan G et al. The Microprocessor complex mediates the genesis of microRNAs. Nature 2004; 432(7014):235-240.
11. Lee Y, Ahn C, Han J et al. The nuclear RNase III Drosha initiates microRNA processing. Nature 2003; 425(6956):415-419.
12. Han J, Lee Y, Yeom K et al. The Drosha-DGCR8 complex in primary microRNA processing. Genes Dev 2004; 18(24):3016-3027.
13. Denli AM, Tops BBJ, Plasterk RHA et al. Processing of primary microRNAs by the Microprocessor complex. Nature 2004; 432(7014):231-235.
14. Han J, Lee Y, Yeom K et al. Molecular basis for the recognition of primary microRNAs by the drosha-DGCR8 complex. Cell 2006; 125(5):887-901.
15. Yi R, Qin Y, Macara IG et al. Exportin-5 mediates the nuclear export of premicroRNAs and short hairpin RNAs. Genes Dev 2003; 17(24):3011-3016.
16. Bohnsack MT, Czaplinski K, Gorlich D. Exportin 5 is a RanGTP-dependent dsRNA-binding protein that mediates nuclear export of premiRNAs. RNA 2004; 10(2):185-191.

17. Lund E, Güttinger S, Calado A et al. Nuclear export of microRNA precursors. Science 2004; 303(5654):95-98.
18. Hammond SM. Dicing and slicing: The core machinery of the RNA interference pathway. FEBS Letters 2005; 579(26):5822-5829.
19. Schwarz DS, Hutvágner G, Du T et al. Asymmetry in the assembly of the RNAi enzyme complex. Cell 2003; 115(2):199-208.
20. Bartel DP. MicroRNAs: Target recognition and regulatory functions. Cell 2009; 136(2):215-233.
21. Filipowicz W, Bhattacharyya SN, Sonenberg N. Mechanisms of posttranscriptional regulation by microRNAs: are the answers in sight? Nat Rev Genet 2008; 9(2):102-114.
22. Wang Y, Baskerville S, Shenoy A et al. Embryonic stem cell-specific microRNAs regulate the G1-S transition and promote rapid proliferation. Nat Genet 2008; 40(12):1478-1483.
23. Bar M, Wyman SK, Fritz BR et al. MicroRNA discovery and profiling in human embryonic stem cells by deep sequencing of small RNA libraries. Stem Cells 2008; 26(10):2496-2505.
24. Suh M, Lee Y, Kim JY et al. Human embryonic stem cells express a unique set of microRNAs. Dev Biol 2004; 270(2):488-498.
25. Kanellopoulou C, Muljo SA, Kung AL et al. Dicer-deficient mouse embryonic stem cells are defective in differentiation and centromeric silencing. Genes Dev 2005; 19(4):489-501.
26. Murchison EP, Partridge JF, Tam OH et al. Characterization of Dicer-deficient murine embryonic stem cells. Proc Natl Acad Sci USA 2005; 102(34):12135-12140.
27. Wang Y, Medvid R, Melton C et al. DGCR8 is essential for microRNA biogenesis and silencing of embryonic stem cell self-renewal. Nat Genet 2007; 39(3):380-385.
28. Savatier P, Huang S, Szekely L et al. Contrasting patterns of retinoblastoma protein expression in mouse embryonic stem cells and embryonic fibroblasts. Oncogene 1994; 9(3):809-818.
29. Planas-Silva MD, Weinberg RA. The restriction point and control of cell proliferation. Curr Opin Cell Biol 1997; 9(6):768-772.
30. Giacinti C, Giordano A. RB and cell cycle progression. Oncogene 2006; 25(38):5220-5227.
31. Savatier P, Lapillonne H, van Grunsven LA et al. Withdrawal of differentiation inhibitory activity/leukemia inhibitory factor up-regulates D-type cyclins and cyclin-dependent kinase inhibitors in mouse embryonic stem cells. Oncogene 1996; 12(2):309-322.
32. Mittnacht S. Control of pRB phosphorylation. Curr Opin Genet Dev 1998; 8(1):21-27.
33. Melton C, Judson R, Blelloch R. Opposing microRNA families regulate self-renewal in mouse embryonic stem cells. Nature 2010; Advance Online.
34. Singh AM, Dalton S. The cell cycle and Myc intersect with mechanisms that regulate pluripotency and reprogramming. Cell Stem Cell 2009; 5(2):141-149.
35. Cartwright P, McLean C, Sheppard A et al. LIF/STAT3 controls ES cell self-renewal and pluripotency by a Myc-dependent mechanism. Development 2005; 132(5):885-896.
36. Lin N, Jackson AL, Guo J et al. Myc-regulated microRNAs attenuate embryonic stem cell differentiation. EMBO J 2009; 28(20):3157-3170.
37. Benetti R, Gonzalo S, Jaco I et al. A mammalian microRNA cluster controls DNA methylation and telomere recombination via Rbl2-dependent regulation of DNA methyltransferases. Nat Struct Mol Biol 2008; 15(3):268-279.
38. Sinkkonen L, Hugenschmidt T, Berninger P et al. MicroRNAs control de novo DNA methylation through regulation of transcriptional repressors in mouse embryonic stem cells. Nat Struct Mol Biol 2008; 15(3):259-267.
39. Xu N, Papagiannakopoulos T, Pan G et al. MicroRNA-145 regulates OCT4, SOX2 and KLF4 and represses pluripotency in human embryonic stem cells. Cell 2009; 137(4):647-658.
40. Tay YM, Tam W, Ang Y et al. MicroRNA-134 modulates the differentiation of mouse embryonic stem cells, where it causes posttranscriptional attenuation of Nanog and LRH1. Stem Cells 2008; 26(1):17-29.
41. Tay Y, Zhang J, Thomson AM et al. MicroRNAs to Nanog, Oct4 and Sox2 coding regions modulate embryonic stem cell differentiation. Nature 2008; 455(7216):1124-1128.
42. Johnson CD, Esquela-Kerscher A, Stefani G et al. The let-7 microRNA represses cell proliferation pathways in human cells. Cancer Res 2007; 67(16):7713-7722.
43. Kumar MS, Erkeland SJ, Pester RE et al. Suppression of nonsmall cell lung tumor development by the let-7 microRNA family. Proc Natl Acad Sci USA 2008; 105(10):3903-3908.
44. Schultz J, Lorenz P, Gross G et al. MicroRNA let-7b targets important cell cycle molecules in malignant melanoma cells and interferes with anchorage-independent growth. Cell Res 2008; 18(5):549-557.
45. Reinhart BJ, Slack FJ, Basson M et al. The 21-nucleotide let-7 RNA regulates developmental timing in Caenorhabditis elegans. Nature 2000; 403(6772):901-906.
46. Pasquinelli AE, Reinhart BJ, Slack F et al. Conservation of the sequence and temporal expression of let-7 heterochronic regulatory RNA. Nature 2000; 408(6808):86-89.
47. Landgraf P, Rusu M, Sheridan R et al. A mammalian microRNA expression atlas based on small RNA library sequencing. Cell 2007; 129(7):1401-1414.
48. Griffiths-Jones S. The microRNA Registry. Nucl Acids Res 2004; 32(suppl_1):D109-111.
49. Griffiths-Jones S, Grocock RJ, van Dongen S et al. miRBase: microRNA sequences, targets and gene nomenclature. Nucl Acids Res 2006; 1;34(Database issue):D140-4.

50. Griffiths-Jones S, Saini HK, van Dongen S et al. miRBase: tools for microRNA genomics. Nucl Acids Res 2008; 36(suppl_1):D154-158.
51. Hagan JP, Piskounova E, Gregory RI. Lin28 recruits the TUTase Zcchc11 to inhibit let-7 maturation in mouse embryonic stem cells. Nat Struct Mol Biol 2009; 16(10):1021-1025.
52. Heo I, Joo C, Cho J et al. Lin28 mediates the terminal uridylation of let-7 precursor MicroRNA. Mol Cell 2008; 32(2):276-84.
53. Heo I, Joo C, Kim Y et al. TUT4 in concert with Lin28 suppresses microRNA biogenesis through premicroRNA uridylation. Cell 2009; 138(4):696-708.
54. Rybak A, Fuchs H, Smirnova L et al. A feedback loop comprising lin-28 and let-7 controls prelet-7 maturation during neural stem-cell commitment. Nat Cell Biol 2008; 10(8):987-93.
55. Thomson JM, Newman M, Parker JS et al. Extensive posttranscriptional regulation of microRNAs and its implications for cancer. Genes Dev 2006; 20(16):2202-7.
56. Viswanathan SR, Daley GQ, Gregory RI. Selective blockade of microRNA processing by Lin28. Science 2008; 320(5872):97-100.
57. Wu L, Belasco JG. Micro-RNA regulation of the mammalian lin-28 gene during neuronal differentiation of embryonal carcinoma cells. Mol Cell Biol 2005; 25(21):9198-9208.
58. Yang D, Moss EG. Temporally regulated expression of Lin-28 in diverse tissues of the developing mouse. Gene Expression Patterns 2003; 3(6):719-726.
59. Marson A, Levine SS, Cole MF et al. Connecting microRNA genes to the core transcriptional regulatory circuitry of embryonic stem cells. Cell 2008; 134(3):521-533.
60. Burdon T, Smith A, Savatier P. Signalling, cell cycle and pluripotency in embryonic stem cells. Trends Cell Biol 2002; 12(9):432-438.
61. Judson RL, Babiarz JE, Venere M et al. Embryonic stem cell-specific microRNAs promote induced pluripotency. Nat Biotechnol 2009; 27(5):459-461.
62. Schwamborn JC, Berezikov E, Knoblich JA. The TRIM-NHL Protein TRIM32 activates microRNAs and prevents self-renewal in mouse neural progenitors. Cell 2009; 136(5):913-925.
63. Maller Schulman BR, Liang X, Stahlhut C et al. The let-7 microRNA target gene, Mlin41/Trim71 is required for mouse embryonic survival and neural tube closure. Cell Cycle 2008; 7(24):3935-3942.
64. Rybak A, Fuchs H, Hadian K et al. The let-7 target gene mouse lin-41 is a stem cell specific E3 ubiquitin ligase for the miRNA pathway protein Ago2. Nat Cell Biol 2009; 11(12):1411-1420.
65. Neumuller RA, Betschinger J, Fischer A et al. Mei-P26 regulates microRNAs and cell growth in the Drosophila ovarian stem cell lineage. Nature 2008; 454(7201):241-245.
66. Hochedlinger K, Plath K. Epigenetic reprogramming and induced pluripotency. Development 2009; 136(4):509-523.
67. Kumar MS, Lu J, Mercer KL et al. Impaired microRNA processing enhances cellular transformation and tumorigenesis. Nat Genet 2007; 39(5):673-677.
68. Ivanovska I, Ball AS, Diaz RL et al. MicroRNAs in the miR-106b Family Regulate p21/CDKN1A and Promote Cell Cycle Progression. Mol Cell Biol 2008; 28(7):2167-2174.
69. Petrocca F, Visone R, Onelli MR et al. E2F1-Regulated microRNAs impair TGF[beta]-dependent cell-cycle arrest and apoptosis in gastric cancer. Cancer Cell 2008; 13(3):272-286.
70. He L, Thomson JM, Hemann MT et al. A microRNA polycistron as a potential human oncogene. Nature 2005; 435(7043):828-833.
71. Mendell JT. miRiad roles for the miR-17-92 cluster in development and disease. Cell 2008; 133(2):217-222.
72. Dews M, Homayouni A, Yu D et al. Augmentation of tumor angiogenesis by a Myc-activated microRNA cluster. Nat Genet 2006; 38(9):1060-1065.
73. Voorhoeve PM, le Sage C, Schrier M et al. A genetic screen implicates miRNA-372 and miRNA-373 as oncogenes in testicular germ cell tumors. Cell 2006; 124(6):1169-1181.
74. Yu F, Yao H, Zhu P et al. let-7 regulates self renewal and tumorigenicity of breast cancer cells. Cell 2007; 131:1109-1123.
75. Trang P, Medina PP, Wiggins JF et al. Regression of murine lung tumors by the let-7 microRNA. Oncogene [Internet] 2009 [cited 2009 Dec 16]; Available from: http://www.ncbi.nlm.nih.gov/pubmed/19966857.
76. Kumar MS, Erkeland SJ, Pester RE et al. Suppression of nonsmall cell lung tumor development by the let-7 microRNA family. Proc Natl Acad Sci USA 2008; 105(10):3903-3908.
77. Esquela-Kerscher A, Trang P, Wiggins JF et al. The let-7 microRNA reduces tumor growth in mouse models of lung cancer. Cell Cycle 2008; 7(6):759-764.
78. Chang T, Zeitels LR, Hwang H et al. Lin-28B transactivation is necessary for Myc-mediated let-7 repression and proliferation. Proc Natl Acad Sci USA 2009; 106(9):3384-3389.
79. Iliopoulos D, Hirsch HA, Struhl K. An epigenetic switch involving NF-kappaB, Lin28, Let-7 MicroRNA and IL6 links inflammation to cell transformation. Cell 2009; 139(4):693-706.
80. Viswanathan SR, Powers JT, Einhorn W et al. Lin28 promotes transformation and is associated with advanced human malignancies. Nat Genet 2009; 41(7):843-848.
81. Dangi-Garimella S, Yun J, Eves EM et al. Raf kinase inhibitory protein suppresses a metastasis signalling cascade involving LIN28 and let-7. EMBO J 2009; 28(4):347-358.
82. Büssing I, Slack FJ, Großhans H. let-7 microRNAs in development, stem cells and cancer. Trends Mol Med 2008; 14(9):400-409.

第9章 成体干细胞和多能胚胎干细胞中的端粒及端粒酶

Rosa M. Marión, Maria A. Blasco*

摘要：端粒酶一般不表达于成体组织之中，但成体干细胞却是一个例外。在某些组织中，成体干细胞的端粒最长。随着机体的老化，成体干细胞的端粒长度不断缩短，直到一个极限的长度，低于该极限就会影响干细胞的迁移和组织再生。p53能够有效地阻止携带过短端粒的成体干细胞成为组织再生的原料，这是p53在体细胞行为调控、维持组织健康及抗肿瘤中的新的作用。有多种方法可以将分化的细胞重编程至多潜能状态，如体细胞核移植，或最近兴起的通过过表达几种转录因子而得到的诱导多能干细胞（iPS）。最近的研究工作证实，在细胞核重编程过程中，端粒染色质被重塑，端粒也被端粒酶延长。这些数据显示端粒染色质结构是动态的，受其表观遗传学程序调控，与细胞分化潜能相关，而且是可以被重编程的。本章内容将着眼于对当前成体干细胞及成体已分化细胞中，通过细胞核重编程形成多潜能胚胎样干细胞过程中端粒及端粒酶的相关知识作简单的介绍。

引言

染色体的天然末端——端粒的逐渐缩短已被认为是细胞衰老的内因之一。伴随衰老的端粒缩短速度不但在男人和女人之间是不一样的，而且还会被加速衰老和导致过早死亡的因素所影响。例如，压力感、吸烟、肥胖等因素都被认为会缩短端粒的长度[1~4]。各种人类衰老相关疾病，如心血管疾病、痴呆、感染[5~11]，以及端粒维持信号通路突变所引起的人类疾病，包括生殖细胞在端粒酶组分中的突变，也都会加速端粒的缩短[12~15]。目前，有模型认为缩短的端粒是一种可以传递给分裂后子细胞的慢性DNA损伤，并且可以通过削弱成体干细胞再生和修复组织的能力来诱导衰老；同样，端粒缩短也可以通过诱导细胞老化和凋亡来引起细胞数量的减少[16~19]。

在发育过程中，细胞逐渐丢失其自我更新和分化的能力。但是，已经分化的成体细胞也可以被逆转回多潜能状态[20]。这种逆转至更原始的发育阶段叫做细胞核的重编程。细胞核的重编程机制包括全基因组的染色体结构和基因表达水平的变化[21~23]，这些通常是由分化潜能相关的表观遗传学状态决定的。

体细胞核移植[24]、分化细胞和胚胎干细胞融合[25]等方法可以实现细胞核重编程。通

---

*Corresponding Author: Maria A. Blasco—Telomeres and Telomerase Group, Molecular Oncology Program, Spanish National Cancer Centre (CNIO), Melchor Fernández Almagro 3, Madrid, E-28029, Spain. Email: mblasco@cnio.es

过体细胞核移植实现的细胞核重编程是将分化的体细胞核移入同物种的去核未受精卵子中（图 9-1）[24]。在这个过程中，体细胞核"回到"合子状态，从而具有了发育成与其具有完全相同遗传背景的完整生物个体的能力。体细胞核移植技术已在诸如小鼠、绵羊、马、猪、兔及猫等多种哺乳动物中获得了成功[24,26~30]。这些研究证实哺乳动物卵母细胞具有通过去分化和重编程细胞核进而获得全能性的能力。此外，该技术在基因操作、物种保存、家畜繁殖等方面具有广泛的应用前景。同时，体细胞核移植技术也可以用于制备患者特异的多潜能干细胞，这种干细胞可以用于研究和治疗人类疾病。

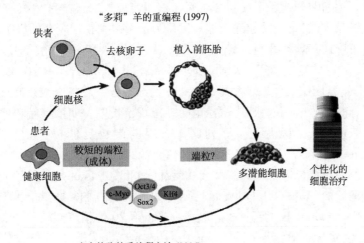

图 9-1 细胞核重编程的两种策略。通过不同的细胞核重编程方法可以从终末分化的成体体细胞产生多潜能干细胞。方案一（上部）：通过体细胞核移植将已分化细胞的细胞核转移入去核卵子中，多潜能细胞就可以从植入前的胚胎中获得。方案二（下部）：通过表达一系列"干性"相关转录因子得到诱导多能干细胞，该方法不需要产生或破坏胚胎。不过，归根到底，成体细胞端粒较短，而干细胞端粒一般较长，因此，人们往往会关注这个问题：在成体细胞重编程为干细胞的过程中，端粒是否被重新延长了

最近，人们惊奇地发现，通过过表达 4 个或更少的转录因子，小鼠和人的体细胞可以被重编程为诱导多能干细胞（induced pluripotent stem cell, iPS）[31,32]。iPS 为患者个性化的干细胞提供了新的来源。而更重要的是，iPS 绕过了体细胞核移植的技术问题，以及包括制造和破坏人类胚胎的伦理学争论（图 9-1）。iPS 的生物学特征还没有被完全认识清楚，最初的研究证实 iPS 和 ES 细胞在基因表达、表观遗传，以及小鼠胚胎发育和形成畸胎瘤等方面具有非常类似的生物学特征[33]。如果要应用于细胞治疗，重编程细胞需要具有足够的能力增殖和维持基因组的完整性，才能够保证其长期功效。从这方面讲，适当的端粒长度和功能对保证重编程细胞染色体稳定性是非常重要的。由于端粒会在细胞分离时缩短且和细胞衰老密切相关，因此，对于重编程细胞，一个重要问题就是，分化细胞相对较短的端粒是否足够形成 ES 样细胞。

端粒及端粒酶的调节

端粒是染色体末端结合有核糖核蛋白的异染色质结构，它可以保护 DNA 不被降解，不被识别为双链 DNA 断裂[34,35]。哺乳动物端粒由 TTAGGG 重复序列和一种被称为"shelterin"的六亚基蛋白复合体组成[35~37]。端粒的功能依赖于一定长度的端粒重复序列，且需要与 shelterin 复合体结合[37]。染色体末端复制的不完全性（所谓的"末端复制问题"）导致了细胞分裂后子细胞端粒逐渐缩短，最终造成严重缩短的/裸露的端粒及细胞周期阻滞和衰老[38,39]。端粒逐渐缩短被认为是引起生物体衰老的机制之一[40]。

端粒长度由端粒酶来维持。端粒酶是一种逆转录酶，它由 *Tert*（telomerase reverse transcriptase，端粒酶逆转录酶）基因和 *Terc*（telomerase RNA component，端粒酶 RNA 组分）基因编码。端粒酶可以在每一次细胞分裂后将端粒重复序列添加到染色体末端[41,42]。除此之外，一些替代途径也能维持端粒长度，如依赖端粒序列或亚端粒（接近端粒的染色质——译者注）序列同源重组的端粒延伸替代途径（alternative lengthening of telomeres，ALT）[43]。目前的研究结果显示，端粒延伸受到端粒染色质表观遗传学状态及端粒结合蛋白的调控[44,45]。端粒区和亚端粒区都是异染色质区，表现出许多异染色特征，如 H3K9（histone H3 lysine 9，组蛋白 H3 第 9 位赖氨酸——译者注）和 H4K20（histone H4 lysine 20，组蛋白 H4 第 20 位赖氨酸——译者注）的三甲基化，与异染色质蛋白 1（heterochromatin protein 1，HP1）结合[46~48]。亚端粒区 DNA 也被高度甲基化[47]。失去这些异染色质标志会引起端粒异常地延长。例如，在缺乏 Suv39h 或 Suv4-20h 组蛋白甲基化转移酶的细胞中，异常长的端粒缺乏 H3K9m3（组蛋白 H3 第 9 位赖氨酸 3 甲基化——译者注）、H4K20m3（组蛋白 H4 第 20 位赖氨酸的 3 甲基化——译者注）和与之结合的 HP1。同样，在缺乏 Dicer 或 DNMT1 和 DNMT3ab DNA 转移酶的细胞中，亚端粒区也缺乏甲基化[47,49]。

在酵母、斑马鱼和哺乳动物中，人们发现端粒是具有转录活性的，它能产生长非编码 RNA，被称为 TelRNA 或 TERRA[44,45,50,51]。TERRA 可以和端粒染色质作用，它们被认为是端粒长度延伸的负调节因子。在体外试验中，TERRA 可以作为端粒酶的潜在抑制剂[44,45,50,51]。因此，在那些需要端粒酶延长端粒的肿瘤形成过程中，TERRA 通常被下调。最后，端粒结合蛋白，如 TRF1、TRF2、Tin2、TANK1 和 TANK2，也可以影响端粒长度[52~55]。以上发现指出，端粒具有受表观遗传调节的高级结构，并且这对于端粒长度的调控是非常重要的（图 9-2）。

端粒酶的表达被限制于胚胎发育中和成体干细胞[16,56~58]。但是，这些组织中的端粒酶活性不足以防止端粒随着衰老而进行的缩短[39,59]。在罹患先天性角化不良、再生障碍性贫血，以及特发性肺纤维化[12~15]等疾病的患者中发现的端粒酶突变会加速端粒缩短，从而导致早衰、组织再生能力丧失等疾病。成年个体内端粒酶水平对机体健康的维持非常重要。端粒酶缺陷的小鼠表现出早衰和成体干细胞增殖能力下降[60~64]。端粒长度和端粒酶在生物体稳态维持和干细胞生物学方面的作用说明了研究细胞核重编程对端粒酶活性、端粒长度及端粒染色质等方面作用的重要性。

第9章　成体干细胞和多能胚胎干细胞中的端粒及端粒酶 · 107 ·

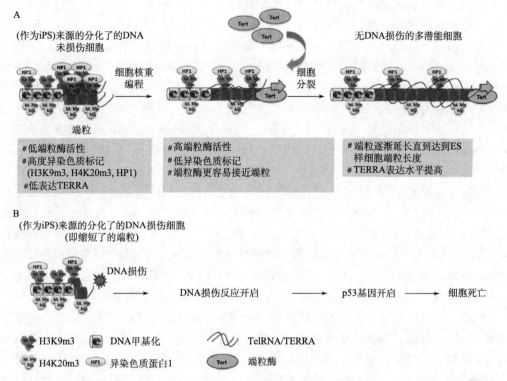

图 9-2　多潜能诱导过程中的端粒的重编程。A. 成体分化细胞中的端粒酶被组织成一种非常紧密的结构，其表现出较高的异染色状态和较低的端粒 RNA TERRA/TelRNA 表达。细胞核重编程会造成 H3K9m3、H4K20m3 和 HP1 异染色质标记的减少，从而使端粒染色质凝集程度降低且易于接近，同时也伴随着端粒酶表达上调。端粒酶有效地延伸了端粒长度，TERRA/TelRNA 表达也被上调。端粒延伸过程一直持续到重编程后，直到其长度达到多潜能细胞所需要的天然端粒长度。B. 一些问题细胞，如端粒较短的细胞，将在重编程的过程中被除去。端粒短于极限长度的细胞，以及具有其他 DNA 损伤的细胞，将在重编程时启动对 DNA 损伤的反应，激活 p53 依赖的凋亡途径，从而阻止具有 DNA 损伤的细胞被重编程为多潜能细胞

端粒和端粒酶在成体干细胞中的作用

端粒酶活性作用的两个重要的生物学进程——肿瘤和衰老，现在越来越多地被认为是干细胞病[40,57,65]。特别地，肿瘤可能经常来源于正常干细胞的恶性转化，而衰老通常伴随着某种干细胞数量和/或功能的逐渐下降[57,65]。

端粒酶活性基本上被限制于干细胞中，这些细胞中端粒酶水平对生物体的健康是非常重要的。关于人类端粒酶突变相关疾病的研究发现，端粒酶在人类衰老和寿命中扮演着限速角色。例如，上述罹患再生障碍性贫血、特发性肺纤维化及先天性角化不良等疾病的患者中可发现端粒酶核心组分、Tert 和 Terc 的突变。这些致命的疾病都是发生于成体，表现为过早地丢失了组织再生能力，同时伴随着端粒缩短[12~15,66~68]。和人类中的

情况相似，端粒酶缺陷的小鼠在第一世代即表现出中位寿命和最长寿命都缩短，同样说明了端粒酶在小鼠中的衰老限速作用[69]。

过去的几年，在许多干细胞中都发现了端粒酶活性，其中包括造血干细胞（hematopoietic stem cell，HSC）[70~73]、表皮干细胞（epidermal stem cell，ESC）[56,57]以及神经干细胞（neural stem cell，NSC）。通过端粒酶缺失和恢复的小鼠模型，人们发现了端粒长度和端粒酶活性在表皮干细胞中的作用[56,57,74]。在Terc缺陷的小鼠中，端粒缩短会导致皮肤表皮干细胞的功能下降[56]。特别地，分裂原诱导的增殖作用下，表皮干细胞从毛囊微环境中的动员（分裂和迁移），在端粒轻微缩短的小鼠（早期代次$Terc^{-/-}$小鼠）中被部分抑制，在端粒严重缩短的小鼠（晚期代次$Terc^{-/-}$小鼠）中被强烈抑制[56]。迁移缺陷直接导致毛囊干细胞和其邻近的短暂扩增细胞增殖速度的降低，继而引起毛发生长缺陷和细胞增殖受阻[56]。这种干细胞功能的缺陷可以完全被重新导入端粒酶活性和延伸缩短的端粒[75]，或者是破坏其p53功能而不延长端粒所恢复[76]。这些结果强调了缩短的端粒以一种需要p53参与的方式引起干细胞的功能异常。除皮肤之外，其他一些需要经常更新细胞的组织，如骨髓、肠、睾丸，在端粒严重缩短的Terc缺陷的小鼠中也表现出形态或功能上的萎缩[61,62]，从而支持了端粒长度在广泛的生物中决定组织稳态的观点。

最后，就像应用体外集落形成试验证实的那样，端粒长度和端粒酶活性在不同的干细胞（胚胎干细胞、造血干细胞或成体神经干细胞）中作用是不同的[56,63]。在Terc缺陷的小鼠模型中，较短的端粒会限制干细胞微环境对植入的野生型干细胞的支持作用[77]。综上所述，随着衰老进程的端粒缩短不但是直接导致干细胞功能异常的内因，而且也影响着干细胞微环境的活力，从而加剧衰老导致的干细胞功能异常。由于端粒酶活性和端粒长度对干细胞功能的影响不但作用于干细胞本身，也作用于干细胞微环境，这说明重新激活端粒酶活性具有潜在的临床应用价值。端粒缩短对组织再生能力的不利影响激发了一个充满诱惑的构想：在成体组织中表达端粒酶可能会增强组织的健康，延缓衰老，达到长寿的目的。其实，过表达端粒酶的小鼠能够抵制肿瘤发生，显著地推迟衰老，延长了40%的中位寿命[78]。这是端粒酶在生物体内第一次被证实具有抗衰老的作用。

在体细胞核移植中端粒及端粒酶的调控

当利用体细胞核移植第一次在成体哺乳动物中获得"多莉"羊时[24]，人们广泛地关注于她的细胞"年龄"。由于"多莉"是从一个成体细胞中克隆而来，她的细胞是否继承了与供者细胞年龄相当的端粒长度；或是在重编程过程中，它的端粒被"再生"为多潜能胚胎干细胞所应该具有的端粒长度？这个问题关系到克隆动物是否会存在发育上的缺陷，关系到所获得的胚胎干细胞的质量，也关系到它们在再生医学上的临床应用价值。

从6岁的"多莉"羊分离细胞进行培养，然后进行端粒长度分析，发现"多莉"的细胞端粒长度比同龄动物短约20%[79]。与之类似的是，从培养的胚胎或胎儿组织细胞进行核移植得到的克隆羊端粒长度也比同龄对照羊短10%~15%[79]。这些发现让人不得不沮丧地承认：作为有丝分裂钟的端粒在重编程过程中并没有被恢复。但是，后续的对多种体细胞核移植克隆动物的研究却得出了不一致的结果,大多数克隆动物表现出和同龄正常动物一致的端粒长度。特别的是,通过体细胞核移植得到的小鼠在6个世代中都保持着稳定的端粒长度[80]。与之类似的是,从成纤维细胞[81,82]、卵丘细胞[82]及颗粒细胞[83]产生的克隆牛都具有正常的端粒长度。其实,从成纤维细胞中得到的克隆羊也恢复了端粒长度[84]。有趣的是,即使以年老的、短端非常粒的细胞作为体细胞核移植的供者细胞,在克隆过程中,端粒长度也能被恢复,甚至增长[81]。这些证据又显示,作为供者的体细胞已经缩短的端粒在重编程的过程中可以被重新延长,即使这种延长的程度有着非常大的差异。这也说明了克隆过程中端粒长度调控的复杂性。到目前为止,仍然不清楚为什么"多莉"的端粒如此地短,但人们认为这种差异可能是由于供者细胞种类、核移植方法及物种的差异造成的[85]。

对于体细胞核移植过程中端粒延长的研究提出了许多新的问题,例如,端粒是在重编程的哪个阶段被延长的?这种延长是否依赖于端粒酶的参与?后者是一个很实际的问题,因为端粒在受精后的早期胚胎分裂中的延长就不是依赖于端粒酶的参与,而是通过一种基于重组的机制[58]。相比之下,在小鼠及牛的胚胎发育中,从桑椹胚到囊胚期转变过程中,端粒以依赖于端粒酶的方式延长[86]。对于由体细胞核移植得到的胚胎而言,不论供者细胞的端粒长度是多少,在桑椹胚时期,其端粒长度与供者细胞相当;在囊胚期时,端粒长度被恢复到正常状态[86]。这些数据显示,在体细胞核移植过程中,可能发生了端粒酶的激活。实际上,供者细胞中没有或只有较低的端粒酶活性,而在克隆得到的胚胎中端粒酶却重新被激活[81~83,87]。在通过克隆得到的胚胎和通过受精得到的胚胎中,这种在发育过程中表现出来的短暂的端粒酶活性是非常相似的:在囊胚期达到最高水平[82,83,87],且和端粒长度的延长相关[86]。

这些结果显示,体细胞核移植克隆动物端粒长度会在胚胎发生过程中得以恢复,其机制很可能是在桑椹胚到囊胚转换期间激活了端粒酶的活性。

在 iPS 产生过程中端粒及端粒酶的调控

当 iPS 细胞第一次被报道时,人们立即对其是否获得了胚胎干细胞一样的端粒、是否延长了其供者细胞端粒长度非常感兴趣。在这一点上,iPS 具有[31,32,89]胚胎干细胞的两个显著的特征:高水平的 TERT 和高端粒酶活性[88]。有报道称,具有端粒酶活性也不一定意味着端粒会得以延长[90]。iPS 产生过程是一个发生在体外细胞培养中的、和体细胞核转移对应的过程,这一过程中端粒是否延长还不能简单地根据其是否表达了端粒酶来确定。最近研究发现,端粒酶依赖的端粒延伸发生在由小鼠胚胎成纤维细胞(mouse embryonic fibroblast,MEF)来源的 iPS 细胞中,在重编程之后,iPS 细胞的端粒持续延长,直到达到胚胎干细胞的端粒长度[91]。即使以年老的小鼠成纤维细胞(比 MEF 细胞端粒短许多)为供者细胞,端粒长度在 iPS 细胞中也被有效地延长。这些结果显示,以端粒较短的供者细胞制备的 iPS 细胞,如具有很短端粒长度的老年个体或患者,只要它们激活了端粒酶信号通路,就会恢复其端粒长度。

有趣的是,与其来源的小鼠胚胎成纤维细胞细胞相比,iPS 细胞也会降低其端粒部位的 H3K9m3 及 H4K20m3 异染色质标志,这一点和胚胎干细胞端粒的低异染色质状态非常类似[91]。这些结果证实,和分化的细胞相比,多潜能的胚胎干细胞具有更高的染色体可塑性[92]。和开放的端粒染色质会伴随着细胞核重编程一样,我们观察到,和小鼠胚胎成纤维细胞细胞相比,胚胎干细胞和 iPS 细胞中端粒重组频率都有所提高[91]。最后,iPS 中比 MEF 更高的 TERRA 水平也支持其具有更长的端粒。一旦 iPS 细胞的端粒达到 ES 细胞的水平,TERRA 的积累又可以作为端粒酶依赖的端粒延伸的负调控因子[91]。由于组蛋白 H3.3 和 ATRX 蛋白(α 地中海贫血/X-连锁精神发育阻滞综合征)都存在于小鼠的胚胎干细胞端粒中[93,94],虽然还没有被证实,它们也非常有可能在 iPS 细胞的端粒中被发现。这些结果显示,在 iPS 细胞产生的过程中,端粒发生了表观遗传的变化,降低异染色质化的组蛋白标志,端粒染色质结构被打开,同时 TERRA 转录活性增强,端粒重组频率增加,随后的端粒被延长至胚胎干细胞的水平。由于 TERRA 具有负调控端粒酶活性的作用,一旦端粒长度达到胚胎干细胞的水平,iPS 中升高的 TERRA 水平就可以阻止端粒酶活性。综上所述,根据细胞不同的分化状态,端粒染色质是动态的,也是可以被重编程的(图 9-2)。

端粒酶活性对于产生"高"质量的 iPS 细胞是必需的

在较高世代的端粒酶缺陷小鼠中,经常发生非常短的端粒和染色体末端融合。这些小鼠的细胞进行重编程的效率非常低,这说明有效的重编程需要一个最小的极限端粒长度[91]。重新导入端粒酶活性会减少短端粒的发生率,从而提高 iPS 细胞产生的效率。因此,损伤的/未包裹的端粒会导致重编程的失败,清除具有未包裹端粒的重编程细胞的"重编程障碍"是的确存在的。由于 p53 在阻止 DNA 受损细胞或具有较短端粒长度的细胞继续增殖的过程中扮演着重要的角色,p53 在重编程障碍中可能也扮演着类似的作用[95]。p53 缺陷的细胞,即使端粒长度较短,或 DNA 受损,也能够被有效地重编程。这说明 p53 在阻止人或小鼠有缺陷的细胞、端粒太短的细胞被重编程为多潜能细胞过程中,扮演着非常重要的作用。其他的一些实验也证实,p53 可以限制 iPS 细胞的产生[96~99]。总之,这些结果显示端粒长度和端粒染色质在体外重编程过程中也被更新了,同时也强调了端粒的生物学作用及在 iPS 细胞产生过程中的动态变化和功能(图 9-2)。

端粒重编程调控

尽管许多研究详细地描述了在细胞核重编程过程中端粒发生的巨大变化,但是端粒重编程是如何被调控的仍然未被完全阐明。由于端粒酶在这个过程中所起到的重要作用,端粒酶活性是如何被上调的就显得非常重要。原癌基因 c-Myc 是因子限定重编程的四个因子之一,它可以在转录水平调控 Tert[100]。这提示 c-Myc 负责激活 iPS 细胞端粒酶活性。但是,利用或者不利用 c-Myc 产生的 iPS 细胞都表现出类似的端粒酶活性[91]。因此,iPS 产生过程中端粒酶活化机制仍然有待进一步研究。此外,端粒酶活性并不是决定端粒长度的唯一因子。端粒酶介导的端粒维持或延长还依赖于端粒本身的结构，而端粒结构又受到表观遗传学修饰和端粒结合蛋白的调节[44,45]。

细胞核重编程需要移除在细胞分化和分裂过程中打上的组织特异的表观遗传学印迹，同时大规模地重新组织染色质的结构和功能，DNA 甲基化及组蛋白乙酰化在全局范围内发生改变，从而形成一种更为开放的染色质状态[23]。这些表观遗传学变化也在重编程过程中改变了端粒染色质[91]。虽然还有待证实，但是在 iPS 细胞中观察到的端粒染色质重编程为更开放的结构可能对端粒酶接近端粒的末端然后延长端粒长度是必需的（图 9-2）。有鉴于此，在通过体细胞核移植获得的克隆胚胎中异常的表观遗传学修饰也许可以解释克隆胚胎中不同长度的端粒可能是由于端粒酶接近端粒的难易程度不同而造成的。在这一点上，更有意思的是克隆胚胎还表现出了不寻常的甲基化状态[22]。从另一方面讲，重编程过程中出现的甲基化改变也可以直接影响端粒酶的表达。

最后，端粒结合蛋白也参与调控端粒长度，它们可以抑制或促进端粒酶与端粒 DNA 结合。虽然还没有证据支持，但这些蛋白质的表达或功能有可能在重编程过程中被调节，以便端粒长度的恢复。因此，研究在端粒重编程过程中染色质修饰活性及端粒结合蛋白所起的作用将引起研究者们极大的兴趣。

结论

随着生物个体的逐渐衰老，成体干细胞的端粒长度会逐渐缩短，正常的干细胞迁移和组织再生需要端粒不能低于一个极限长度，而且端粒也可能是生物个体寿命的一个决定因素。因而，不管是基于体细胞核移植，还是基于表达几种特定的转录因子的方法获得的细胞核重编程，都能增加端粒酶活性，恢复端粒长度到健全的状态。此外，至少在通过体外重编程诱导的多能干细胞中，端粒染色质也被重编程至胚胎干细胞类似的状态，形成一个开放的结构，增加 TERRA 的表达。目前，所有的数据都显示在分化和重编程过程中端粒进行着复杂的调控，端粒染色质结构发生着动态的变化，这些变化由细胞类型特异的表观遗传学修饰决定，而且在重编程过程中是可逆的。

（王有为　韩忠朝　译）

参 考 文 献

1. Epel ES, Blackburn EH, Lin J et al. Accelerated telomere shortening in response to life stress. Proc Natl Acad Sci USA 2004; 101:17312-17315.
2. Valdes AM, Andrew T, Gardner JP et al. Obesity, cigarette smoking and telomere length in women. Lancet 2005; 366:662-664.
3. Canela A, Vera E, Klatt P et al. High-thoughput telomere length quantification by FISH and its application to human population studies. Proc Natl Acad Sci USA 2007; 104:5300-5305.
4. Cherkas LF, Aviv A, Valdes AM et al. The effects of social status on biological aging as measured by white-blood-cell telomere length. Aging Cell 2006; 5:361-365.
5. Oh H, Wang SC, Prahash A et al. Telomere attrition and Chk2 activation in human heart failure. Proc Natl Acad Sci USA 2003; 100:5378-5383.
6. O'Sullivan JN, Bronner MP, Brentnall TA et al. Chromosomal instability in ulcerative colitis is related to telomere shortening. Nat Genet 2002; 32:280-284.
7. Wiemann SU, Satyanarayana A, Tsahuridu M et al. Hepatocyte telomer shortening and senescence are general markers of human liver cirrhosis. FASEB J 2002; 16:935-942.

8. Samani NJ, Boultby R, Butler R et al. Telomere shortening in atherosclerosis. Lancet 2001; 358:472-473.
9. Wolthers KC, Bea G, Wisman A et al. T-cell telomere length in HIV-1 infection: no evidence for increased CD4$^+$ T-cell turnover. Science 1996; 274:1543-1547.
10. Cawthon RM, Smith KR, O'Brien E et al. Association between telomere length in blood and mortality in people aged 60 years or older. Lancet 2003; 361:393-395.
11. Honig LS, Schupf N, Lee JH et al. Shorter telomeres are associated with mortality in those with APOE epsilon4 and dementia. Ann Neurol 2006; 60:181-187.
12. Mitchell JR, Wood E, Collins K. A telomerase component is defective in the human disease dyskeratosis congenital. Nature 1999; 402:551-555.
13. Yamaguchi H, Calado RT, Ly H et al. Mutations in TERT, the gene for telomerase reverse transcriptase, in aplastic anemia. N Engl J Med 2005; 352:1413-1424.
14. Tsakiri KD, Cronkhite JT, Kuan PJ et al. Adult-onset pulmonary fibrosis caused by mutations in telomerase. Proc Natl Acad Sci USA 2007; 104:7552-7557.
15. Armanios MY, Chen JJ, Cogan JD et al. Telomerase mutations in families with idiopathic pulmonary fibrosis. N Engl J Med 2007; 356:1317-1326.
16. Blasco MA. Telomere length, stem cells and aging. Nat Chem Biol 2007; 3:640-649.
17. Serrano M, Blasco MA. Cancer and ageing: convergent and divergent mechanisms. Nat Rev Mol Cell Biol 2007; 9:715-722.
18. Finkel T, Serrano M, Blasco MA. The common biology of cancer and ageing. Nature 2007; 448:767-774.
19. Collado M, Blasco MA, Serrano M. Cellular senescence in cancer and aging. Cell 2007; 130:223-233.
20. Gurdon JB, Melton DA. Nuclear reprogramming in cells. Science 2008; 322:1811-1815.
21. Rideout WM 3rd, Eggan K, Jaenisch R. Nuclear cloning and epigenetic reprogramming of the genome. Science 2001; 293:1093-1098.
22. Yang X, Smith SL, Tian XC et al. Nuclear reprogramming of cloned embryos and its implications for therapeutic cloning. Nat Genet 2007; 39:295-302.
23. Hochedlinger K, Plath K. Epigenetic reprogramming and induced pluripotency. Development 2009; 136:509-523.
24. Wilmut I, Schnieke AE, McWhir J et al. Viable offspring derived from fetal and adult mammalian cells. Nature 1997; 385:810-813.
25. Tada M, Takahama Y, Abe K et al. Nuclear reprogramming of somatic cells by in vitro hybridization with ES cells. Curr Biol 2001; 11:1553-1558.
26. Wakayama T, Perry AC, Zuccotti M et al. Full-term development of mice from enucleated oocytes injected with cumulus cell nuclei. Nature 1998; 394:369-374.
27. Kato Y, Tani T, Sotomaru Y et al. Eight calves cloned from somatic cells of a single adult. Science 1998; 282:2095-2098.
28. Polejaeva IA, Chen SH, Vaught TD et al. Cloned pigs produced by nuclear transfer from adult somatic cells. Nature 2000; 407:86-90.
29. Chesné P, Adenot PG, Viglietta C et al. Cloned rabbits produced by nuclear transfer from adult somatic cells. Nat Biotechnol 2002; 20:366-369.
30. Shin T, Kraemer D, Pryor J et al. A cat cloned by nuclear transplantation. Nature 2002; 415:859.
31. Takahashi K, Yamanaka S. Induction of pluripotent stem cells from mouse embryonic and adult fibroblast cultures by defined factors. Cell 2006; 126:663-676.
32. Takahashi K, Tanabe K, Ohnuki M et al. Induction of pluripotent stem cells from adult human fibroblasts by defined factors. Cell 2007; 131:861-872.
33. Amabile G, Meissner A. Induced pluripotent stem cells: current progress and potential for regenerative medicine. Trends Mol Med 2009; 15:59-68.
34. Chan SR, Blackburn EH. Telomeres and telomerase. Philos Trans R Soc Lond B Biol Sci 2004; 359:109-121.
35. Palm W, de Lange T. How shelterin protects mammalian telomeres. Annu Rev Genet 2008; 42:301-334.
36. Blackburn EH. Switching and signaling at the telomere. Cell 2001; 106:661-673.
37. de Lange T. Shelterin: the protein complex that shapes and safeguards human telomeres. Genes Dev 2005; 19:2100-2110.
38. d'Adda di Fagagna F, Reaper PM, Clay-Farrace L et al. A DNA damage checkpoint response in telomere-initiated senescence. Nature 2003; 426:194-198.
39. Harley CB, Futcher AB, Greider CW. Telomeres shorten during ageing of human fibroblasts. Nature 1990; 345:458-460.
40. Blasco MA. Telomeres and human disease: ageing, cancer and beyond. Nat Rev Genet 2005; 6:611-622.
41. Chan SW, Blackburn EH. New ways not to make ends meet: telomerase, DNA damage proteins and heterochromatin. Oncogene 2002; 21:553-563.
42. Collins K, Mitchell JR. Telomerase in the human organism. Oncogene 2002; 21:564-579.
43. Dunham MA, Neumann AA, Fasching CL et al. Telomere maintenance by recombination in human cells. Nature Genet 2000; 26:447-450.
44. Schoeftner S, Blasco MA. A 'higher order' of telomere regulation: telomere heterochromatin and telomeric RNAs. EMBO J 2009; 28:2323-2336.

45. Schoeftner S, Blasco MA. Chromatin regulation and noncoding RNAs at mammalian telomeres. Semin Cell Dev Biol 2009; doi:10.1016/j.semcdb.2009.09.015.
46. García-Cao M, O'Sullivan R, Peters AH et al. Epigenetic regulation of telomere length in mammalian cells by the Suv39h1 and Suv39h2 histone methyltransferases. Nat Genet 2004; 36:94-99.
47. Gonzalo S, Jaco I, Fraga MF et al. DNA methyltransferases control telomere length and telomere recombination in mammalian cells. Nat Cell Biol 2006; 8:416-424.
48. Benetti R, Gonzalo S, Jaco I et al. Suv4-20h deficiency results in telomere elongation and derepression of telomere recombination. J Cell Biol 2007; 178:925-936.
49. Benetti R, Gonzalo S, Jaco I et al. A mammalian microRNA cluster controls DNA methylation and telomere recombination via Rbl2-dependent regulation of DNA methyltransferases. Nat Struct Mol Biol 2008; 15:268-279.
50. Azzalin CM, Reichenbach P, Khoriauli L et al. Telomeric repeat containing RNA and RNA surveillance factors at mammalian chromosome ends. Science 2007; 318:798-801.
51. Schoeftner S, Blasco MA. Developmentally regulated transcription of mammalian telomeres by DNA-dependent RNA polymerase II. Nat Cell Biol 2008; 10:228-236.
52. van Steensel B, de Lange T. Control of telomere length by the human telomeric protein TRF1. Nature 1997; 385:740-743.
53. Muñoz P, Blanco R, Flores JM et al. XPF nuclease-dependent telomere loss and increased DNA damage in mice overexpressing TRF2 result in premature aging and cancer. Nat Genet 2005; 37:1063-1071.
54. Ye JZ, de Lange T. TIN2 is a tankyrase 1 PARP modulator in the TRF1 telomere length control complex. Nat Genet 2004; 36:618-623.
55. Smith S, de Lange T. Tankyrase promotes telomere elongation in human cells. Curr Biol 2000; 10:1299-1302.
56. Flores I, Cayuela ML, Blasco MA. Effects of telomerase and telomere length on epidermal stem cell behavior. Science 2005; 309:1253-1256.
57. Flores I, Benetti R, Blasco MA. Telomerase regulation and stem cell behavior. Curr Opin Cell Biol 2006; 18:254-260.
58. Liu L, Bailey SM, Okuka M et al. Telomere lengthening early in development. Nat Cell Biol 2007; 9:1436-1441.
59. Flores I, Canela A, Vera E et al. The longest telomeres: A general signature of adult stem cell compartments. Genes Dev 2008; 22:654-667.
60. Blasco MA, Lee HW, Hande MP et al. Telomere shortening and tumor formation by mouse cells lacking telomerase RNA. Cell 1997; 91:25-34.
61. Lee HW, Blasco MA, Gottlieb GJ et al. Essential role of mouse telomerase in highly proliferative organs. Nature 1998; 392:569-574.
62. Herrera E, Samper E, Martín-Caballero J et al. Disease states associated with telomerase deficiency appear earlier in mice with short telomeres. EMBO J 1999; 18:2950-2960.
63. Samper E, Fernández P, Eguía R et al. Long-term repopulating ability of telomerase-deficient murine hematopoietic stem cells. Blood 2002; 99:2767-2775.
64. Ferrón S, Mira H, Franco S et al. Telomere shortening and chromosomal instability abrogates proliferation of adult but not embryonic neural stem cells. Development 2004; 131:4059-4070.
65. Bell DR, Van Zant G. Stem cells, aging and cancer: inevitabilities and outcomes. Oncogene 2004; 23:7290-7296.
66. Vulliamy T, Marrone A, Goldman F et al. The RNA component of telomerase is mutated in autosomal dominant dyskeratosis congenita. Nature 2001; 413:432-435.
67. Vulliamy T, Marrone A, Szydlo R et al. Disease anticipation is associated with progressive telomere shortening in families with dyskeratosis congenita due to mutations in TERC. Nat Genet 2004; 36:447-449.
68. Marrone A, Stevens D, Vulliamy T et al. Heterozygous telomerase RNA mutations found in dyskeratosis congenita and aplastic anemia reduce telomerase activity via haploinsufficiency. Blood 2004; 104:3936-3942.
69. García-Cao I, García-Cao M, Tomás-Loba A et al. Increased p53 activity does not accelerate telomere-driven ageing. EMBO Rep 2006; 7:546-552.
70. Vaziri H, Dragowska W, Allsopp RC et al. Evidence for a mitotic clock in human hematopoietic stem cells: loss of telomeric DNA with age. Proc Natl Acad Sci USA 1994; 91:9857-9860.
71. Allsopp RC, Cheshier S, Weissman IL. Telomere shortening accompanies increased cell cycle activity during serial transplantation of hematopoietic stem cells. J Exp Med 2001; 193:917-924.
72. Allsopp RC, Morin GB, DePinho R et al. Telomerase is required to slow telomere shortening and extend replicative lifespan of HSCs during serial transplantation. Blood 2003; 102:517-520.
73. Allsopp RC, Morin GB, Horner JW et al. Effect of TERT over-expression on the long-term transplantation capacity of hematopoietic stem cells. Nat Med 2003; 9:369-371.
74. Sarin KY, Cheung P, Gilison D et al. Conditional telomerase induction causes proliferation of hair follicle stem cells. Nature 2005; 436:1048-1052.
75. Siegl-Cachedenier I, Flores I, Klatt P et al. Telomerase reverses epidermal hair follicle stem cell defects and loss of long-term survival associated with critically short telomeres. J Cell Biol 2007; 179:277-290.

76. Flores I, Blasco MA. A p53-dependent response limits epidermal stem cell functionality and organismal size in mice with short telomeres. PLoS One 2009; 4(3):e4934. doi:10.1371/journal.pone.0004934
77. Ju Z, Jiang H, Jaworski M et al. Telomere dysfunction induces environmental alterations limiting hematopoietic stem cell function and engraftment. Nat Med 2007; 3:742-747.
78. Tomás-Loba A, Flores I, Fernández-Marcos PJ et al. Telomerase reverse transcriptase delays aging in cancer-resistant mice. Cell 2008; 135:609-622.
79. Shiels PG, Kind AJ, Campbell KH et al. Analysis of telomere lengths in cloned sheep. Nature 1999; 399:316-317.
80. Wakayama T, Shinkai Y, Tamashiro KL et al. Cloning of mice to six generations. Nature 2000; 407:318-319.
81. Lanza RP, Cibelli JB, Blackwell C et al. Extension of cell life-span and telomere length in animals cloned from senescent somatic cells. Science 2000; 288:665-669.
82. Tian XC, Xu J, Yang X. Normal telomere lengths found in cloned cattle. Nat Genet 2000; 26:272-273.
83. Betts D, Bordignon V, Hill J et al. Reprogramming of telomerase activity and rebuilding of telomere length in cloned cattle. Proc Natl Acad Sci USA 2001; 98:1077-1082.
84. Clark AJ, Ferrier P, Aslam S et al. Proliferative lifespan is conserved after nuclear transfer. Nat Cell Biol 2003; 5:535-538.
85. Miyashita N, Shiga K, Yonai M et al. Remarkable differences in telomere lengths among cloned cattle derived from different cell types. Biol Reprod 2002; 66:1649-1655.
86. Schaetzlein S, Lucas-Hahn A, Lemme E et al. Telomere length is reset during early mammalian embryogenesis. Proc Natl Acad Sci USA 2004; 101:8034-8038.
87. Xu J, Yang X. Telomerase activity in early bovine embryos derived from parthenogenetic activation and nuclear transfer. Biol Reprod 2001; 64:770-774.
88. Thomson JA, Itskovitz-Eldor J, Shapiro SS et al. Embryonic stem cell lines derived from human blastocysts. Science 1998; 282:1145-1147.
89. Stadtfeld M, Maherali N, Breault DT et al. Defining molecular cornerstones during fibroblast to iPS cell reprogramming in mouse. Cell Stem Cell 2008; 2:230-240.
90. Zhu J, Wang H, Bishop JM et al. Telomerase extends the lifespan of virus-transformed human cells without net telomere lengthening. Proc Natl Acad Sci USA 1999; 96:3723-3728.
91. Marion RM, Strati K, Li H et al. Telomeres acquire embryonic stem cell characteristics in induced pluripotent stem cells. Cell Stem Cell 2009; 4:141-154.
92. Meshorer E, Yellajoshula D, George E et al. Hyperdynamic plasticity of chromatin proteins in pluripotent embryonic stem cells. Dev Cell 2006; 10:105-116.
93. Wong LH, Ren H, Williams E et al. Histone H3.3 incorporation provides a unique and functionally essential telomeric chromatin in embryonic stem cells. Genome Res 2009; 19:404-414.
94. Wong LH, McGhie JD, Sim M. ATRX interacts with H3.3 in maintaining telomere structural integrity in pluripotent embryonic stem cells. Genome Res 2010; doi 10.1101/gr.101477.109.
95. Marión RM, Strati K, Li H et al. A p53-mediated DNA damage response limits reprogramming to ensure iPS cell genomic integrity. Nature 2009; 460:1149-1153.
96. Li H, Collado M, Villasante A et al. The Ink4/Arf locus is a barrier for iPS cell reprogramming. Nature 2009; 460:1136-1139.
97. Hong H, Takahashi K, Ichisaka T et al. Suppression of induced pluripotent stem cell generation by the p53-p21 pathway. Nature 2009; 460:1132-1135.
98. Kawamura T, Suzuki J, Wang YV et al. Linking the p53 tumour suppressor pathway to somatic cell reprogramming. Nature 2009; 460:1140-1144.
99. Utikal J, Polo JM, Stadtfeld M et al. Immortalization eliminates a roadblock during cellular reprogramming into iPS cells. Nature 2009; 460:1145-1148.
100. Wang J, Xie LY, Allan S et al. Myc activates telomerase. Genes Dev 1998; 12:1769-1774.

第10章 X染色体失活与胚胎干细胞

Tahsin Stefan Barakat，Joost Gribnau*

摘要：X染色体失活（XCI）是平衡男性与女性细胞中由X染色体编码的基因表达量所必需的一种过程。X染色体失活发生于女性胚胎发育的早期或女性胚胎干细胞（ES）的分化起始时期，该过程可导致每一个女性体细胞中的一条X染色体发生失活。参与调节X染色体失活的各种因子在维持ES细胞数量及分化方面也发挥重要作用，因此，研究X染色体失活过程也成为对ES细胞生物学研究的极佳模型。在这一章我们将会阐述X染色体失活中重要的顺式及反式调节因子并介绍和解释女性细胞中特异诱发X染色体失活的模型。我们还将讨论参与形成失活X染色体的蛋白质，以及与失活过程相关的不同核染色质修饰。最后，我们将阐述以小鼠和人ES细胞，以及诱导多能干细胞（iPS）作为研究X染色体失活过程的模型系统潜力。

引言

在许多物种中，个体的性别通常是由性染色体上的基因决定的[1]。哺乳动物为异形配子生物，雌性的细胞核内含有两条X染色体，而雄性的细胞核内则含有一条X染色体和一条Y染色体。性染色体起源于一对常染色体，关键的雄性性别决定基因SRY由位于前Y染色体上的原始*SOX3*基因进化而来，它的出现启动了这些常染色体向前X染色体及前Y染色体的转变[2~5]。在随后的过程中，前Y染色体获得了具有雄性生育能力的基因，使其具有了一个与X染色体非同源的基因组区域。人们认为，正是同源性的丧失导致了Y染色体的退化。现在，Y染色体只包含有数量有限的基因，它们中大部分与雄性生育能力及性别决定有关[6,7]。而X染色体仍然可在雌性生殖系中发生重组，这种重组抑制了降解，从而维持其成为包含有超过1000种基因的大染色体，其包含的基因具有调节从大脑发育到新陈代谢及生殖能力的多种生物学功能[1,8]。

由于两性都具有相同数量的常染色体，故需要相同剂量的X连锁基因来实现细胞的生物学功能。在胎盘哺乳动物中，两性间的X连锁基因的数量补偿是通过雌性两条X染色体中的一条发生失活来实现的，这种过程即称为X染色体失活（XCI）[9~11]。X染色体失活发生于雌性发育的早期，进而产生了功能性的hetero-chromatinization，使X染色体沉默，并且在贯穿终生的细胞分裂过程中得以维持[12]。X染色体失活的调控使得在两性中，都只有一条X染色体具有功能活性。与常染色体基因表达相比，这条未失

---

* Corresponding Author：Joost Gribnau—Department of Reproduction and Development，Room Ee 09-71，Erasmus MC，3015 GE Rotterdam，The Netherlands. Email：j.gribnau@erasmusmc.nl

活 X 染色体上编码基因的表达，以及雄性细胞中单一 X 染色体编码基因的表达都双倍上调，从而可对 X 染色体编码基因的表达剂量补偿发挥进一步作用[13~16]。在小鼠体内，X 染色体失活表现为两种形式。在胚外组织，X 染色体失活通过发生印记（imprinting）使所有细胞中的父源 X 染色体（Xp）失活[17]。这一过程在 2~8 细胞阶段的发育早期发生[18,19]，并在胚胎胚外组织（包括胎盘组织）发育过程中保持。与此相反，父源 X 染色体 Xp 在内细胞团（ICM）中发生再活化使胚体生长，而在随后发育到 5.5 天左右时会发生随机的 X 染色体失活(图 10-1)。

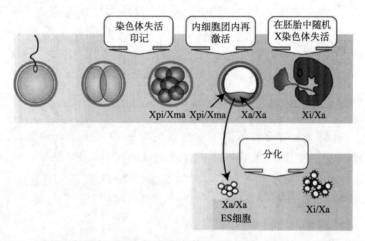

图 10-1 雌性小鼠发育过程中的 X 染色体失活。在小鼠发育早期的 2~8 细胞阶段，启动了 X 染色体的印记失活，这一过程导致父源 X 染色体（Xpi，红色细胞）的失活。印记失活的 X 染色体在滋养外胚层（发育成未来的胎盘）及原始内胚层中得以维持。但在胚囊的内细胞团中，印记的 X 染色体发生失活逆转，两条 X 染色体均获得转录活性（XaXa，灰色）。就在胚胎子宫着床发生前后，在早期胚胎的 ICM 发育而来的胚胎前体细胞中发生随机的 X 染色体失活（XiXa，蓝色细胞父源 X 染色体有活性，红色细胞母源 X 有活性），因而 X 染色体失活与个体发育及细胞分化相关联。胚胎干细胞来源于胚囊的 ICM，雌性的 ES 细胞在分化过程中可发生随机 X 染色体失活，因此可作为进行 X 染色体失活的体外研究模型。该图的彩版见 www.landesbioscience.com/curie

胚胎干细胞（ES）来自于胚囊的内细胞团（ICM）。它们的特征是具有自我更新能力，以及通过分化形成胚体和成熟有机体所有类型细胞的多潜能性[20,21]。除了在再生医学方面的应用潜力，ES 细胞也是研究哺乳动物由着床前时期开始的发育过程的理想体系[22,23]。小鼠的雌性 ES 细胞具有两条活性 X 染色体（Xa），当这些细胞开始分化后，细胞中将发生随机的 X 染色体失活，这使它们成为用于研究 X 染色体失活的最普遍的模型系统[24,25]。除了模拟早期发育，ES 细胞中的 X 染色体失活对于实现其自身特有细胞功能及发育潜能都具有重要的意义。最近对于诱导多能干细胞（iPS）的发现再次强调了 X 染色体失活的重要性。诱导多能干细胞（iPS）是通过特定多能性因子对体细胞的重编程衍化而来，iPS 细胞具有与 ES 细胞相同的特征[26~31]。ES 细胞及 iPS 细胞中的活化态 X 染色体可以作为一种多潜能标记，因为在 iPS 重编程过程中，在体细胞中失活状态的 X 染色体（Xi）可以重新活化变为 iPS 细胞中的活化状态[32,33]。

在这一章中，我们将先描述在 X 染色体失活中起主要作用的基因，进而讨论用于解释在 X 染色体失活中 X 染色体的计数和选择，以及在随后的细胞分裂过程中建立及维持失活 X 染色体的不同模型。最后，我们将讨论小鼠及人类 ES 细胞 X 染色体失活的不同点。

X 染色体失活中的顺势作用因子

几十年来，生物学家们一直对发育过程中整个染色体的转录沉寂非常感兴趣，而最近几年我们已经获得了帮助我们理解 X 染色体失活中所涉及的分子机制的一系列重要知识。对于小鼠及人类 X 染色体-常染色体易位的遗传学研究显示，X 连锁控制位点，即 X 染色体失活中心（在小鼠为 Xic，在人类为 XIC）对于 X 染色体失活的发生是必不可少的[34~37]。在小鼠 X 染色体中，长度超过 1Mb 的 Xic[38] 已经被证实包含有至少 4 个与 X 染色体失活相关的基因（图 10-2）。这其中的 *Xist*、*Tsix* 及 *Xite* 三个基因为非编码基因，它们代表了参与 X 染色体沉寂的顺式调节过程中最主要的调节位点。第四个基因 *Rnf12* 编码了参与 X 染色体失活反式调节的泛素连接酶 E3，这将在后续的章节中介绍。

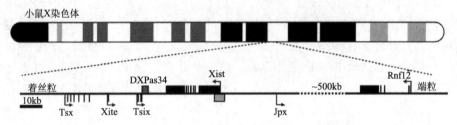

图 10-2 X 染色体上的 X 失活中心。小鼠的 X 染色体及部分 X 失活中心位置的示意图。X 染色体失活中心含有参与 X 染色体失活调节的 *Xist*、*Tsix*、*Xite* 及 *Rnf12* 等多个基因

X-失活特异转录产物（在小鼠为 *Xist*，在人类为 *XIST*）是第一个被发现与 XIC/Xic 相关的基因[37,39,40]。*Xist* 是已知的唯一由失活的 X 染色体特异表达的基因。*Xist* 是一个非编码基因，在小鼠中包括 7 个外显子，在人类中则为 8 个，它可以产生一个可发生选择性剪接的多聚腺苷酸 RNA 分子（在人类中大小为 17kb，在小鼠中为 15kb）[41,42]。*Xist* RNA 与失活的 X 染色体结合[43,44]，在 X 染色体失活的顺式调节中发挥不可替代的作用。研究表明，在雌性 ES 细胞及小鼠中敲除 *Xist* 基因将会使突变的 X 染色体无法失活[45,47]。在 X 染色体失活发生过程中，将选择失活的 X 染色体上的 *Xist* 的表达首先上调[48,49]，然后 *Xist* RNA 分子从 X 染色体失活中心顺式扩展，最后覆盖了整个失活的 X 染色体。在 *Xist* RNA 分子扩展的过程中，*Xist* RNA 通过吸引与基因沉默相关的染色质修饰蛋白使 X 染色体发生异染色质化[50~52]。

另一个定位于 X 染色体失活中心的非编码基因为 $Tsix$，其产物为 $Xist$ 的反义转录本[53]。$Tsix$ 包含 4 个外显子及至少 2 个转录起始点，可产生大小为 40kb 的转录产物，经 RNA-FISH 实验证实其只定位于 X 染色体失活中心。在小鼠中，$Tsix$ 基因与 $Xist$ 基因完全重合，并且 $Tsix$ 对 $Xist$ 的表达具有负性调节作用。$Tsix$ 基因的缺失会导致 $Xist$ 表达产物上调，并可专一性失活雌性细胞中发生缺失突变的 X 染色体[54,55]。在 X 染色体失活发生之前，两条 X 染色体均表达 $Tsix$，且其表达量超过 $Xist$ 的 10～100 倍，在 X 染色体失活发生后，$Tsix$ 仅在活性 X 染色体上继续短暂表达[56]。在 X 染色体失活过程中，$Tsix$ 的表达关闭与 $Tsix$ 启动子的染色质改变相伴随[57]。$Tsix$ 介导的 $Xist$ 沉默的精确机制目前还未知。由于发现 $Tsix$ 的转录水平沿 $Xist$ 基因呈现梯度下降，即相对于 $Tsix$ 基因的 3′端，5′端合成更多的转录产物，人们提出了转录干扰（transcriptional interference）作为一种 $Xist$ 抑制机制[49,56,58]。另一种 $Tsix$ 抑制 $Xist$ 转录的可能机制为经 RNA 介导的沉寂。已有证据证明 $Tsix$ 可通过甲基转移酶 3a（DNMT3A）调节 $Xist$ 启动子的甲基化并进一步调节其转录活性[49]。同时在具有 $Tsix$ 基因缺陷的细胞中，顺式的 $Xist$ 基因启动子区拥有更多的活性染色质标记，而相对应的抑制性染色质标记减少[59,60]。当 Tsix 的长度削减至原长度的 93% 后，其将不能诱导 $Xist$ 沉寂，提示产生经过 $Xist$ 启动子自身的反转录本似乎对于抑制性染色质标记的出现具有决定性作用[61]。DXPas34 是位于 $Tsix$ 主要启动子下游的一个 CpG 岛，可启动反义转录，同样的，去除 DXPas34 元件也使 $Xist$ 的顺式沉寂无法实现，这进一步强调了在 $Tsix$ 介导的 $Xist$ 沉寂过程中反义转录的重要性[60,62,63]。此外，这一 CpG 岛的甲基化状态也与 $Xist$ 的反义转录起始完美契合[64,65]。由于 $Xist$ 与 $Tsix$ 的转录产物有部分重叠，故 RNAi 介导的 X 染色体失活调节机制可能也参与其中[66]。大小为 25～42 核苷酸的小 xiRNA 已经确实在 $Xist$ 基因的不同区域检测到，且 Dicer 内切核酸酶的突变可导致 xiRNA 结构的丧失及 $Xist$ 的去甲基化，这些都提示了 Dicer 在 X 染色体失活中的可能作用。但这一观点也受到争议，一些研究者已经发现缺乏 Dicer 的 ES 细胞也表现出正常的 X 染色体失活，而且对于活性 X 染色体（Xa）的影响也是由新合成甲基转移酶的活性降低介导的，而不是 Dicer 的直接作用[67,68]。因此，小 RNA 在 X 染色体失活启动过程中的确切功能现在仍不明确。与此同时，在缺乏内源性 $Tsix$ 表达的细胞系中顺式（in cis）过表达 $Tsix$ cDNA 并不能恢复由 $Tsix$ 介导的 $Xist$ 顺式沉默，这一结果也不支持 RNAi 介导过程[69]。

参与 X 染色体失活的第三个非编码基因为 $Xite$，是 X 染色体基因间转录产物元件[70]，其位于 $Tsix$ 上游大约 10kb 的位置，它在 X 染色体失活过程中的表达及甲基化模式均与 $Tsix$ 相似。$Xite$ 被认为是 $Tsix$ 的正向调节因子[71]。$Xite$ 的缺失会导致 $Xist$ 位点反义转录水平下降，这暗示了 $Xite$ 在抑制 $Xist$ 表达中与 $Tsix$ 及 DXPas34 有相似的作用[65,70]。

X 染色体失活中的反式作用因子

$Xist$、$Tsix$ 及 $Xite$ 又是如何被调节的？在参与 $Tsix$ 调节过程的蛋白质中，在

$DXpas34$ 区域及 $Xite$ 启动子上已经识别了绝缘蛋白（insulator protein）CTCF 及转录因子 YY1（yin yang1）的一些串联结构结合位点[72]。对于 $Yy1$ 的敲除研究或经 RNAi 介导的 $Yy1$ 及 $Ctcf$ 基因的部分去除研究显示了二者可下调 $Tsix$ 表达并伴随 $Xist$ 的上调表达，这支持了 YY1 及 CTCF 在 $Tsix$ 表达中发挥作用的观点[73]。最近发现，多潜能转录因子 OCT4 可结合于 DXpas34 元件及 $Xite$ 基因启动子，并可能与 CTCF 协作对 $Tsix$ 转录发挥调节作用[74]。另一项研究则表明多潜能因子 NANOG、OCT4 及 SOX2 可与 $Xist$ 的内含子 1 的某区域结合发挥抑制 $Xist$ 转录的作用[79]。$Nanog$ 缺陷的 ES 细胞表现出早于 $Tsix$ 表达下调的 $Xist$ 上调表达，支持了在 $Xist$ 表达调控中还有一种非 $Tsix$ 依赖的多潜能因子机制。有趣的是，在 ICM 中对失活的父源 X 染色体的重新活化也需要 $Nanog$ 的表达，这一调控过程很可能是通过抑制 $Xist$ 表达实现的[80]。因此，看起来这些已经在多潜能性研究中被广泛研究的主要调控因子可以通过不同途径参与到分化前对 X 染色体失活的抑制过程。

一项在 ES 细胞中利用针对 X 染色体失活中心——Xic 部分区域的 BAC 转基因筛查发现了由 X 染色体编码的 $Rnf12$ 基因对 X 染色体失活的激活作用[81]。$Rnf12$ 位于距离 $Xist$ 端粒方向 500kb 处一个最初定义为 Xic 的结构区域，这一区域编码一个 E3 泛素连接酶，此泛素连接酶已经被证实参与调节 LIM-同源结构域转录因子、雌激素依赖的转录活化，以及维持端粒长度的稳定性[82~84]。额外的 $Rnf12$ 基因拷贝可介导雄性 ES 细胞中唯一的 X 染色体，以及雌性细胞中两条 X 染色体大部分诱导的异常 X 染色体失活[81]。与野生型雌性 ES 细胞相比，含有杂合体 $Rnf12$ 的雌性 ES 细胞中 X 染色体失活的起始发生率降低。这些发现表明，虽然目前对 RNF12 介导的 X 染色体失活的分子机制仍然不明，但 RNF12 在激活 X 染色体失活过程中具有重要的剂量依赖性。除此以外，如何解释在杂合体 $Rnf12$ 敲除的雌性 ES 细胞中启动 X 染色体失活过程还有待于发现更多未知的 X 染色体失活激活分子。

数量及选择

在 X 染色体失活启动调节中必须有适当数量的 X 染色体被失活，而被失活的 X 染色体数量及这些 X 染色体是如何被选择的机制是 X 染色体失活研究中一个长期存在的问题。细胞是如何感知细胞核中存在的 X 染色体数量并知道它们其中有多少需要被失活呢？针对具有异常 X 染色体数量的患者的临床观察给这个问题带来了启示。拥有多余 X 染色体的患者，如具有 46，XXX 的超级女性或具有 46，XXY 的 Klinefelter 综合征的患者中除了一条 X 染色体，其余的均被失活[85~87]。在 Turner 综合征的患者中，这些女性的唯一 X 染色体并未发生 X 染色体的失活，然而在四倍体女性胚胎中却发生了两条 X 染色体的失活[88,89]。通过这些可以推断出 X 染色体失活的一般规则，即每两条染色体中会保存一条活化的 X 染色体[90]。

复杂而又机制交叉的模型被用来解释这些观察到的现象（图 10-3）。阻抑因子（BF）模型假设存在一种常染色体编码的因子，并以单一因子形式存在于二倍体细胞核中[35,36]。人们认为这种 BF 可与 DNA 上的一个元件结合后发挥作用，这个元件被称为计

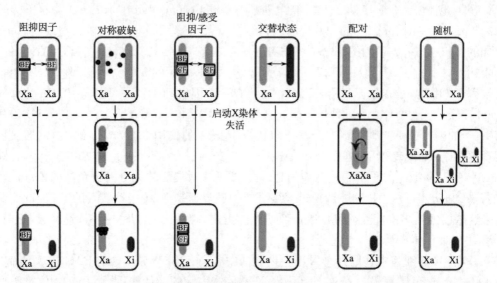

图 10-3 X 染色体失活中染色体计数及选择的几个模型。其中一些模型只用于解释雌性细胞中的 X 染色体失活启动。BF 模型假设细胞核中存在由常染色体编码的、称为阻抑因子（BF）的单一蛋白质或实体，这一因子决定了一条 X 染色体的失活。对称性阻抑模型假设阻抑因子（BF）由多个蛋白质组成并可在保持活性的 X 染色体上完成自我组装。阻抑因子/互补因子（BF/CF）模型加入由 X 染色体编码的互补性因子（CF）。其中一个 CF 因子拷贝对于在无 BF 保护的 X 染色体上启动 X 染色体失活是必需的，而另一个拷贝则被有活性的 X 染色体上的 BF 中和。在可替换态模型中，X 染色体失活由两条 X 染色体的不同状态预先决定。雌性特异的 X 染色体失活启动还可以通过一种配对模型进行解释，在这种模型中雌性细胞的两个 X 染色体失活中心（Xic）彼此紧密靠近，引起相互信息交换并调节 X 染色体失活的启动。最后，随机模型也被用来解释 X 染色体失活。在这一模型中，X 染色体均具有启动 X 染色体失活的可能性，而只有具有正确活化 X 染色体数的细胞才能存活

数元件，其只与一条 X 染色体相互作用，使这条染色体免于 X 染色体失活的顺式作用。由于在二倍体细胞核中，BF 的量只够阻止一条 X 染色体发生 X 染色体失活，故其他多余的 X 染色体都会发生沉寂。对称抑制相关模型认为 BF 不仅仅是一种单一分子，而是由多个常染色体编码分子组成，这些分子可以在活化的 X 染色体上整合并抑制该染色体发生失活[91,92]。由于 X 染色体失活中心（Xic）对于 X 染色体失活的发生是必需的，X 染色体失活只发生于拥有超过一个 Xic 的二倍体细胞核中，因而使得假设的计数元件定位于 X 染色体失活中心。一些研究试图利用转基因及缺失方法来确定 X 染色体计数元件，这些研究的设计基础为，当一个额外的 X 染色体计数元件被引入到雄性 ES 细胞中时，这一计数元件将能够中和掉限量的 BF，从而诱导计数过程，导致内源性 X 染色体由于缺乏 BF 保护而发生 X 染色体失活。

事实上，已经通过向常染色体中整合入含有 *Xist* 及 *Tsix*，或仅含有 *Xist* 的基因得到一系列雄性转基因 ES 细胞系，并在这些细胞系中观察到唯一的内源性 X 染色体会发生异常 X 染色体失活[93~98]。同时在这些细胞系中还观察到常染色体 *Xist* 的表达及在接

近整合位点处的其他常染色体基因沉默。研究所使用的转基因载体多种多样，从包含超过 500kb 插入片段的巨大的 YAC 到只含有 $Xist$ 及其两侧相邻序列仅 35kb 的黏粒[94]，研究结果表明，参与计数过程的因子可能位于被转基因所覆盖的序列。然而，使用相似转基因的其他相关研究却没有观察到计数现象[81,99,100]，或仅观察到只有多拷贝的转基因才能诱导计数现象[101]。我们已经知道在雌性 ES 细胞中，$Xist$ 对于 X 染色体失活的顺式发生具有重要作用，有趣的是，对于含有一条被去除 $Xist$ 的 X 染色体的雌性 ES 细胞的研究显示，X 染色体失活仍然可在野生型 X 染色体上正常发生，表明 $Xist$ 的转录及 $Xist$ 基因被去除的部分并不参与计数过程[45~47,102~104]。与此相反的是，不同的 $Tsix$ 突变雄性 ES 细胞（部分在 $Tsix$ 被发现之前就已经得到）却在单一的 X 染色体上出现了 X 染色体失活，这说明缺失序列在计数过程中也可能发挥了作用[55,58,60,63,105]。然而，相同的突变却不能抑制在雌性细胞的计数过程，但可导致突变染色体的选择性失活，这证明了 $Tsix$ 介导的顺式 $Xist$ 沉寂[54]。这些发现可通过以下机制来解释，即这些突变破坏了计数元件，抑制了 BF 的结合。可是，雌性 ES 细胞及小鼠的 $Xist$、$Tsix$ 和 $Xite$（ΔXTX）区域的杂合缺失并未干扰计数过程，野生型的 X 染色体一样发生了失活[106]。虽然对于 $Xist$ 转基因的研究显示，被转的序列在 X 染色体计数及失活起始过程中可能发挥额外作用，但上述研究结果说明，$Xist$、$Tsix$ 和 $Xite$ 对于计数过程并不是必需的，并且提示了可能的计数元件并不在被去除的区域。

与之前的所有关于 $Tsix$ 突变的报道相比，有一发生在 DXPas34 区域的 $Tsix$ 缺失突变（ΔCpG）并没有导致雄性细胞的异常 X 染色体失活[107]。有趣的是，带有纯合子 ΔCpG $Tsix$ 突变的雌性细胞可表现出无序的 X 染色体失活，许多细胞的两条 X 染色体都同时发生了 X 染色体失活。基于这项发现，一种 X 编码的获能性因子（competence factor，CF）也被认为参与了 X 染色体失活的起始。一种结合了 BF 及 CF 功能的模型认为，大量存在的获能性因子 CF 可使得除 BF 结合的 X 染色体外所有的 X 染色体失活[103]。另一种假设则认为 X 编码的 CF 是限量的，并且其作用可被常染色体编码的 BF 抵消，而常染色体编码的 BF 只足以中和一条 X 染色体编码的 CF[107]。当二倍体细胞中出现了一条以上 X 染色体时，多余的 CF 将不能被 BF 中和，从而导致不受保护的 X 染色体失活。

选择态模型认为，XX 细胞中的两条 X 染色体在 X 染色体失活发生前就已有区别[108]。这一模型可被以下事实支持：在未分化细胞中，两条染色体的姐妹染色单体结合时会受到不同调节。当然，不同的甲基化水平及染色单体状态也会对此产生影响。因此，可能在 X 染色体失活发生之前，从遗传学角度来看，完全相同的两条染色体可能就已发生了内源的表观遗传学差异。但该模型仍需进一步的实验证实。

还有一种不同的模型用来解释 X 染色体失活的计数及选择。在雌性二倍体细胞核中，两条 X 染色体失活中心发生了配对及短暂的染色体互换（transient transvection）[109,110]。已观察到，在 ES 细胞分化早期，细胞核中存在一种非随机的 X 染色体失活中心（Xic）空间限制分布，即在 X 染色体失活发生之前 X 染色体失活中心移动，相互靠近。这种短暂的配对现象可能在计数及选择过程中发挥一定作用。这一配对受到 $Tsix$ 及 $Xite$ 序列的辅助，并且似乎也依赖于 CTCF 的作用[111]。干细胞转录因子 $Oct4$

及RNA聚合酶Ⅱ介导的转录对于配对过程也发挥重要作用[74,111]。位于$Xist$端粒侧250~350kb，包含有部分$Slc16A2$基因的染色体区域也被认为介导了X染色体失活起始时的X染色体配对[112]，并在X染色体失活激活过程中发挥作用。到现在，我们还不清楚配对在X染色体失活过程中是起到调控的作用，还是仅仅是由于转录活化导致的X染色体失活中心（Xic）在细胞核中的重定位的结果。有趣的是，对于X染色体失活的启动而言，配对似乎并不是必需的。在$XX^{\Delta 65kb}$ES细胞中发生了$Tsix$及$Xite$基因缺失，从而消除了X染色体失活配对的发生，然而细胞中仍可发生X染色体失活起始[109,110]。因此，需要更多的研究来阐明配对环节在X染色体失活过程中所发挥的作用。

在上面我们所讨论的很多模型中，我们都假设X染色体发生失活过程是明确的，是互相排斥的。在这一过程中，雌性细胞中总是有正确数量的X染色体发生失活。然而，在体外对二倍体或四倍体ES细胞或ICM细胞所进行的研究显示，在相当多比例的细胞中发生了过多或过少失活X染色体（Xi），提示指导X染色体失活过程中具有一个随机机制，这种机制对于每条X染色体所发生的X染色体失活都具有相对独立的概率[106]。在对比不同的二倍体、三倍体及四倍体ES细胞发生X染色体失活的细胞数量时发现，X染色体与常染色体的比例决定了一条X染色体发生失活的可能性[114]。这种可能性是不同因子作用的结果，包括X编码的X染色体失活激活因子和常染色体编码的X染色体失活抑制因子，这两种因子分别促进或抑制$Xist$的产生。随着发育及分化，X染色体失活激活因子的浓度将会上升，同时伴随X染色体失活抑制因子浓度的下降。在雌性细胞中，这将使$Xist$在特定时间能够达到充足浓度并开始顺式扩散（图10-4）。X染色体失活抑制因子设定了$Xist$开始积累的阈值。由于X染色体失活激活因子基因是X染色体连锁的，故而$Xist$的扩散将会通过顺式作用下调X染色体失活激活因子基因表达，使得第二条X染色体免于失活。在这种模型中，$Xist$的扩散启动发生是随机的，所以任何一条X染色体上X染色体失活激活因子基因的沉默概率都是相同的。在雄性细胞中，X染色体失活激活因子的浓度不足以打破这个临界值，所以无法启动X染色体失活。因此，雌性特异的X染色体失活启动是通过性别决定X染色体编码的X染色体失活激活因子剂量差异从而可以促进$Xist$积聚来实现的。通过对具有$Xist$或$Tsix$突变的细胞系及小鼠的研究表明，$Xist$及$Tsix$在X染色体失活的可能性的选择中发挥主要作用，而且X染色体失活激活因子及抑制因子可能正是通过这些基因发挥作用的。

一些X染色体失活抑制因子已经被人们所认识，它们中的一些，如$Yy1$、$Ctcf$及$Oct4$，通过激活$Tsix$的表达发挥作用；或者通过直接抑制$Xist$发挥作用，如$Nanog$、$Oct4$及$Sox2$[72~74,79]。$Rnf12$是一个X编码X染色体失活激活因子[81]。观察到的由$Rnf12$转基因诱导的异常X染色体失活与转基因的$Rnf12$/RNF12表达有关，这提供了RNF12剂量在X染色体失活活化过程中作用的依据。与此相一致的是，$Rnf12$敲除的杂合雌性ES细胞只能在较少的细胞中引发X染色体失活。在X染色体失活起始之前或在其起始时间附近，野生型的ES细胞中的RNF12表达量发生上调，与雄性ES细胞相比，雌性细胞中检测到的RNF12蛋白是雄性的两倍。在X染色体失活过程起始后，$Rnf12$基因迅速被沉默，这可以阻止雌性细胞中的第二条X染色体发生X染色体失活。

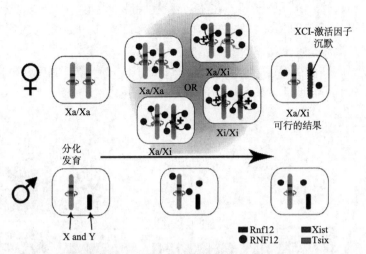

图 10-4　X 染色体失活的随机模型。X 染色体失活随机模型认为 X 染色体失活的启动受 X 染色体编码的 X 染色体失活激活因子及常染色体编码的 X 染色体失活抑制因子共同调控。随着分化的进行或在发育过程中，细胞核内的多个 X 染色体失活激活因子如 RNF12（蓝色）浓度不断上升，同时，X 染色体失活抑制因子的浓度则不断下降，如 NANOG、OCT4 及 SOX2（未标出）；或维持不变，如 YY1 及 CTCF（未标出）。X 染色体失活激活因子由 X 染色体编码，因此只有在雌性细胞核中，这些因子浓度间的平衡才足以产生独立而持续性的 X 染色体失活（灰色及黑色箭头）。不同的因子间浓度比例可导致以下几种情况：细胞不发生 X 染色体失活；在一条或两条 X 染色体上发生 X 染色体失活。X 染色体失活导致 Xist 的扩散及 X 染色体失活激活因子基因的顺式沉寂，后者将下调 X 染色体失活激活因子的浓度，这一下调会抑制对剩余的活性 X 染色体的失活启动。本图的彩版见 www.landesbioscience.com/curie

但 $Rnf12$ 是如何调节 X 染色体失活的仍不可知，未来的研究将致力于 $Rnf12$ 介导的 X 染色体失活活化的分子机制。

最近对于 X 染色体失活抑制因子及活化因子的了解为 X 染色体失活建立了一种随机模型。这种模型认为，X 染色体失活过程并没有一种选择机制，而是由一种起始反馈程序控制，这种程序受 X 染色体失活抑制因子及活化因子间精细复杂的平衡关系调节。

沉默与沉默的维持

X 染色体失活起始后，将发生一系列事件引发活性 X 染色体中的常染色质转变为失活 X 染色体（Xi）中紧密折叠的非活性异染色质，这一变化可以通过雌性体细胞中的 Barr 小体识别[115,116]。这一系列级联事件的第一步就是将失活 X 染色体上 $Xist$ 基因的转录上调和扩散[37,39]（图 10-5）。$Xist$ 的转录产物包括多个重复序列，其中 A 重复序列位于 $Xist$ 的 5' 端，参与沉默过程[117]。$Xist$ RNA 在前 Xi 上的扩散导致了 RNA 聚合酶 Ⅱ 及结合的转录因子的快速脱落，这使得在这个染色体上基因转录减少并产生沉默核区

（silent nuclear compartment）[118,119]。在 *Xist* 积累后最早发生的是特异性染色质的改变，包括：活性染色质标记的丢失，如 H3K9 乙酰化和 H3K4 单或双甲基化[19,120,121]；沉默染色质标记的增加，如 H3K27 三甲基化（H3K27-me3）[50,51,57,122]、H3K9 双甲基化（H3K9-me2）[121,123-126] 和 H4K20 单甲基化（H4K20-me1）[127] 等；之后是 macroH2A[128] 的异化组蛋白的嵌入，以及 DNA 的甲基化和复制时间的改变[129~131]。

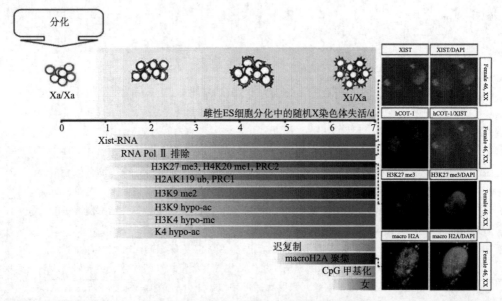

图 10-5　X 染色体失活（Xi）的表观遗传学改变。雌性 ES 细胞的分化诱导了 X 染色体的随机失活。在即将失活的 X 染色体（Xi）上最初的改变是 *Xist* RNA 的顺式扩散（红色），接下来是 RNA 聚合酶Ⅱ排除（棕色），以及活性染色体标记的丢失（蓝色）、失活染色体的标记获取或特异性组蛋白变异体的聚集（紫色）。这些染色体的改变伴随着蛋白质和蛋白质复合体的聚集（绿色）及其他表观遗传学改变，比如 S 期 DNA 复制期后延和 CpG 甲基化（橙色）。右边栏图示人的雌性纤维母细胞的 *XIST* RNA FISH、用于检测 RNA 聚合酶Ⅱ转录重复区域的 Cot 探针的 RNA FISH、H3K27 me3 抗体和抗 macroH2A 抗体的免疫染色。每个细胞都有一个 *XIST* RNA 信号包被的 Xi，同一 Xi 上是 Cot RNA 阴性，并富含 H3K27me3 及杂合 macroH2A。彩图可参见 www.landesbioscience.com/curie

为了启动染色质改变及沉默，*Xist* 需要发生积累并在 X 染色体上的扩散。*Xist* 的扩散机制还不清楚。针对 X-常染色体易位和常染色体整合的 *Xist* 转基因研究显示，*Xist* 在常染色质区域的扩增并不完整，这表示 X 染色体的序列对于 *Xist* 有效扩增是必需的[35,95,131~135]。增强子（booster element）或中间点（way station）已经被认为对于 X 染色体上的 *Xist* 扩增具有增强作用[136]，而 LINE-1 转座子则被认为是最可能的增强子[137~140]。LINE-1 转座子是哺乳动物基因组中最常见的重复序列，并且与常染色体相比，在 X 染色体上更为丰富[141~144]。LINE-1 转座子在 X 染色体上发生 X 染色体失活的区域与不发生失活基因的区域相比具有更高密度[8,142,143,145~147]，且在 X-常染色体易位区域的融合染色质的 LINE-1 密度与 *Xist* 扩散程度呈正相关[133~135,148~153]。然而，在其他研究中并没有发现 LINE-1 密度与扩散之间的关系，或者说没有重要作用，因而存在其

他对于 $Xist$ 扩增也很重要的序列[57,146,154,~156]。目前还不知道 $Xist$ 在扩增过程中是如何与这些序列相互作用的,以及是否还存在其他分子对这种相互作用产生直接或间接的影响。

在 $Xist$ 积聚之后,在失活 X 染色体（Xi）上最早被检测到的组蛋白修饰是 H3K27-me3。这一修饰依赖于 Ploycomb 抑制复合体 PRC2。这种复合体由 EED 蛋白、RbAp46/48、SUZ12 和 EZH2 组成并在失活 X 染色体上积累。其中 EZH2 是一种甲基转移酶,参与 H3K27 的三甲基化[51,157,158]。最近有关 EZH2 与 $Xist$ 的 A 重复序列间的直接相互作用的报道,对于在雌性 ES 细胞中的 EZH2 敲除研究表明,PRC2 对随机 X 染色体失活时失活 X 染色体（Xi）形成发挥作用[52]。有趣的是,EED 基因的敲除导致了 PRC2 复合体的功能丧失,且 H3K27-me3 的缺失发现了在小鼠胚胎外组织中具有失活 X 染色体的复活,提示了 X 染色体印记失活的缺陷[159]。尽管由于 EED 的缺失导致胚胎发育过程中 X 染色体失活的异常,但在胚胎内的随机 X 染色体失活并没有受到影响[160],这与体外敲除 $Ezh2$ 后获得的结果形成对比[52]。另一项使用常染色质整合的诱导性 $Xist$ 转基因进行的研究表明,缺乏 EED 导致一些 PRC1 复合体组分如 MPH1 和 MPH2 不再分布于有 $Xist$ 覆盖的常染色质,然而对于复合体 PRC1 的组分——RING1B 的募集及 $Xist$ 介导的常染色体序列沉默并没被影响[161]。在失活 X 染色体上的 H2AK119ub1（119 位赖氨酸单泛素化的组蛋白 H2A）积聚依赖于 RING1B[162,163],但纯合子 $Ring1b$ 的突变并不影响 $Xist$ 介导的、对带有 $Xist$ 转基因插入的常染色体沉默[161]。这些发现提示,PRC1 与 PRC2 并非随机 X 染色体失活所必需的,但也不能排除这些复合体对于失活 X 染色体的建立及维持过程具有冗余调控作用。

其他蛋白质也可能对失活 X 染色体（Xi）的建立和维持起作用。一种由点型（speckle-type）POZ 蛋白 SPOP 和 CULLIN3 组成的 E3 泛素化连接酶复合体与失活 X 染色体中 macroH2A 的沉积调节有关[164]。SPOP 和 CULLIN3 均可泛素化组成 PRC1 复合体的 BMI1 分子和 macroH2A。macroH2A 的泛素化对于该分子在失活 X 染色体上的募集具有重要作用。由 RNAi 介导的 SPOP 或 CULLIN3 敲除将导致失活 X 染色体上 macroH2A 的消失。RNAi 介导的 macroH2A 或 SPOP/CULLIN3 表达敲除结合去甲基化和脱乙酰化抑制作用物的作用导致失活 X 染色体上连锁的报告基因的转录活化。仅用去甲基化和脱乙酰化抑制作用物并不能导致基因的活化,表明了 macroH2A 分子对于维持失活 X 染色体沉默状态的重要作用。macroH2A 的沉默功能可能通过募集 Poly（ADP-核糖）聚合酶 1（PARP-1）而间接作用[165]。PARP-1 是一种参与染色质结构调节的核酶,可以感知 DNA 损伤并调节基因表达[166~168]。MacroH2A 可将 PARP-1 募集到失活 X 染色体上并抑制 PARP-1 的催化活性。失去酶活性的 PARP-1 可结合核小体并抑制转录[169]。联合去甲基化和脱乙酰化抑制处理的 PARP-1 去除实验可导致失活 X 染色体-连锁报告基因的活化。因此,macroH2A 可能通过调节 PARP-1 酶活性继而在染色体沉默发挥作用。失活 X 染色体上还富含核支架蛋白 SAF-A。SAF-A 被认为是一种 DNA/RNA 结合蛋白,并且被认为可能是调节基因表达及 DNA 复制相关的核支架的组成成分。SAF-A 的富集取决于 SAF-A 分子上的 RNA 结合功能区,这个功能区与 $Xist$ 介导的失活 X 染色体分子募集有关[170]。BRCA1 可协助失活 X 染色体对 $Xist$ 的募集。

BRCA1 涉及许多通路调控，包括检查点的激活和 DNA 的修复。而 BRCA1 与失活 X 染色体的结合只存在于一小部分细胞，实验表明 BRCA1 在 Xist/XIST 染色体定位中发挥作用[171,172]。但是，这种说法存在争议，在 X 染色体失活中 BRCA1 的作用仍未确定[173,174]。SMCHD1 也与失活 X 染色体有关。该分子参与受 X 染色体失活调节的基因的 CpG 岛甲基化[175]。SMCHD1 包含一个 SMC-铰链结构，这种结构在参与粘连及染色质浓缩的蛋白质中可见。SmcHD1 敲除小鼠表现出对胚胎及外胚组织中失活 X 染色体的维持缺陷。对 Atrx 敲除的雌性杂合体小鼠的分析发现，ATRX 在印记 X 染色体失活中发挥作用（ATRX 是位于 X 染色体上与 α-地中海贫血及智力低下相关的一种染色体重塑蛋白）[176]。从雌性胚系中遗传有突变等位基因的雌性小鼠不能使其外胚组织中的父系 X 染色体失活，说明失败可能发生在 X 染色体印记失活上。有趣的是，在 ES 细胞分化的稍后阶段，ATRX 仍与失活 X 染色体相伴，提示其在失活 X 染色体维持上的作用[177]。SatB1 和 SatB2 同样与失活 X 染色体的建立有关。这些肿瘤关联基因编码的核蛋白发挥着基因组组织者（genome organizer）和基因调节因子的作用[178]。对 SatB1 和 SatB2 敲除研究显示，分化中的雌性 ES 细胞的 X 染色体沉默具有部分缺陷，提示这些蛋白质在失活 X 染色体的异染色质化过程中的作用[179]。

有趣的是，当失活 X 染色体已经建立后，维持失活状态不再需要 Xist 的表达，因为 Xist 的条件性敲除并不会导致沉默基因的复活[47]。因此，失活 X 染色体的建立与修复涉及多个层次功能冗余的蛋白质和表观遗传调控。这些因素的联合作用确保了失活 X 染色体保持稳定的沉寂状态，实现恒定的基因剂量补偿。在小鼠中只有有限的 X-连锁基因可以逃脱 X 染色体失活介导的沉默，人类中则发现了更多的逃脱者[180~183]。因此，物种不同，沉默普遍性可能不同，这可能是由于物种间在进化上对于剂量补偿上的差异要求导致的。那些逃脱了剂量补偿并且在 Y 染色体上没有等位基因的基因，将在雄性与雌性间出现表达剂量差异，这些可能对于建立两性差异产生作用。

总之，在 X 染色体基因沉默及沉默维持中，涉及许多层次的表观遗传层调控，以及仅为人们了解了一小部分的多种复合物作用。尽管已有许多研究关注于这些机制，但对于 X 染色体失活沉寂及维持的了解目前尚处于非常初级的阶段。

X 染色体失活与人 ES 细胞

研究人类 X 染色体失活是很有挑战性的一项工程。出于人类伦理道德因素，早期人类胚胎细胞的研究是受到严格限制的。因此，许多关于人的 X 染色体失活的既有知识是来源于对不同模型系统研究的结果，包括人鼠细胞杂交、人类胚胎肿瘤细胞、肿瘤细胞系[184,185]和人类基因插入鼠 ES 细胞基因组[81,97,186~188]。人 ES 细胞系的建立为研究人 X 染色体失活提供了可与鼠 ES 细胞相比拟的研究模型。然而，在人类 ES 细胞中，对 X 染色体失活发生的不同研究，其结果是多样的，甚至是自相矛盾的[189~194]。首先，大多数未分化的雌性人 ES 细胞系存在 X 染色体失活及失活 X 染色体（Xi）上的 XIST 表达、包被（coating）及异染色质的积累标记。其他一些未分化人 ES 细胞系则与鼠 ES 细胞相似，存在活性 X 染色体，且在分化过程中存在 X 染色体的失活（表 10-1）。

有趣的是,某些细胞系在不同的实验室存在差异。一些亚克隆表现为在分化过程中随机发生 X 染色体失活,然而其他的则表现在分化前便可检测到失活 X 染色体的标记。在一项对 11 株确定的 hES 细胞系的研究中,Silva 等根据 X 染色体失活的不同确认了三群 hES 细胞系[191]。第一群细胞系只在分化后出现 X 染色体失活特征。第二群与第三群细胞系则表现为在未分化状态已发生 X 染色体的失活,但第三群细胞系还伴随 XIST 的表达缺失。在第三群细胞系中,尽管细胞不再表达 XIST,但其他 X 染色体失活标记仍然存在,如对于 X 染色体的 Cot-1 RNA 的排异。有趣的是,在这群细胞系中 H3K27 三甲基化也丢失了,这表明 H3K27 三甲基化在 X 染色体失活的维持过程中并非必需。这些结果支持了之前发现的关于鼠与人类的 H3K27 三甲基化依赖 XIST 的结论[50]。

表 10-1 鼠与人类多能干细胞的特征

	鼠			人	
	ES 细胞	iPS 细胞	EpiSC	ES 细胞	iPS
XaXa	+	+	—	+/−	−[1]
XaXi	—	—	+	+/−	+[1]
Xist/XIST	—	—	+	+/−	+[1]
H3K27 me3,Cot exclusion	—	—	+	+/−	+[1]
bFGF,Activin/Nodal sign.	—	—	+	+	+
Lif/Stat3 signalling	+	+	—	—	—
diff. in extra embr. tissue	—	—	—	+	?

注:对比不同报道中人类与鼠雌性多能干细胞 X 染色体失活的特征与培养条件(1,只测试了 1 个雌性 iPS 细胞系)。失活 X 染色体(Xi)及与 X 染色体失活相关的不同表观遗传学的改变用存在(+)与缺失(—)标识。用于维持干细胞或其向外胚层组织分化潜能的组织培养条件也用(+)标识出。

如何解释人类与鼠 ES 细胞在 X 染色体失活上的不同呢？小鼠的 ICM 细胞在分化之前表现为两个活化的 X 染色体(Xa)。不同物种在 X 染色体失活发生的时间窗上的差异性也许可以用来解释,同样是来自 ICM,为什么人 ES 细胞在分化前就表现出 X 染色体失活的特征。是否在人的胚胎更早期就发生一个随机的 X 染色体失活,因而人胚胎干细胞中已存在一个失活 X 染色体(Xi)？另一种可能的解释是,在人类胚胎发育早期卵裂时的失活的 X 染色体再活化过程晚于小鼠胚胎。在小鼠,印记 X 染色体失活发生于胚胎着床之前并引发父系 X 染色体(Xp)的失活[17~19,195,196]。印记失活的 X 染色体外胚组织被保存了,但是在 ICM 细胞中印记失活的 X 染色体却被再活化(图 10-1)。这一过程很可能是通过在外胚层细胞中表达的 Nanog 的诱导[80]。尽管在人类,X 染色体失活过程也发生于早期卵裂期,但不清楚是否人类失活 X 染色体也发生印记[198~207]。另外,如果存在的话,失活 X 染色体在哪个时期重活化也不清楚。因而在未分化的 hES 细胞中存在的 X 染色体失活有可能在着床前的胚胎中就已存在,而不是由提前发生的随机 X 染色体失活导致。在人类女性 hES 细胞中发现了两条 X 染色体中存在其中一条更倾向于失活的现象,作为对这一可能性的支持证据,这种高度偏态的 X 染色体失活模式正好支持了后一种观点[190,208]。但遗憾的是，这种失活 X 染色体的父系起源并没有确定。

除了不同物种 X 染色体失活发生时间的差异外，人类和鼠未分化 ES 细胞中 X 染色体失活的显著差异还可以通过两种细胞不同来解释[209,210]。人类和鼠的 ES 细胞在形态学、克隆生成、分子形态和培养基要求上都有很明显的不同。例如，人 ES 细胞的自我更新需要 bFGF 和 Activin/Nodal 信号，而鼠 ES 细胞则依赖于 LIF/Stat3 信号[211,212]，而且两种细胞多能性启动子的活性在基因组水平具有广泛性差异[213]。尽管两种细胞都具有在体外分化成所有胚层细胞及形成畸胎瘤的潜能，但只对鼠 ES 细胞进行过形成生殖细胞系及产生完整个体的测试。对于 hES 细胞，由于伦理道德因素，不能进行嵌合体形成的测试。因此，现在仍不清楚是否人类 ES 细胞等同于鼠 ES 细胞。目前，利用含 bFGF 及激活素而缺乏 LIF 的培养基已从鼠的植入后外胚层内分离出来一个新的小鼠多能干细胞群[214,215]。这些被称为 EPiSC（植入后外胚层生发细胞）的细胞可表达多能因子如 *OCT4*、*Sox2* 和 *Nanog*，但其与鼠的 ES 细胞相比从形态学及转录产物上都具有差异，这使得它们与人的 ES 细胞更具有可比性[216]。尽管这类细胞可在体外分化成全胚层细胞类型和外胚滋养层，但形成杂合体的效率很低，并且没有报道表示它们能实现生殖细胞传递（germ line transmission）。有趣的是，雌性 EpiSC 也经历 X 染色体失活，并保留一条活性 X 染色体。因此，在人 ES 细胞中观察到的 X 染色体失活可能反映的是人类与鼠 ES 细胞的不同，人 ES 细胞实际上就是鼠 EpiSC 细胞在人类的对应细胞[217]。基于这一观点，鼠 EpiSC 和人 ES 细胞都有在体外分化成为外胚层组织的能力[218]，而鼠 ES 细胞只能分化成三种胚层细胞。有实验表明 EpiSC 可由 ES 细胞培养分化而来，这提示了 EpiSC 细胞由衍生 ES 细胞[219]。同时，过表达多能分化因子 *Klf4*[219]、*Nanog*[80]，或在含有 LIF[220] 的培养基中培养 EpiSC，可诱导重编程为 ES，这一过程伴随着 X 染色体失活后的复活。尽管人 ES 细胞像鼠 ES 细胞一样衍生自 ICM，为什么人 ES 细胞代表的是人的分化度较高的细胞类型呢？可能是用于培养人 ES 细胞的培养基适合于基态人 ES 细胞向具有 EpiSC 特征的人 ES 细胞分化，并在体外抑制人 ICM 多能状态的稳定性[221]。此外，在早期分化中，啮齿类与人类有所不同，在小鼠中存在诸如卵筒（egg cylinder），或者被称为小鼠特有的胚胎发育的暂时停止的滞育现象。这些差异也许可以解释为什么人类与小鼠相比仅有一个相对较短的窗口期可以获得基态的 ES 细胞。因而在实际获得 ES 细胞的培养过程中，更多的可能性是获得已经产生一定分化的细胞，从而可以解释观察到的在这些 ES 细胞中的 X 染色体失活现象。

导致在未分化的人 ES 细胞中有一条 X 染色体失活也可以解释为是对两条 X 染色体间的负选择性压力。事实上，在发育过程中，早期胚胎细胞具有两条活性 X 染色体的时间窗很短。由于这些细胞并不是要永久停留在 ICM 细胞状态，故这些细胞可能为适应 ES 细胞衍生的培养条件启动 X 染色体失活，这一过程有可能使它们具有成长和生存优势。对两条活性 X 染色体的负选择同样出现在许多鼠 ES 细胞系中。雌性鼠 ES 细胞表现出基因组水平的低甲基化[223]，这可能导致基因的不稳定性。这一现象可以解释为什么许多近交鼠 ES 细胞在扩增时丢失两条 X 染色体中的一条，而只有杂交 ES 细胞可以稳定存在两条 X 染色体。因为人 ES 细胞并没有发现这种 X 染色体丢失趋势的特征，所以必须借用X染色体失活来阻止两条X染色体的活化。

总而言之，对于在未分化人ES细胞中就已经出现X染色体的活化可有许多不同的解释，而这些机制的综合作用可能最终会成为在人ES细胞中观察到的表观遗传学流动性现象的原因。今后的研究将需要针对这些观察到的现象的起因，因为关于人ES细胞表观遗传学稳定性，以及这些细胞的分化潜能的所有疑义需要在人ES细胞引入临床应用前得到阐明[191,224~226]。

最近发现的人iPS细胞是否可以为我们研究X染色体失活提供一个更好的模型？对于鼠iPS细胞而言，在体细胞重编程阶段沉寂X染色体的复活已经证实了[32,33]，而且这些细胞通过分化重新启动X染色体失活过程。尽管直到今天还没有一项研究确定在人iPS细胞中X染色体失活的发生情况，我们实验室的一些初步研究结果已经观察到在人iPS细胞中早期 XIST 积累（T.S.B和J.G未发表的研究结果）。而这些观察到的结果是由于培养条件产生的，或由于人ES细胞及人iPS细胞本身多能性状态较之于鼠有根本差异，还是由于重编程的不完全所引起的，仍待进一步调查研究来澄清。

结论

因X染色体失活随雌性ES细胞的分化而开始，所以鼠ES细胞可为研究X染色体失活提供有力的模型系统。最近的研究表明，在雌性ES细胞中X染色体失活受到 *Nanog*、*Sox2* 和 *Oct4* 等因子的抑制作用，这些因子也作为维持ES细胞的多能基础状态一个复杂细胞因子网络的一部分发挥关键作用。X染色体失活的活化需要下调这些因子，以及上调一系列X-编码的X染色体失活激活因子，这些因子中包括了涉及分化过程 *Rnf12*。这表明X染色体失活受参与细胞多能性及分化调控的相同细胞因子的调节。尽管如此，许多调节X染色体失活的分子及其下游靶点仍未确定。多能性与X染色体失活之间的紧密联系，使得X染色体的表观遗传学状态测定为研究雌鼠细胞的多潜能性及核重编码过程提供了重要指标。不尽如人意的是，对于人类雌性ES细胞而言，这一过程的调节更为复杂，是否可用女性ES细胞作为可靠的模型系统研究X染色体失活过程仍不明确，为解决这些问题我们还要进行更多的研究。

致谢

感谢实验室里所有给予帮助及参与激烈讨论的成员。这项工作获得NWO TOP项目资助（授予J.G）并得到丹麦政府研究项目的资助（BSIK programme 03038，SC-DD）。

（胡　晓　阮　峥　刘文静　译）

参 考 文 献

1. Ohno S. Sex chromosomes and Sex-linked genes. 1967, Berlin: Springer.
2. Charlesworth B. The evolution of sex chromosomes. Science 1991; 251(4997):1030-3.
3. Graves JA. The origin and function of the mammalian Y chromosome and Y-borne genes—an evolving understanding. Bioessays 1995; 17(4):311-20.
4. Graves JA, Koina E, Sankovic N. How the gene content of human sex chromosomes evolved. Curr Opin Genet Dev 2006; 16(3):219-24.
5. Graves JA. Sex chromosome specialization and degeneration in mammals. Cell 2006; 124(5):901-14.
6. Skaletsky H, Kuroda-Kawaguchi T, Minx PJ et al. The male-specific region of the human Y chromosome is a mosaic of discrete sequence classes. Nature 2003; 423(6942):825-37.
7. Koopman P, Gubbay J, Vivian N et al. Male development of chromosomally female mice transgenic for Sry. Nature 1991; 351(6322):117-21.
8. Ross MT, Grafham DV, Coffey AJ et al. The DNA sequence of the human X chromosome. Nature 2005; 434(7031):325-37.
9. Lyon MF. Gene action in the X-chromosome of the mouse (Mus musculus L.). Nature 1961; 190:372-3.
10. Lyon MF, Phillips RJ, Searle AG. A test for mutagenicity of caffeine in mice. Z Vererbungsl 1962; 93:7-13.
11. Lyon MF. Sex chromatin and gene action in the mammalian X-chromosome. Am J Hum Genet 1962; 14:135-48.
12. Davidson RG, Nitowsky HM, Childs B. Demonstration of two populations of cells in the human female heterozygous for glucose-6-phosphate dehydrogenase variants. Proc Natl Acad Sci USA 1963; 50:481-5.
13. Adler DA, Rugarli EI, Lingenfelter PA et al. Evidence of evolutionary up-regulation of the single active X chromosome in mammals based on Clc4 expression levels in Mus spretus and Mus musculus. Proc Natl Acad Sci USA 1997; 94(17):9244-8.
14. Birchler JA, Fernandez HR, Kavi HH. Commonalities in compensation. Bioessays 2006; 28(6):565-8.
15. Lin H, Gupta V, Vermilyea MD et al. Dosage compensation in the mouse balances up-regulation and silencing of X-linked genes. PLoS Biol 2007; 5(12):e326.
16. Nguyen DK. Disteche CM. Dosage compensation of the active X chromosome in mammals. Nat Genet 2006; 38(1):47-53.
17. Takagi N, Sasaki M. Preferential inactivation of the paternally derived X chromosome in the extraembryonic membranes of the mouse. Nature 1975; 256(5519):640-2.
18. Huynh KD, Lee JT. Inheritance of a pre-inactivated paternal X chromosome in early mouse embryos. Nature 2003; 426(6968):857-62.
19. Okamoto I, Otte AP, Allis CD et al. Epigenetic dynamics of imprinted X inactivation during early mouse development. Science 2004; 303(5658):644-9.
20. Evans MJ, Kaufman MH. Establishment in culture of pluripotential cells from mouse embryos. Nature 1981; 292(5819):154-6.
21. Martin GR. Isolation of a pluripotent cell line from early mouse embryos cultured in medium conditioned by teratocarcinoma stem cells. Proc Natl Acad Sci USA 1981; 78(12):7634-8.
22. Thomson JA, Itskovitz-Eldor J, Shapiro SS et al. Embryonic stem cell lines derived from human blastocysts. Science 1998; 282(5391):1145-7.
23. Keller G. Embryonic stem cell differentiation: emergence of a new era in biology and medicine. Genes Dev 2005; 19(10):1129-55.
24. Leahy A, Xiong JW, Kuhnert F et al. Use of developmental marker genes to define temporal and spatial patterns of differentiation during embryoid body formation. J Exp Zool 1999; 284(1):67-81.
25. Monk M. A stem-line model for cellular and chromosomal differentiation in early mouse-development. Differentiation 1981; 19(2):71-6.
26. Takahashi K, Yamanaka S. Induction of pluripotent stem cells from mouse embryonic and adult fibroblast cultures by defined factors. Cell 2006; 126(4):663-76.
27. Nakagawa M, Koyanagi M, Tanabe K et al. Generation of induced pluripotent stem cells without Myc from mouse and human fibroblasts. Nat Biotechnol 2008; 26(1):101-6.
28. Wernig M, Meissner A, Foreman R et al. In vitro reprogramming of fibroblasts into a pluripotent ES-cell-like state. Nature 2007; 448(7151):318-24.
29. Yu J, Vodyanik MA, Smuga-Otto K et al. Induced pluripotent stem cell lines derived from human somatic cells. Science 2007; 318(5858):1917-20.
30. Takahashi K, Tanabe K, Ohnuki M et al. Induction of pluripotent stem cells from adult human fibroblasts by defined factors. Cell 2007; 131(5):861-72.
31. Park IH, Zhao R, West JA et al. Reprogramming of human somatic cells to pluripotency with defined factors. Nature 2008; 451(7175):141-6.
32. Maherali N, Sridharan R, Xie W et al. Directly reprogrammed fibroblasts show global epigenetic remodeling and widespread tissue contribution. Cell Stem Cell 2007; 1(1):55-70.
33. Stadtfeld M, Maherali N, Breault DT et al. Defining molecular cornerstones during fibroblast to iPS cell reprogramming in mouse. Cell Stem Cell 2008; 2(3):230-40.

34. Russell LB. Mammalian X-chromosome action: inactivation limited in spread and region of origin. Science 1963; 140:976-8.
35. Rastan S. Non-random X-chromosome inactivation in mouse X-autosome translocation embryos—location of the inactivation centre. J Embryol Exp Morphol 1983; 78:1-22.
36. Rastan S, Robertson EJ. X-chromosome deletions in embryo-derived (EK) cell lines associated with lack of X-chromosome inactivation. J Embryol Exp Morphol 1985; 90:379-88.
37. Brown CJ, Ballabio A, Rupert JL et al. A gene from the region of the human X inactivation centre is expressed exclusively from the inactive X chromosome. Nature 1991; 349(6304):38-44.
38. Heard E, Avner P. Role play in X-inactivation. Hum Mol Genet 1994; 3 Spec No:1481-5.
39. Borsani G, Tonlorenzi R, Simmler MC et al. Characterization of a murine gene expressed from the inactive X chromosome. Nature 1991; 351(6324):325-9.
40. Brockdorff N, Ashworth A, Kay GF et al. Conservation of position and exclusive expression of mouse Xist from the inactive X chromosome. Nature 1991; 351(6324):329-31.
41. Brockdorff N, Ashworth A, Kay GF et al. The product of the mouse Xist gene is a 15 kb inactive X-specific transcript containing no conserved ORF and located in the nucleus. Cell 1992; 71(3):515-26.
42. Brown CJ, Hendrich BD, Rupert JL et al. The human XIST gene: analysis of a 17 kb inactive X-specific RNA that contains conserved repeats and is highly localized within the nucleus. Cell 1992; 71(3):527-42.
43. Clemson CM, McNeil JA, Willard HF et al. XIST RNA paints the inactive X chromosome at interphase: evidence for a novel RNA involved in nuclear/chromosome structure. J Cell Biol 1996; 132(3):259-75.
44. Jonkers I, Monkhorst K, Rentmeester E et al. Xist RNA is confined to the nuclear territory of the silenced X chromosome throughout the cell cycle. Mol Cell Biol 2008; 28(18):5583-94.
45. Penny GD, Kay GF, Sheardown SA et al. Requirement for Xist in X chromosome inactivation. Nature 1996; 379(6561):131-7.
46. Marahrens Y, Panning B, Dausman J et al. Xist-deficient mice are defective in dosage compensation but not spermatogenesis. Genes Dev 1997; 11(2):156-66.
47. Csankovszki G, Panning B, Bates B et al. Conditional deletion of Xist disrupts histone macroH2A localization but not maintenance of X inactivation. Nat Genet 1999; 22(4):323-4.
48. Kay GF, Penny GD, Patel D et al. Expression of Xist during mouse development suggests a role in the initiation of X chromosome inactivation. Cell 1993; 72(2):171-82.
49. Sun BK, Deaton AM, Lee JT. A transient heterochromatic state in Xist preempts X inactivation choice without RNA stabilization. Mol Cell 2006; 21(5):617-28.
50. Plath K, Fang J, Mlynarczyk-Evans SK et al. Role of histone H3 lysine 27 methylation in X inactivation. Science 2003; 300(5616):131-5.
51. Silva J, Mak W, Zvetkova I et al. Establishment of histone h3 methylation on the inactive X chromosome requires transient recruitment of Eed-Enx1 polycomb group complexes. Dev Cell 2003; 4(4):481-95.
52. Zhao J, Sun BK, Erwin JA et al. Polycomb proteins targeted by a short repeat RNA to the mouse X chromosome. Science 2008; 322(5902):750-6.
53. Lee JT, Davidow LS, Warshawsky D. Tsix, a gene antisense to Xist at the X-inactivation centre. Nat Genet 1999; 21(4):400-4.
54. Lee JT, Lu N. Targeted mutagenesis of Tsix leads to nonrandom X inactivation. Cell 1999; 99(1):47-57.
55. Clerc P, Avner P. Role of the region 3' to Xist exon 6 in the counting process of X-chromosome inactivation. Nat Genet 1998; 19(3):249-53.
56. Shibata S, Lee JT. Characterization and quantitation of differential Tsix transcripts: implications for Tsix function. Hum Mol Genet 2003; 12(2):125-36.
57. Marks H, Chow JC, Denissov S et al. High-resolution analysis of epigenetic changes associated with X inactivation. Genome Res 2009; 19(8):1361-73.
58. Luikenhuis S, Wutz and A, Jaenisch R. Antisense transcription through the Xist locus mediates Tsix function in embryonic stem cells. Mol Cell Biol 2001; 21(24):8512-20.
59. Navarro P, Pichard S, Ciaudo C et al. Tsix transcription across the Xist gene alters chromatin conformation without affecting Xist transcription: implications for X-chromosome inactivation. Genes Dev 2005; 19(12):1474-84.
60. Vigneau S, Augui S, Navarro P et al. An essential role for the DXPas34 tandem repeat and Tsix transcription in the counting process of X chromosome inactivation. Proc Natl Acad Sci USA 2006; 103(19):7390-5.
61. Ohhata T, Hoki Y, Sasaki H et al. Crucial role of antisense transcription across the Xist promoter in Tsix-mediated Xist chromatin modification. Development 2008; 135(2):227-35.
62. Cohen DE, Davidow LS, Erwin JA et al. The DXPas34 repeat regulates random and imprinted X inactivation. Dev Cell 2007; 12(1):57-71.
63. Debrand E, Chureau C, Arnaud D et al. Functional analysis of the DXPas34 locus, a 3' regulator of Xist expression. Mol Cell Biol 1999; 19(12):8513-25.
64. Prissette M, El-Maarri O, Arnaud D et al. Methylation profiles of DXPas34 during the onset of X-inactivation. Hum Mol Genet 2001; 10(1):31-8.
65. Boumil RM, Ogawa Y, Sun BK et al. Differential methylation of Xite and CTCF sites in Tsix mirrors the pattern of X-inactivation choice in mice. Mol Cell Biol 2006; 26(6):2109-17.

66. Ogawa Y, Sun BK, Lee JT. Intersection of the RNA interference and X-inactivation pathways. Science 2008; 320(5881):1336-41.
67. Nesterova TB, Popova BC, Cobb BS et al. Dicer regulates Xist promoter methylation in ES cells indirectly through transcriptional control of Dnmt3a. Epigenetics Chromatin 2008; 1(1):2.
68. Kanellopoulou C, Muljo SA, Dimitrov SD et al. X chromosome inactivation in the absence of Dicer. Proc Natl Acad Sci USA 2009; 106(4):1122-7.
69. Shibata S Lee JT. Tsix transcription- versus RNA-based mechanisms in Xist repression and epigenetic choice. Curr Biol 2004; 14(19):1747-54.
70. Ogawa Y, Lee JT. Xite, X-inactivation intergenic transcription elements that regulate the probability of choice. Mol Cell 2003; 11(3):731-43.
71. Stavropoulos N, Rowntree RK, Lee JT. Identification of developmentally specific enhancers for Tsix in the regulation of X chromosome inactivation. Mol Cell Biol 2005; 25(7):2757-69.
72. Chao W, Huynh KD, Spencer RJ et al. CTCF, a candidate trans-acting factor for X-inactivation choice. Science 2002; 295(5553):345-7.
73. Donohoe ME, Zhang LF, Xu N et al. Identification of a Ctcf cofactor, Yy1, for the X chromosome binary switch. Mol Cell 2007; 25(1):43-56.
74. Donohoe ME, Silva SS, Pinter SF et al. The pluripotency factor Oct4 interacts with Ctcf and also controls X-chromosome pairing and counting. Nature 2009; 460(7251):128-32.
75. Chambers I, Colby D, Robertson M et al. Functional expression cloning of Nanog, a pluripotency sustaining factor in embryonic stem cells. Cell 2003; 113(5):643-55.
76. Mitsui K, Tokuzawa Y, Itoh H et al. The homeoprotein Nanog is required for maintenance of pluripotency in mouse epiblast and ES cells. Cell 2003; 113(5):631-42.
77. Nichols J, Zevnik B, Anastassiadis K et al. Formation of pluripotent stem cells in the mammalian embryo depends on the POU transcription factor Oct4. Cell 1998; 95(3):379-91.
78. Avilion AA, Nicolis SK, Pevny LH et al. Multipotent cell lineages in early mouse development depend on SOX2 function. Genes Dev 2003; 17(1):126-40.
79. Navarro P, Chambers I, Karwacki-Neisius V et al. Molecular coupling of Xist regulation and pluripotency. Science 2008; 321(5896):1693-5.
80. Silva J, Nichols J, Theunissen TW et al. Nanog is the gateway to the pluripotent ground state. Cell 2009; 138(4):722-37.
81. Jonkers I, Barakat TS, Achame EM et al. RNF12 is an X-Encoded dose-dependent activator of X chromosome inactivation. Cell 2009; 139(5):999-1011.
82. Bach I, Rodriguez-Esteban C, Carrière C et al. RLIM inhibits functional activity of LIM homeodomain transcription factors via recruitment of the histone deacetylase complex. Nat Genet 1999; 22(4):394-9.
83. Her YR. Chung IK. Ubiquitin Ligase RLIM Modulates Telomere Length Homeostasis through a Proteolysis of TRF1. J Biol Chem 2009; 284(13):8557-66.
84. JJohnsen SA, Güngör C, Prenzel T et al. Regulation of estrogen-dependent transcription by the LIM cofactors CLIM and RLIM in breast cancer. Cancer Res 2009; 69(1):128-36.
85. Jacobs Pa, Baikie Ag, Brown Wm et al. Evidence for the existence of the human "super female". Lancet 1959; 2(7100):423-5.
86. Maclean N, Mitchell JM. A survey of sex-chromosome abnormalities among 4514 mental defectives. Lancet 1962; 1(7224):293-6.
87. Grumbach MM, Morishima A, Taylor JH. Human Sex Chromosome Abnormalities in Relation to DNA Replication and Heterochromatinization. Proc Natl Acad Sci USA 1963; 49(5):581-9.
88. Carr DH. Chromosome studies in selected spontaneous abortions. 1. Conception after oral contraceptives. Can Med Assoc J 1970; 103(4):343-8.
89. Webb S, de Vries TJ, Kaufman MH. The differential staining pattern of the X chromosome in the embryonic and extraembryonic tissues of postimplantation homozygous tetraploid mouse embryos. Genet Res 1992; 59(3):205-14.
90. Harnden DG. Nuclear sex in triploid XXY human cells. Lancet 1961; 2(7200):488.
91. Nicodemi M, Prisco A. Symmetry-breaking model for X-chromosome inactivation. Phys Rev Lett 2007; 98(10):108104.
92. Nicodemi M, Prisco A. Self-assembly and DNA binding of the blocking factor in x chromosome inactivation. PLoS Comput Biol 2007; 3(11):e210.
93. Lee JT, Strauss WM, Dausman JA et al. A 450 kb transgene displays properties of the mammalian X-inactivation center. Cell 1996; 86(1):83-94.
94. Herzing LB, Romer JT, Horn JM et al. Xist has properties of the X-chromosome inactivation centre. Nature 1997; 386(6622):272-5.
95. Lee JT, Jaenisch R. Long-range cis effects of ectopic X-inactivation centres on a mouse autosome. Nature 1997; 386(6622):275-9.
96. Lee JT, Lu N, Han Y. Genetic analysis of the mouse X inactivation center defines an 80-kb multifunction domain. Proc Natl Acad Sci USA 1999; 96(7):3836-41.
97. Migeon BR, Kazi E, Haisley-Royster C et al. Human X inactivation center induces random X chromosome inactivation in male transgenic mice. Genomics 1999; 59(2):113-21.

98. Migeon BR, Winter H, Kazi E et al. Low-copy-number human transgene is recognized as an X inactivation center in mouse ES cells, but fails to induce cis-inactivation in chimeric mice. Genomics 2001; 71(2):156-62.
99. Heard E, Kress C, Mongelard F et al. Transgenic mice carrying an Xist-containing YAC. Hum Mol Genet 1996; 5(4):441-50.
100. Matsuura S, Episkopou V, Hamvas R et al. Xist expression from an Xist YAC transgene carried on the mouse Y chromosome. Hum Mol Genet 1996; 5(4):451-9.
101. Heard E, Mongelard F, Arnaud D et al. Xist yeast artificial chromosome transgenes function as X-inactivation centers only in multicopy arrays and not as single copies. Mol Cell Biol 1999; 19(4):3156-66.
102. Gribnau J et al. X chromosome choice occurs independently of asynchronous replication timing. J Cell Biol 2005; 168(3):365-73.
103. Marahrens Y, Loring J, Jaenisch R. Role of the Xist gene in X chromosome choosing. Cell 1998; 92(5):657-64.
104. Nesterova TB, Johnston CM, Appanah R et al. Skewing X chromosome choice by modulating sense transcription across the Xist locus. Genes Dev 2003; 17(17):2177-90.
105. Sado T, Li E, Sasaki H. Effect of TSIX disruption on XIST expression in male ES cells. Cytogenet Genome Res 2002; 99(1-4):115-8.
106. Monkhorst K, Jonkers I, Rentmeester E et al. X inactivation counting and choice is a stochastic process: evidence for involvement of an X-linked activator. Cell 2008; 132(3):410-21.
107. Lee JT. Homozygous Tsix mutant mice reveal a sex-ratio distortion and revert to random X-inactivation. Nat Genet 2002; 32(1):195-200.
108. Mlynarczyk-Evans S, Royce-Tolland M, Alexander MK et al. X chromosomes alternate between two states prior to random X-inactivation. PLoS Biol 2006; 4(6):e159.
109. Bacher CP, Guggiari M, Brors B et al. Transient colocalization of X-inactivation centres accompanies the initiation of X inactivation. Nat Cell Biol 2006; 8(3):293-9.
110. Xu N, Tsai CL, Lee JT. Transient homologous chromosome pairing marks the onset of X inactivation. Science 2006; 311(5764):1149-52.
111. Xu N, Donohoe ME, Silva SS et al. Evidence that homologous X-chromosome pairing requires transcription and Ctcf protein. Nat Genet 2007; 39(11):1390-6.
112. Augui S, Filion GJ, Huart S et al. Sensing X chromosome pairs before X inactivation via a novel X-pairing region of the Xic. Science 2007; 318(5856):1632-6.
113. Takagi N. Variable X chromosome inactivation patterns in near-tetraploid murine EC x somatic cell hybrid cells differentiated in vitro. Genetica 1993; 88(2-3):107-17.
114. Monkhorst K, de Hoon B, Jonkers I et al. The probability to initiate X chromosome inactivation is determined by the X to autosomal ratio and X chromosome specific allelic properties. PLoS One 2009; 4(5):e5616.
115. Barr ML, Bertram EG. A morphological distinction between neurones of the male and female and the behaviour of the nucleolar satellite during accelerated nucleoprotein synthesis. Nature 1949; 163(4148):676.
116. Ohno S, Hauschka TS. Allocycly of the X-chromosome in tumors and normal tissues. Cancer Res 1960; 20:541-5.
117. Wutz A, Rasmussen TP, Jaenisch R. Chromosomal silencing and localization are mediated by different domains of Xist RNA. Nat Genet 2002; 30(2):167-74.
118. Chaumeil J, Le Baccon P, Wutz A et al. A novel role for Xist RNA in the formation of a repressive nuclear compartment into which genes are recruited when silenced. Genes Dev 2006; 20(16):2223-37.
119. Clemson CM, Hall LL, Byron M et al. The X chromosome is organized into a gene-rich outer rim and an internal core containing silenced nongenic sequences. Proc Natl Acad Sci USA 2006; 103(20):7688-93.
120. Goto Y, Gomez M, Brockdorff N et al. Differential patterns of histone methylation and acetylation distinguish active and repressed alleles at X-linked genes. Cytogenet Genome Res 2002; 99(1-4):66-74.
121. Heard E, Rougeulle C, Arnaud D et al. Methylation of histone H3 at Lys-9 is an early mark on the X chromosome during X inactivation. Cell 2001; 107(6):727-38.
122. Mak W, Baxter J, Silva J et al. Mitotically stable association of polycomb group proteins eed and enx1 with the inactive x chromosome in trophoblast stem cells. Curr Biol 2002; 12(12):1016-20.
123. Boggs BA, Cheung P, Heard E et al. Differentially methylated forms of histone H3 show unique association patterns with inactive human X chromosomes. Nat Genet 2002; 30(1):73-6.
124. Mermoud JE, Popova B, Peters AH et al. Histone H3 lysine 9 methylation occurs rapidly at the onset of random X chromosome inactivation. Curr Biol 2002; 12(3):247-51.
125. Peters AH, Mermoud JE, O'Carroll D et al. Histone H3 lysine 9 methylation is an epigenetic imprint of facultative heterochromatin. Nat Genet 2002; 30(1):77-80.
126. Rougeulle C, Chaumeil J, Sarma K et al. Differential histone H3 Lys-9 and Lys-27 methylation profiles on the X chromosome. Mol Cell Biol 2004; 24(12):5475-84.
127. Kohlmaier A, Savarese F, Lachner M et al. A chromosomal memory triggered by Xist regulates histone methylation in X inactivation. PLoS Biol 2004; 2(7):E171.
128. Costanzi C, Pehrson JR. Histone macroH2A1 is concentrated in the inactive X chromosome of female mammals. Nature 1998; 393(6685):599-601.
129. Lock LF, Takagi N, Martin GR. Methylation of the Hprt gene on the inactive X occurs after chromosome inactivation. Cell 1987; 48(1):39-46.

130. Norris DP, Brockdorff N, Rastan S. Methylation status of CpG-rich islands on active and inactive mouse X chromosomes. Mamm Genome 1991; 1(2):78-83.
131. Mohandas T, Sparkes RS, Shapiro LJ. Reactivation of an inactive human X chromosome: evidence for X inactivation by DNA methylation. Science 1981; 211(4480):393-6.
132. Russell LB, Montgomery CS. Comparative studies on X-autosome translocations in the mouse. II. Inactivation of autosomal loci, segregation and mapping of autosomal breakpoints in five T (X;1) S. Genetics 1970; 64(2):281-312.
133. Duthie SM, Nesterova TB, Formstone EJ et al. Xist RNA exhibits a banded localization on the inactive X chromosome and is excluded from autosomal material in cis. Hum Mol Genet 1999; 8(2):195-204.
134. Keohane AM et al. H4 acetylation, XIST RNA and replication timing are coincident and define x; autosome boundaries in two abnormal X chromosomes. Hum Mol Genet 1999; 8(2):377-83.
135. Popova BC, Tada T, Takagi N et al. Attenuated spread of X-inactivation in an X; autosome translocation. Proc Natl Acad Sci USA 2006; 103(20):7706-11.
136. Gartler SM, Riggs AD. Mammalian X-chromosome inactivation. Annu Rev Genet 1983; 17:155-90.
137. Lyon MF. X-chromosome inactivation: a repeat hypothesis. Cytogenet Cell Genet 1998; 80(1-4):133-7.
138. Lyon MF. LINE-1 elements and X chromosome inactivation: a function for "junk" DNA? Proc Natl Acad Sci USA 2000; 97(12):6248-9.
139. Lyon MF. Do LINEs Have a Role in X-Chromosome Inactivation? J Biomed Biotechnol 2006; 2006(1):59746.
140. Lyon MF. No longer 'all-or-none'. Eur J Hum Genet 2005; 13(7):796-7.
141. Furano AV. The biological properties and evolutionary dynamics of mammalian LINE-1 retrotransposons. Prog Nucleic Acid Res Mol Biol 2000; 64:255-94.
142. Waters PD, Dobigny G, Pardini AT et al. LINE-1 distribution in Afrotheria and Xenarthra: implications for understanding the evolution of LINE-1 in eutherian genomes. Chromosoma 2004; 113(3):137-44.
143. Parish DA, Vise P, Wichman HA et al. Distribution of LINEs and other repetitive elements in the karyotype of the bat Carollia: implications for X-chromosome inactivation. Cytogenet Genome Res 2002; 96(1-4):191-7.
144. Wichman HA, Van den Bussche RA, Hamilton MJ et al. Transposable elements and the evolution of genome organization in mammals. Genetica 1992; 86(1-3):287-93.
145. Bailey JA, Carrel L, Chakravarti A et al. Molecular evidence for a relationship between LINE-1 elements and X chromosome inactivation: the Lyon repeat hypothesis. Proc Natl Acad Sci USA 2000; 97(12):6634-9.
146. Carrel L, Willard HF. X-inactivation profile reveals extensive variability in X-linked gene expression in females. Nature 2005; 434(7031):400-4.
147. Wang Z, Willard HF, Mukherjee S et al. Evidence of influence of genomic DNA sequence on human X chromosome inactivation. PLoS Comput Biol 2006; 2(9):e113.
148. Hall LL, Clemson CM, Byron M et al. Unbalanced X; autosome translocations provide evidence for sequence specificity in the association of XIST RNA with chromatin. Hum Mol Genet 2002; 11(25):3157-65.
149. Sharp AJ, Spotswood HT, Robinson DO et al. Molecular and cytogenetic analysis of the spreading of X inactivation in X; autosome translocations. Hum Mol Genet 2002; 11(25):3145-56.
150. Sharp A, Robinson DO, Jacobs P. Absence of correlation between late-replication and spreading of X inactivation in an X; autosome translocation. Hum Genet 2001; 109(3):295-302.
151. White WM, Willard HF, Van Dyke DL et al. The spreading of X inactivation into autosomal material of an x; autosome translocation: evidence for a difference between autosomal and X-chromosomal DNA. Am J Hum Genet 1998; 63(1):20-8.
152. Solari AJ, Rahn IM, Ferreyra ME et al. The behavior of sex chromosomes in two human X-autosome translocations: failure of extensive X-inactivation spreading. Biocell 2001; 25(2):155-66.
153. Dobigny G, Ozouf-Costaz C, Bonillo C et al. Viability of X-autosome translocations in mammals: an epigenomic hypothesis from a rodent case-study. Chromosoma 2004; 113(1):34-41.
154. Chureau C, Prissette M, Bourdet A et al. Comparative sequence analysis of the X-inactivation center region in mouse, human and bovine. Genome Res 2002; 12(6):894-908.
155. Cantrell MA, Carstens BC, Wichman HA. X chromosome inactivation and Xist evolution in a rodent lacking LINE-1 activity. PLoS One 2009; 4(7):e6252.
156. Ke X, Collins A. CpG islands in human X-inactivation. Ann Hum Genet 2003; 67(Pt 3):242-9.
157. Cao R, Zhang Y. SUZ12 is required for both the histone methyltransferase activity and the silencing function of the EED-EZH2 complex. Mol Cell 2004; 15(1):57-67.
158. de la Cruz CC, Fang J, Plath K et al. Developmental regulation of Suz 12 localization. Chromosoma 2005; 114(3):183-92.
159. Wang J, Mager J, Chen Y et al. Imprinted X inactivation maintained by a mouse Polycomb group gene. Nat Genet 2001; 28(4):371-5.
160. Kalantry S, Magnuson T. The Polycomb group protein EED is dispensable for the initiation of random X-chromosome inactivation. PLoS Genet 2006; 2(5):e66.
161. Schoeftner S, Sengupta AK, Kubicek S et al. Recruitment of PRC1 function at the initiation of X inactivation independent of PRC2 and silencing. EMBO J 2006; 25(13):3110-22.
162. de Napoles M, Mermoud JE, Wakao R et al. Polycomb group proteins Ring1A/B link ubiquitylation of histone H2A to heritable gene silencing and X inactivation. Dev Cell 2004; 7(5):663-76.

163. Fang J, Chen T, Chadwick B et al. Ring1b-mediated H2A ubiquitination associates with inactive X chromosomes and is involved in initiation of X inactivation. J Biol Chem 2004; 279(51):52812-5.
164. Hernández-Muñoz I, Lund AH, van der Stoop P et al. Stable X chromosome inactivation involves the PRC1 Polycomb complex and requires histone MACROH2A1 and the CULLIN3/SPOP ubiquitin E3 ligase. Proc Natl Acad Sci USA 2005; 102(21):7635-40.
165. Nusinow DA, Hernández-Muñoz I, Fazzio TG et al. Poly(ADP-ribose) polymerase 1 is inhibited by a histone H2A variant, MacroH2A and contributes to silencing of the inactive X chromosome. J Biol Chem 2007; 282(17):12851-9.
166. Kim MY, Zhang T, Kraus WL. Poly(ADP-ribosyl)ation by PARP-1: 'PAR-laying' NAD+ into a nuclear signal. Genes Dev 2005; 19(17):1951-67.
167. Rouleau M, Aubin RA, Poirier GG. Poly(ADP-ribosyl)ated chromatin domains: access granted. J Cell Sci 2004; 117(Pt 6):815-25.
168. Hassa PO, Haenni SS, Elser M et al. Nuclear ADP-ribosylation reactions in mammalian cells: where are we today and where are we going? Microbiol Mol Biol Rev 2006; 70(3):789-829.
169. Kim MY, Mauro S, Gévry N et al. NAD+-dependent modulation of chromatin structure and transcription by nucleosome binding properties of PARP-1. Cell 2004; 119(6):803-14.
170. Helbig R, Fackelmayer FO. Scaffold attachment factor A (SAF-A) is concentrated in inactive X chromosome territories through its RGG domain. Chromosoma 2003; 112(4):173-82.
171. Ganesan S, Silver DP, Greenberg RA et al. BRCA1 supports XIST RNA concentration on the inactive X chromosome. Cell 2002; 111(3):393-405.
172. Silver DP, Dimitrov SD, Feunteun J et al. Further evidence for BRCA1 communication with the inactive X chromosome. Cell 2007; 128(5):991-1002.
173. Vincent-Salomon A, Ganem-Elbaz C, Manié E et al. X inactive-specific transcript RNA coating and genetic instability of the X chromosome in BRCA1 breast tumors. Cancer Res 2007; 67(11):5134-40.
174. Xiao C, Sharp JA, Kawahara M, Davalos AR et al. The XIST noncoding RNA functions independently of BRCA1 in X inactivation. Cell 2007; 128(5):977-89.
175. Blewitt ME, Gendrel AV, Pang Z et al. SmcHD1, containing a structural-maintenance-of-chromosomes hinge domain, has a critical role in X inactivation. Nat Genet 2008; 40(5):663-9.
176. Garrick D, Sharpe JA, Arkell R et al. Loss of Atrx affects trophoblast development and the pattern of X-inactivation in extraembryonic tissues. PLoS Genet 2006; 2(4):e58.
177. Baumann C, De La Fuente R. ATRX marks the inactive X chromosome (Xi) in somatic cells and during imprinted X chromosome inactivation in trophoblast stem cells. Chromosoma 2009; 118(2):209-22.
178. Han HJ, Russo J, Kohwi Y et al. SATB1 reprogrammes gene expression to promote breast tumour growth and metastasis. Nature 2008; 452(7184):187-93.
179. Agrelo R, Souabni A, Novatchkova M et al. SATB1 defines the developmental context for gene silencing by Xist in lymphoma and embryonic cells. Dev Cell 2009; 16(4):507-16.
180. Disteche CM. Escape from X inactivation in human and mouse. Trends Genet 1995; 11(1):17-22.
181. Disteche CM. Escapees on the X chromosome. Proc Natl Acad Sci USA 1999; 96(25):14180-2.
182. Disteche CM, Filippova GN, Tsuchiya KD. Escape from X inactivation. Cytogenet Genome Res 2002; 99(1-4):36-43.
183. Brown CJ, Greally JM. A stain upon the silence: genes escaping X inactivation. Trends Genet 2003; 19(8):432-8.
184. Looijenga LH, Gillis AJ, van Gurp RJ et al. X inactivation in human testicular tumors. XIST expression and androgen receptor methylation status. Am J Pathol 1997; 151(2):581-90.
185. Chow JC, Hall LL, Clemson CM et al. Characterization of expression at the human XIST locus in somatic, embryonal carcinoma and transgenic cell lines. Genomics 2003; 82(3):309-22.
186. Migeon BR, Chowdhury AK, Dunston JA et al. Identification of TSIX, encoding an RNA antisense to human XIST, reveals differences from its murine counterpart: implications for X inactivation. Am J Hum Genet 2001; 69(5):951-60.
187. Chow JC, Hall LL, Lawrence JB et al. Ectopic XIST transcripts in human somatic cells show variable expression and localization. Cytogenet Genome Res 2002; 99(1-4):92-8.
188. Hall LL, Byron M, Sakai K et al. An ectopic human XIST gene can induce chromosome inactivation in postdifferentiation human HT-1080 cells. Proc Natl Acad Sci USA 2002; 99(13):8677-82.
189. Hall LL, Byron M, Butler J et al. X-inactivation reveals epigenetic anomalies in most hESC but identifies sublines that initiate as expected. J Cell Physiol 2008; 216(2):445-52.
190. Shen Y, Matsuno Y, Fouse SD et al. X-inactivation in female human embryonic stem cells is in a nonrandom pattern and prone to epigenetic alterations. Proc Natl Acad Sci USA 2008; 105(12):4709-14.
191. Silva SS, Rowntree RK, Mekhoubad S et al. X-chromosome inactivation and epigenetic fluidity in human embryonic stem cells. Proc Natl Acad Sci USA 2008; 105(12):4820-5.
192. Enver T, Soneji S, Joshi C et al. Cellular differentiation hierarchies in normal and culture-adapted human embryonic stem cells. Hum Mol Genet 2005; 14(21):3129-40.
193. Hoffman LM, Hall L, Batten JL et al. X-inactivation status varies in human embryonic stem cell lines. Stem Cells 2005; 23(10):1468-78.

194. Dhara SK, Benvenisty N. Gene trap as a tool for genome annotation and analysis of X chromosome inactivation in human embryonic stem cells. Nucleic Acids Res 2004; 32(13):3995-4002.
195. Takagi N. Imprinted X-chromosome inactivation: enlightenment from embryos in vivo. Semin Cell Dev Biol 2003; 14(6):319-29.
196. West JD, Papaioannou VE, Frels WI et al. Preferential expression of the maternally derived X chromosome in the mouse yolk sac. Cell 1977; 12(4):873-82.
197. van den Berg IM, Laven JS, Stevens M et al. X chromosome inactivation is initiated in human preimplantation embryos. Am J Hum Genet 2009; 84(6):771-9.
198. Looijenga LH, Gillis AJ, Verkerk AJ et al. Heterogeneous X inactivation in trophoblastic cells of human full-term female placentas. Am J Hum Genet 1999; 64(5):1445-52.
199. Ropers HH, Wolff G, Hitzeroth HW. Preferential X inactivation in human placenta membranes: is the paternal X inactive in early embryonic development of female mammals? Hum Genet 1978; 43(3):265-73.
200. Daniels R, Zuccotti M, Kinis T et al. XIST expression in human oocytes and preimplantation embryos. Am J Hum Genet 1997; 61(1):33-9.
201. Ray PF, Winston RM, Handyside AH. XIST expression from the maternal X chromosome in human male preimplantation embryos at the blastocyst stage. Hum Mol Genet 1997; 6(8):1323-7.
202. Bamforth F, Machin G, Innes M. X-chromosome inactivation is mostly random in placental tissues of female monozygotic twins and triplets. Am J Med Genet 1996; 61(3):209-15.
203. Migeon BR, Wolf SF, Axelman J et al. Incomplete X chromosome dosage compensation in chorionic villi of human placenta. Proc Natl Acad Sci USA 1985; 82(10):3390-4.
204. Migeon BR, Axelman J, Jeppesen P. Differential X reactivation in human placental cells: implications for reversal of X inactivation. Am J Hum Genet 2005; 77(3):355-64.
205. Goto T, Wright E, Monk M. Paternal X-chromosome inactivation in human trophoblastic cells. Mol Hum Reprod 1997; 3(1):77-80.
206. Harrison KB. X-chromosome inactivation in the human cytotrophoblast. Cytogenet Cell Genet 1989; 52(1-2):37-41.
207. Harrison KB, Warburton D. Preferential X-chromosome activity in human female placental tissues. Cytogenet Cell Genet 1986; 41(3):163-8.
208. Liu W, Sun X. Skewed X chromosome inactivation in diploid and triploid female human embryonic stem cells. Hum Reprod 2009; 24(8):1834-43.
209. Sato N, Sanjuan IM, Heke M et al. Molecular signature of human embryonic stem cells and its comparison with the mouse. Dev Biol 2003; 260(2):404-13.
210. Ginis I, Luo Y, Miura T et al. Differences between human and mouse embryonic stem cells. Dev Biol 2004; 269(2):360-80.
211. Xu RH, Peck RM, Li DS et al. Basic FGF and suppression of BMP signaling sustain undifferentiated proliferation of human ES cells. Nat Methods 2005; 2(3):185-90.
212. Ying QL, Nichols J, Chambers I et al. BMP induction of Id proteins suppresses differentiation and sustains embryonic stem cell self-renewal in collaboration with STAT3. Cell 2003; 115(3):281-92.
213. Boyer LA, Lee TI, Cole MF et al. Core transcriptional regulatory circuitry in human embryonic stem cells. Cell 2005; 122(6):947-56.
214. Brons IG, Smithers LE, Trotter MW et al. Derivation of pluripotent epiblast stem cells from mammalian embryos. Nature 2007; 448(7150):191-5.
215. Tesar PJ, Chenoweth JG, Brook FA et al. New cell lines from mouse epiblast share defining features with human embryonic stem cells. Nature 2007; 448(7150):196-9.
216. Rossant J. Stem cells and early lineage development. Cell 2008; 132(4):527-31.
217. Lovell-Badge R. Many ways to pluripotency. Nat Biotechnol 2007; 25(10):1114-6.
218. Xu RH, Chen X, Li DS et al. BMP4 initiates human embryonic stem cell differentiation to trophoblast. Nat Biotechnol 2002; 20(12):1261-4.
219. Guo G, Yang J, Nichols J et al. Klf4 reverts developmentally programmed restriction of ground state pluripotency. Development 2009; 136(7):1063-9.
220. Bao S, Tang F, Li X et al. Epigenetic reversion of post-implantation epiblast to pluripotent embryonic stem cells. Nature 2009; 461(7268):1292-5.
221. Hanna J, Markoulaki S, Mitalipova M et al. Metastable pluripotent states in NOD-mouse-derived ESCs. Cell Stem Cell 2009; 4(6):513-24.
222. Nichols J, Smith A. Naive and primed pluripotent states. Cell Stem Cell 2009; 4(6):487-92.
223. Zvetkova I, Apedaile A, Ramsahoye B et al. Global hypomethylation of the genome in XX embryonic stem cells. Nat Genet 2005; 37(11):1274-9.
224. Lagarkova MA, Volchkov PY, Lyakisheva AV et al. Diverse epigenetic profile of novel human embryonic stem cell lines. Cell Cycle 2006; 5(4):416-20.
225. Rugg-Gunn PJ, Ferguson-Smith AC, Pedersen RA. Epigenetic status of human embryonic stem cells. Nat Genet 2005; 37(6):585-7.
226. Allegrucci C, Wu YZ, Thurston A et al. Restriction landmark genome scanning identifies culture-induced DNA methylation instability in the human embryonic stem cell epigenome. Hum Mol Genet 2007; 16(10):1253-68.

第11章 成体干细胞及其细胞龛

Francesca Ferraro, Cristina Lo Celso, David Scadden*

摘要： 干细胞参与决定发育事件结局和机体应激的动态生理过程。这些细胞是组织维持和修复的基础，因此其所接受的信号在机体的整体性方面具有重要的作用。过去，多数研究集中于干细胞的鉴定及与其调控相关的分子通路。但是，我们对于这些通路对干细胞生理功能应答的影响知之甚少。本章将回顾在干细胞龛功能异常、肿瘤发生和衰老条件下，我们对干细胞的了解。

"龛"的概念、定义与历史

体内干细胞龛指干细胞定居并接受决定其命运刺激的微环境。因此，不应简单认为"龛"是干细胞所定居的物理位置，这里是与外源性信号发生相互作用的场所，并且最终将影响干细胞的行为。这些刺激因素包括细胞与细胞间或细胞与细胞基质间相互作用，以及活化和/或抑制基因和转录程序的信号（分子）。这一相互作用的直接结果就是干细胞维持休眠状态、诱发自我更新或转变为分化状态。

1978年，Schoefield首先提出干细胞特定的微环境假说[1]。他认为，干细胞龛具有明确的解剖部位，并且将干细胞移出细胞龛后将导致细胞分化。首先在非脊椎动物模型黑腹果蝇和秀丽隐杆线虫的生殖腺证明细胞龛成分及其特征[2,3]。研究这些特征较为简单的动物，对于进一步了解更为复杂的哺乳动物细胞龛结构具有重要的作用。结果表明，细胞龛微环境中基本解剖成分和分子通路在各物种间具有高度保守性，虽然这些物质的相应作用在各细胞龛中完全不同。因此，人们认为可能会发现功能相似的干细胞龛共有成分（图11-1）。

一般细胞龛模型包括定居干细胞和细胞龛中各类细胞（细胞龛细胞）的相互作用。但是，仅多种不同类型的异质性细胞是不足的，还应包括细胞外基质（或其他非细胞成分），这些成分亦可能对干细胞龛起决定性作用。需要说明的是，即使在没有干细胞的情况下（如通过照射治疗去除干细胞后），细胞龛微环境仍可保留其关键作用与特性，允许招募外源性干细胞并使其归巢于原有干细胞龛。

细胞龛中的保守成分包括以下几种。

（1）基质，支持细胞，包括细胞与细胞间黏附分子及分泌的可溶性因子，存在于干

---

*David Scadden—Harvard Stem Cell Institute, Massachusetts General Hospital, Department of Stem Cell and Regenerative Biology, Harvard University, 185 Cambridge Street, Boston, Massachusettes, USA. Email: dscadden@mgh.harvard.edu

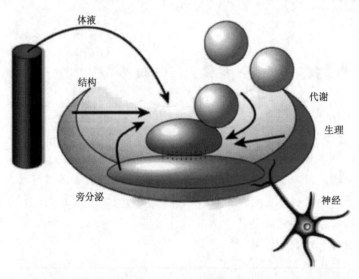

图 11-1 干细胞龛示例图：干细胞龛指体液、神经、局部（旁分泌）、体位（生理）及代谢之间相互作用调控干细胞的部位。（改编自 Scadden DT. Nature 441：1075-1079）

细胞附近。

（2）细胞外基质蛋白是干细胞的"锚定部位"，其组成机械支架单元用于细胞信号传递。

（3）血管，其可向干细胞龛输送营养并传递来自其他器官的全身信号，参与干细胞招募至细胞龛的过程。

（4）输入神经，其有助于干细胞向细胞龛外动员并整合来自不同器官的信号。研究表明，神经在造血干细胞的运输中具有重要的作用[4]。

根据细胞龛微环境对干细胞行为的重要作用，越来越多的研究开始探索细胞龛紊乱对干细胞功能障碍的影响，如常见的衰老或肿瘤转化[5~9]。

干细胞龛成分

在无脊柱动物模型果蝇的卵巢中，生殖干细胞（germinal stem cell，GSC）位于卵巢，其与冠细胞（cap cell）和终丝细胞（terminal filaments cell）存在物理相互接触。在不对称分裂的过程中，通过 E 选择素连接与冠细胞发生物理接触的 GSC 细胞保持干性，但是那些失去与冠细胞接触的细胞分化为成熟的卵泡细胞。在果蝇的睾丸中存在相似的极化信号系统，在这里，两套干细胞，即生殖干细胞（GSC）和成体干细胞（somatic stem cell，SSC）与睾丸中心细胞的顶端结构相关。与中心细胞相接触的子细胞启动分化程序分化为精原细胞和精囊细胞。在秀丽隐杆线虫中，225 个生殖细胞与远端细胞（DTC）相互作用，通过来自远端细胞的信号保持干性。

在许多哺乳动物组织，如造血系统、皮肤、小肠、脑和骨骼肌内亦发现了一些细龛（图11-2）。

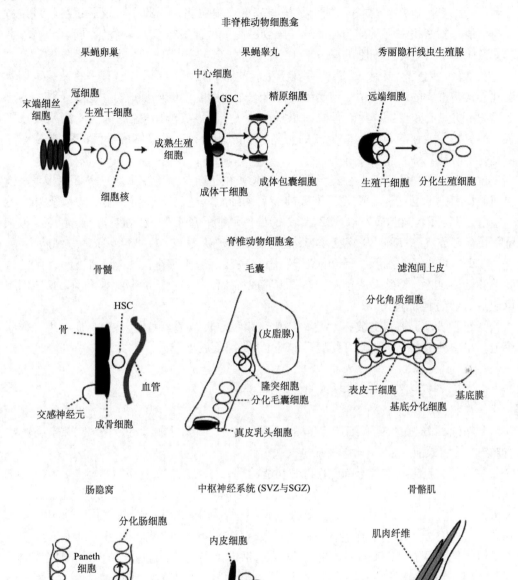

图11-2 干细胞龛。本图显示了脊椎和非脊椎动物的干细胞龛及其相应成分。GSC，生殖干细胞；HSC，造血干细胞；ISC，小肠干细胞；SVZ，前脑室管膜下层；SGZ，海马齿状回颗粒细胞下层；GFAP，胶质纤维酸性蛋白

在小梁骨骨髓中，造血干细胞及前体细胞（HSPC）沿骨内膜表面分布，邻近成骨细胞[10,11]并靠近血管[12,13]。"龛"的定义是调节干细胞功能，因此不可仅通过在物理位置上与干细胞紧密相邻判断其为龛成分。至今已有一些研究证实成骨细胞具有调节作用，但是目前尚不清楚内皮细胞的调节作用[10,11,14~16]。

在皮肤，内皮干细胞（ESC）存在于毛囊的膨出区域[17]。虽然已有一些证据表明真皮乳头具有重要的调节作用，但是目前尚未确定皮肤龛中的确切成分。这些干细胞对于毛囊的再生具有重要的作用，而黏附于基底膜（分割表皮与真皮的基底膜、基底角质细胞）的散在干细胞参与滤泡间上皮的更新[18]。皮脂腺由介于腺体底部间的细胞维持[19]，但是尚不清楚其细胞龛。

在成人中枢神经系统，神经干细胞（NSC）位于前脑室管膜下层（SVZ）和海马齿状回颗粒细胞下层（SGZ）[20~22]。该区神经干系细胞表达星形胶质细胞标志胶质纤维酸性蛋白（GFAP）。在 SVZ 和 SGZ 区域 GFAP 阳性的星形胶质细胞可以产生神经母细胞并最终形成成熟神经元。邻近 NSC 的内皮细胞被认为是中枢神经系统的龛细胞[22]。

在小肠，小肠干细胞（ISC）位于肠隐窝底部、潘氏细胞间[23]。隐窝周围区域富含肠神经元和血管。特殊基质细胞，亦称为肌成纤维细胞，位于隐窝细胞周围，这些细胞可能是 ISC 的龛细胞[24]。

在骨骼肌，干细胞也被称为卫星细胞，其沿肌纤维束黏附围绕在每一肌纤维束的质膜上。在这种情况下，基底纤维层可能是卫星细胞的龛[25,26]。

与龛功能相关的分子通路

龛细胞可分泌决定干细胞命运的分子，这些分子可能通过旁分泌或神经输出（神经-内分泌效应）发挥作用。

在不同细胞龛，维持干细胞稳态和功能的重要分子存在重复作用。但是，在特定细胞龛中其具有不同的作用。这些信号通路包括 Wnt/β-catenin、骨形成蛋白（BMP）、Notch、血管生成素 1（Ang1）及多种生长因子，如成纤维母细胞生长因子（FGF）、胰岛素生长因子（IGF）、血管内皮生长因子（VEGF）、转化成长因子 α（TGFα）及血小板衍化生长因子（PDGF）。在这些信号通路中，无论是在非脊椎动物还是哺乳动物，BMP 和 Wnt 信号通路在控制自我更新和分化方面具有高度的保守性。

根据组织不同，Wnt/β-catenin 信号可具有不同的作用。例如，在造血系统[27]和小肠[28,29]，Wnt/β-catenin 是干细胞自我更新和增殖的重要调节因素，但在皮肤，其促进毛囊前体细胞的分化[30~32]。在哺乳动物大脑，通过 Wnt 信号过量表达 β-catenin 可使神经干细胞群扩增[33]。

在黑腹果蝇中，BMP 信号通过抑制 bam 的表达调控 GCS 扩增[34]，该信号通路是包囊细胞分化的调节子。在果蝇睾丸中心细胞中亦存在 BMP 信号，并且其为 GSC 自我更新调控所必需的。在造血系统，BMP 在控制造血干细胞数量方面也起着重要的作用[11]；而在皮肤，BMP 信号与 Wnt 信号的作用相反，其抑制毛囊干细胞活化并反映了表皮细胞的最终结局[35,36]。在中枢神经系统，BMP 有益于 NSC 分化为星形细胞，但是 BMP 抑制剂（Noggin）促进神经的分化[37]。在这些系统中，Notch 信号通路对于维持干细胞未分化态是必需的，但是其刺激表皮前体细胞的分化。

除如上所述的分泌蛋白，其他分子，如离子、氧气和活性氧（ROS）也作用于干细胞并影响其行为。例如，在骨髓，较高的钙离子浓度位于邻近骨内膜表面、富含破骨细胞和成骨细胞的骨重建活跃区域。HSC 正常表达钙传感受体（CaR），这一受体缺失使 HSC 功能异常并导致骨移植受损[38]。进一步研究表明 HSC 趋向于根据氧浓度梯度定位，并且 ROS 可使不成熟 HSC 衰老[39]。这些成分小结见表 11-1。

表 11-1 已知干细胞的龛细胞与分子成分

组织	干细胞	支持细胞	信号通路	黏附
秀丽隐杆线虫性腺	GSC	远端细胞	Notch	NI
黑腹果蝇睾丸	GSC	中心细胞	JAK-STAT	DE-钙黏素 β-catenin
黑腹果蝇卵巢	GSC	冠细胞，ESC	DPP-BMP	DE-钙黏素 β-catenin
黑腹果蝇睾丸	CPC	中心细胞	JAK-STAT	DE-钙黏素 β-catenin
黑腹果蝇卵巢	ESC	NI	JAK-STAT	NI
黑腹果蝇卵巢	FSC	NI	Hedgehog	DE-钙黏素 β-catenin
小鼠骨骼肌	卫星细胞	NI	Notch	β-1 整合素
小鼠骨髓	HSC	成骨细胞，内皮细胞	Wnt, Notch, ANG1, OPN	β-1 整合素
小鼠小肠	CBC	隐窝成纤维细胞，Paneth 细胞	Wnt, BMP	β-catenin
小鼠皮肤	滤泡间细胞角化细胞	NI	Wnt, Shh, Notch	E-钙黏素 β-catenin
小鼠皮肤	隆突干细胞	表皮成纤维细胞	Wnt, BMP	β-catenin β-1 整合素
小鼠大脑（侧脑室）	SVZ 干细胞	血管细胞，星形细胞	Shh, BMP	N-钙黏素 β-catenin
大鼠大脑（海马回）	SVZ 干细胞	血管细胞，星形细胞	Shh, Wnt	N-钙黏素 β-catenin

注：ANG1，血管生成素 1；BMP，骨形态蛋白；CBC，隐窝基底柱状细胞；*C. elegans*，*Caenorabditis elegans*；CPC，隐窝前体细胞（成体干细胞）；DPP，同源序列；*D. melanogaster*，*Drosophia melanogaster*；ECM，细胞外基质；ESC，护卫干细胞；FSC，滤泡干细胞；GSC，生殖干细胞；HSC，造血干细胞；ISC，小肠干细胞；JAK，Janus 激酶；NI，未鉴定；OPN，骨桥蛋白；SGZ，亚颗粒区；Shh，音猬因子；STAT，细胞信号转导和转录活化蛋白；SVZ，脑室下区。

细胞外基质与细胞-细胞间相互作用

ECM 作为干细胞的支持系统，含有基质细胞和分子信号。其作用是滞留干细胞于特定部位、集中信号并且建立梯度指导干细胞进行自我更新与分化。这些作用的典型例子就是 β1 整合素，该分子表达于各种细胞，如造血干细胞[40]、皮肤[41~43]及骨骼肌干细胞，其调节干细胞与基质成分的黏附、调控干细胞的维持。但是，去除 β 整合素不引起骨髓基质细胞丢失[45]，这表明在 HSC 的定位中存在其他因子[46,47]。肌腱蛋白 C (tenascin-C)是另一种表达于脑组织[48]和骨髓[49]基质成分的蛋白质。在脑组织，该蛋白质可增加 NSC 对成纤维生长因子 2（FGF2）和骨形态蛋白 4（BMP4）的敏感性[48]。在骨髓，骨桥蛋白（OPN）是另一种基质糖蛋白，其与表达于 HSC 表面的黏附分子相互作用，如 CD44 及其他整合素[50,51]，从而有利于 HSC 滞留于细胞龛内。

近期研究表明，ECM 的多种机械特性（如刚性和弹性）通过影响家族谱系的定向分化来参与干细胞的分化[52]。这在组织损伤后可能尤为重要，在损伤部位形成的瘢痕可能对干细胞损伤修复能力具有不良影响。

早先对非脊椎动物模型的研究表明，通过钙黏素的细胞-细胞间相互作用和纺锤细胞定向可代表明确的系统，细胞可通过这一系统进行对称细胞分裂（自我更新）或非对称细胞分裂（定向/分化）[53,54]。研究表明，虽然尚有一些问题仍未解决，但哺乳动物细胞龛中亦存在这样的活动。例如，前期研究表明 N-钙黏素有益于造血干细胞和成骨细胞间相互作用[11]，并且 M-钙黏蛋白与骨骼肌干细胞和骨骼肌纤维的相互作用相关[55]。但是，在造血系统或骨骼肌中缺失 N-钙黏蛋白和 M-钙黏蛋白不会导致这些系统的干细胞出现明显的功能缺陷[56~58]。

干细胞龛动态性

干细胞龛收集和调节来自外周损伤应激后组织修复所必需的信号，因此干细胞的功能是多样且可变的。

在果蝇，卵巢龛细胞可以诱导成体干细胞进入细胞龛并替代丢失的种系干细胞[59]，并且在睾丸细胞龛中，如果细胞需要增加新细胞，其可诱导精原细胞去分化[60]。

在哺乳动物表皮，毛囊隆突部干细胞可以在损伤后向上迁移并再生成所有表皮成分（虽然这一过程仅为暂时性过程）；在稳态间，它们仅与毛囊维持相关[61]。在表皮基底层，人为诱导 Wnt 信号活化可导致新真皮乳头和异位毛囊形成，从而产生新的干细胞龛[62]。

在果蝇小肠和哺乳动物骨骼肌，研究表明，应对损伤时，不同细胞均可反馈诱导干细胞增殖[63,64]。

哺乳动物造血干细胞龛是细胞龛应对不同刺激动态变化的典型例子。例如，研究表明 HSC 动员因子 G-CSF 首先在成骨细胞中发挥作用，诱导其增殖。成骨细胞增殖后，HSC 增殖，并且成骨细胞增殖的数量与 HSC 释放出细胞龛和动员至外周血的细胞数相关[65]。而且，HSC 细胞龛在接受致死剂量的辐射后仍能存活，吸引新鲜移植 HSC 并使其再生，整个造血分化是成骨细胞、巨核细胞、内皮细胞及血管周围细胞分子相互作用的结果[15,66]。在 HSC 细胞龛再生过程中，趋化因子 CXCL12（亦称为 SDF1）具有重要的作用，其参与移植 HSC 的招募与滞留并调节新生血管产生、巨核细胞和成骨细胞前体细胞的存活[67~69]。

损伤修复检测方法的建立，以及大量转基因和基因敲除模型使我们可以了解干细胞龛的多样性并研究组织再生的分子机制。近来，由于体内成像技术的发展，可以直接观察移植造血干细胞到达骨髓并启动定植过程[70~73]。联合体内细胞示踪及精确的干细胞和谱系标志鉴定，是目前评价成人干细胞龛对组织损伤动态应答反应最有前景的方案[74]。并且，生理应力设置可以观察新细胞龛的形成。在造血过程中，异常细胞侵入骨髓（骨髓纤维化）或造血干细胞生成需求过高（如血红蛋白病）均可导致异位造血，如淋巴结、脾脏和肝脏。

干细胞龛衰老

衰老过程不但影响干细胞，亦对其微环境造成影响。在多种组织中，衰老的干细胞具有较高增殖能力，但在自我更新和产生子代方面能力较差（见果蝇小肠干细胞[75]及哺乳动物造血干细胞）[76]。根据对果蝇和小鼠生殖腺的研究表明，干细胞并不是由于自身原因衰老，而是由于衰老的细胞龛支持能力不足所致[77~80]。实际上，当将小鼠生殖干细胞无菌移植至幼龄小鼠的睾丸时，这些生殖细胞可以维持功能很多年[80]。当将年轻的造血干细胞移植至衰老的受体小鼠，这些干细胞至少在瞬时表现出衰老表型。而且，当将年轻的干细胞在体外与衰老小鼠的成骨细胞培养后，在移植测试时，其表现出衰老HSC特性[9]。

局部信号通路失调及应激诱导损伤增加，包括活性氧（ROS）是衰老过程中引起干细胞及细胞龛功能丧失的主要原因。衰老的细胞龛持续表达维持损伤应答的通路和机制，而年轻的细胞龛仅偶尔活化这些信号应对损伤。在这种意义下，细胞龛衰老可以认为是一种能够将灵活应对损伤转变为持续、低效应对损伤的转化过程，这最终耗尽干细胞的再生能力[75,83~85]。调节年轻干细胞与细胞龛相互作用的分子信号仍存在于衰老的细胞龛中，但在后者，这些调节信号失调，如在果蝇小肠的Notch通路[84]及哺乳动物骨骼肌细胞的Wnt信号通路[6]。

不仅局部因素，系统因素亦在干细胞和细胞龛老化中发挥作用。虽然很难确定这些因素，但是通过实验，首次利用这些因素使干细胞和细胞龛年轻化。利用联体共生实验（通过手术方法将两个动物连在一起，并且建立共用的循环系统），当老年小鼠接触年轻系统的因子时出现骨骼肌、造血干细胞和细胞龛的年轻化[6,9]。

恶性干细胞龛

与微环境对干细胞功能具有重要的调节作用相似，细胞龛与细胞龛的相互作用亦在肿瘤的发生中起重要作用。一百多年前，人们就提出微环境可能是肿瘤发生的一个因素[86]。很久之前，研究者即认识到局部血管对肿瘤的生长具有支持作用[87,88]，并且抗血管分子的研发一直是基础和转化研究（旨于制定有效的抗肿瘤治疗方案）的热门领域[89]。对基底细胞、肿瘤细胞和基质细胞的转录组学分析表明，肿瘤成纤维细胞表达高水平的Gremlin1，可对抗BMP2和BMP4，抑制表皮细胞分化并促进局部增殖[90]。而且在乳腺癌中，肿瘤相关成纤维细胞可以促进肿瘤进展[91]。免疫细胞在肿瘤进展的调控中具有双重调节作用。一方面，它们可以提供免疫监视，并且通过活化移除组织中的转化细胞；另一方面，慢性炎症使局部分子信号发生转换，转变为慢性应激状态，导致细胞损伤和提前衰老，所有这些均导致微环境发生改变，从而支持肿瘤生长[92,93]。

此外，干细胞亦被认为是某些恶性细胞的来源，很可能由于其微环境的缺陷促进其恶性表型发生。在造血干细胞中，很多例子表明细胞龛/微环境可诱导细胞恶化，某些转基因或基因敲除小鼠会出现骨髓增殖或骨髓增生异常综合征，这种疾病无法移植，但是当这类小鼠接受野生型小鼠骨髓移植时，这些症状仍会出现[94~96]。例如，在间充质细胞（成骨细胞前体细胞）的一个选择性亚群中去除 *Dicer1* 基因后出现骨髓增生异常综合征和明显的急性白血病。白血病存在继发遗传改变，但是具有正常的 Dicer1。仅在受体骨髓成骨前体细胞亦缺失 Dicer1 时，再次的白血病移植方会成功。因此，这一细胞龛可能是肿瘤中致瘤事件的始动因素，并可能是肿瘤维持所必需的[97]。

近期，将干细胞模型用于肿瘤的研究使人们质疑是否存在肿瘤干细胞并由特定的微环境调控，并且是否它们与正常干细胞相互竞争。问题的答案是复杂的，分析的肿瘤不同，其结果不同。根据 SDF1 的表达类型，人类 B 细胞白血病细胞移植入小鼠后定位于骨髓，因此，其表现与正常 HSC 相似[13]。但是，在慢性髓系白血病小鼠模型中，恶性、非正常 HSC 依赖 CD44 功能定位于骨髓并引起白血病[98]，这表明在微环境中存在白血病特异性相互作用机制。而且，进展期疾病开始不依赖于细胞龛的支持。即使白血病可不受微环境的影响而独立产生，某些基质细胞产生的特定细胞因子和其他分泌因子也会导致化疗干预抵抗[99,100]。细胞龛对正常细胞和恶性细胞具有不同的支持作用，这代表了一个新的有前景的研究方向。若这些细胞对细胞龛信号具有不同的敏感性，那么这一对细胞龛信号不同的敏感性可作为治疗新方案的研究方向。

通过损伤正常干细胞，肿瘤进展可以改变现有细胞龛。例如，在白血病进展期，骨髓中的白血病细胞可以抑制正常移植的 HSC 定位于细胞龛，并分泌 SCF 形成一个新的抑制性细胞龛[101]。骨髓瘤细胞通过分泌 Wnt 抑制剂 DKK1 破坏骨内膜 HSC 细胞龛[102]。

侵袭性原发肿瘤分泌的可溶性因子可作用于远端形成转移前细胞龛，细胞龛一旦形成转而招募肿瘤细胞并支持转移生长[103]。例如，骨髓瘤细胞产生的细胞因子可活化肝星形细胞，使其转变为与肌原纤维相似的细胞，并支持骨髓瘤细胞转移[104,105]。最后，正常细胞龛可以吸引与细胞龛中正常干细胞特征相似的恶性细胞。在骨髓中即是如此，像乳腺癌、前列腺癌和神经母细胞癌这类嗜骨肿瘤均基于 SDF1-CXCR4 信号轴的转移[103]。

结论

自从 Schofield 提出干细胞龛这一概念以来，已有多项实验证实干细胞龛是存在的并且极为复杂。干细胞内不同细胞间的相互交流为以这些细胞交流网络为靶点提供机会，并且调整正常干细胞的动力学可增强其应对损伤应答能力，并提升其与恶性细胞的竞争力。仍需进一步研究确定细胞龛是否适于作为一个干预点，这将为再生医学和抗肿瘤药物的研发提供新的机会。

（李 雪 韩忠朝 译）

参 考 文 献

1. Schofield R. The relationship between the spleen colony-forming cell and the haemopoietic stem cell. Blood Cells 1978; 4(1-2):7-25.
2. Xie T, Spradling AC. decapentaplegic is essential for the maintenance and division of germline stem cells in the Drosophila ovary. Cell 1998; 94(2):251-260.
3. Kimble JE, White JG. On the control of germ cell development in Caenorhabditis elegans. Dev Biol 1981; 81(2):208-219.
4. Katayama Y, Battista M, Kao WM et al. Signals from the sympathetic nervous system regulate hematopoietic stem cell egress from bone marrow. Cell 2006; 124(2):407-421.
5. Conboy IM, Conboy MJ, Wagers AJ et al. Rejuvenation of aged progenitor cells by exposure to a young systemic environment. Nature 2005; 433(7027):760-764.
6. Brack AS, Conboy MJ, Roy S et al. Increased Wnt signaling during aging alters muscle stem cell fate and increases fibrosis. Science 2007; 317(5839):807-810.
7. Zhu Y, Ghosh P, Charnay P et al. Neurofibromas in NF1: Schwann cell origin and role of tumor environment. Science 2002; 296(5569):920-922.
8. Trimboli AJ, Cantemir-Stone CZ, Li F et al. Pten in stromal fibroblasts suppresses mammary epithelial tumours. Nature 2009; 461(7267):1084-1091.
9. Mayack SR, Shadrach JL, Kim FS et al. Systemic signals regulate ageing and rejuvenation of blood stem cell niches. Nature 2010; 463(7280):495-500.
10. Calvi LM, Adams GB, Weibrecht KW et al. Osteoblastic cells regulate the haematopoietic stem cell niche. Nature 2003; 425(6960):841-846.
11. Zhang J, Niu C, Ye L et al. Identification of the haematopoietic stem cell niche and control of the niche size. Nature 2003; 425(6960):836-841.
12. Kiel MJ, Yilmaz OH, Iwashita T et al. SLAM family receptors distinguish hematopoietic stem and progenitor cells and reveal endothelial niches for stem cells. Cell 2005; 121(7):1109-1121.
13. Sipkins DA, Wei X, Wu JW et al. In vivo imaging of specialized bone marrow endothelial microdomains for tumour engraftment. Nature 2005; 435(7044):969-973.
14. Visnjic D, Kalajzic Z, Rowe DW et al. Hematopoiesis is severely altered in mice with an induced osteoblast deficiency. Blood 2004; 103(9):3258-3264.
15. Hooper AT, Butler JM, Nolan DJ et al. Engraftment and reconstitution of hematopoiesis is dependent on VEGFR2-mediated regeneration of sinusoidal endothelial cells. Cell Stem Cell 2009; 4(3):263-274.
16. Jung Y, Wang J, Song J et al. Annexin II expressed by osteoblasts and endothelial cells regulates stem cell adhesion, homing and engraftment following transplantation. Blood 2007; 110(1):82-90.
17. Cotsarelis G, Sun TT, Lavker RM. Label-retaining cells reside in the bulge area of pilosebaceous unit: implications for follicular stem cells, hair cycle and skin carcinogenesis. Cell 1990; 61(7):1329-1337.
18. Levy V, Lindon C, Harfe BD, et al. Distinct stem cell populations regenerate the follicle and interfollicular epidermis. Dev Cell 2005; 9(6):855-861.
19. Blanpain C, Fuchs E. Epidermal homeostasis: a balancing act of stem cells in the skin. Nat Rev Mol Cell Biol 2009; 10(3):207-217.
20. Doetsch F, Caille I, Lim DA, et al. Subventricular zone astrocytes are neural stem cells in the adult mammalian brain. Cell 1999; 97(6):703-716.
21. Palmer TD, Takahashi J, Gage FH. The adult rat hippocampus contains primordial neural stem cells. Mol Cell Neurosci 1997; 8(6):389-404.
22. Shen Q, Goderie SK, Jin L et al. Endothelial cells stimulate self-renewal and expand neurogenesis of neural stem cells. Science 2004; 304(5675):1338-1340.
23. Barker N, van Es JH, Kuipers J et al. Identification of stem cells in small intestine and colon by marker gene Lgr5. Nature 2007; 449(7165):1003-1007.
24. Mills JC, Gordon JI. The intestinal stem cell niche: there grows the neighborhood. Proc Natl Acad Sci USA 2001; 98(22):12334-12336.
25. Mauro A. Satellite cell of skeletal muscle fibers. J Biophys Biochem Cytol 1961; 9:493-495.
26. Kuang S, Kuroda K, Le Grand F et al. Asymmetric self-renewal and commitment of satellite stem cells in muscle. Cell 2007; 129(5):999-1010.
27. Reya T, Duncan AW, Ailles L et al. A role for Wnt signalling in self-renewal of haematopoietic stem cells. Nature 2003; 423(6938):409-414.
28. Brittan M, Wright NA. Gastrointestinal stem cells. J Pathol 2002; 197(4):492-509.
29. Brittingham J, Phiel C, Trzyna WC et al. Identification of distinct molecular phenotypes in cultured gastrointestinal smooth muscle cells. Gastroenterology 1998; 115(3):605-617.
30. Huelsken J, Vogel R, Erdmann B et al. beta-Catenin controls hair follicle morphogenesis and stem cell differentiation in the skin. Cell 2001; 105(4):533-545.
31. Merrill BJ, Gat U, DasGupta R et al. Tcf3 and Lef1 regulate lineage differentiation of multipotent stem cells in skin. Genes Dev 2001; 15(13):1688-1705.

32. Niemann C, Owens DM, Hulsken J et al. Expression of DeltaNLef1 in mouse epidermis results in differentiation of hair follicles into squamous epidermal cysts and formation of skin tumours. Development 2002; 129(1):95-109.
33. Chenn A, Walsh CA. Regulation of cerebral cortical size by control of cell cycle exit in neural precursors. Science 2002; 297(5580):365-369.
34. Chen D, McKearin D. Dpp signaling silences bam transcription directly to establish asymmetric divisions of germline stem cells. Curr Biol 2003; 13(20):1786-1791.
35. Botchkarev VA, Botchkareva NV, Nakamura M et al. Noggin is required for induction of the hair follicle growth phase in postnatal skin. FASEB J 2001; 15(12):2205-2214.
36. Kulessa H, Turk G, Hogan BL. Inhibition of Bmp signaling affects growth and differentiation in the anagen hair follicle. EMBO J 2000; 19(24):6664-6674.
37. Temple S. The development of neural stem cells. Nature 2001; 414(6859):112-117.
38. Adams GB, Chabner KT, Alley IR et al. Stem cell engraftment at the endosteal niche is specified by the calcium-sensing receptor. Nature 2006; 439(7076):599-603.
39. Ito K, Hirao A, Arai F et al. Regulation of oxidative stress by ATM is required for self-renewal of haematopoietic stem cells. Nature 2004; 431(7011):997-1002.
40. Wagers AJ, Allsopp RC, Weissman IL. Changes in integrin expression are associated with altered homing properties of Lin(-/lo)Thy1.1(lo)Sca-1(+)c-kit(+) hematopoietic stem cells following mobilization by cyclophosphamide/granulocyte colony-stimulating factor. Exp Hematol 2002; 30(2):176-185.
41. Brakebusch C, Grose R, Quondamatteo F et al. Skin and hair follicle integrity is crucially dependent on beta 1 integrin expression on keratinocytes. EMBO J 2000; 19(15):3990-4003.
42. Jones PH, Watt FM. Separation of human epidermal stem cells from transit amplifying cells on the basis of differences in integrin function and expression. Cell 1993; 73(4):713-724.
43. Jensen UB, Lowell S, Watt FM. The spatial relationship between stem cells and their progeny in the basal layer of human epidermis: a new view based on whole-mount labelling and lineage analysis. Development 1999; 126(11):2409-2418.
44. Sherwood RI, Christensen JL, Conboy IM et al. Isolation of adult mouse myogenic progenitors: functional heterogeneity of cells within and engrafting skeletal muscle. Cell 2004; 119(4):543-554.
45. Bungartz G, Stiller S, Bauer M et al. Adult murine hematopoiesis can proceed without beta1 and beta7 integrins. Blood 2006; 108(6):1857-1864.
46. Fleming WH, Alpern EJ, Uchida N et al. Steel factor influences the distribution and activity of murine hematopoietic stem cells in vivo. Proc Natl Acad Sci USA 1993; 90(8):3760-3764.
47. Sugiyama T, Kohara H, Noda M et al. Maintenance of the hematopoietic stem cell pool by CXCL12-CXCR4 chemokine signaling in bone marrow stromal cell niches. Immunity 2006; 25(6):977-988.
48. Garcion E, Halilagic A, Faissner A et al. Generation of an environmental niche for neural stem cell development by the extracellular matrix molecule tenascin C. Development 2004; 131(14):3423-3432.
49. Ohta M, Sakai T, Saga Y et al. Suppression of hematopoietic activity in tenascin-C-deficient mice. Blood 1998; 91(11):4074-4083.
50. Stier S, Ko Y, Forkert R et al. Osteopontin is a hematopoietic stem cell niche component that negatively regulates stem cell pool size. J Exp Med 2005; 201(11):1781-1791.
51. Nilsson SK, Johnston HM, Whitty GA et al. Osteopontin, a key component of the hematopoietic stem cell niche and regulator of primitive hematopoietic progenitor cells. Blood 2005; 106(4):1232-1239.
52. Engler AJ, Sen S, Sweeney HL et al. Matrix elasticity directs stem cell lineage specification. Cell 2006; 126(4):677-689.
53. Deng W, Lin H. Spectrosomes and fusomes anchor mitotic spindles during asymmetric germ cell divisions and facilitate the formation of a polarized microtubule array for oocyte specification in Drosophila. Dev Biol 1997; 189(1):79-94.
54. Yamashita YM, Jones DL, Fuller MT. Orientation of asymmetric stem cell division by the APC tumor suppressor and centrosome. Science 2003; 301(5639):1547-1550.
55. Irintchev A, Zeschnigk M, Starzinski-Powitz A et al. Expression pattern of M-cadherin in normal, denervated and regenerating mouse muscles. Dev Dyn 1994; 199(4):326-337.
56. Hollnagel A, Grund C, Franke WW et al. The cell adhesion molecule M-cadherin is not essential for muscle development and regeneration. Mol Cell Biol 2002; 22(13):4760-4770.
57. Hooper AT, Butler J, Petit I et al. Does N-cadherin regulate interaction of hematopoietic stem cells with their niches? Cell Stem Cell 2007; 1(2):127-129.
58. Kiel MJ, Radice GL, Morrison SJ. Lack of evidence that hematopoietic stem cells depend on N-cadherin-mediated adhesion to osteoblasts for their maintenance. Cell Stem Cell 2007; 1(2):204-217.
59. Kai T, Spradling A. An empty Drosophila stem cell niche reactivates the proliferation of ectopic cells. Proc Natl Acad Sci USA 2003; 100(8):4633-4638.
60. Brawley C, Matunis E. Regeneration of male germline stem cells by spermatogonial dedifferentiation in vivo. Science 2004; 304(5675):1331-1334.
61. Ito M, Liu Y, Yang Z et al. Stem cells in the hair follicle bulge contribute to wound repair but not to homeostasis of the epidermis. Nat Med 2005; 11(12):1351-1354.

62. Silva-Vargas V, Lo Celso C, Giangreco A et al. Beta-catenin and Hedgehog signal strength can specify number and location of hair follicles in adult epidermis without recruitment of bulge stem cells. Dev Cell 2005; 9(1):121-131.
63. Jiang H, Patel PH, Kohlmaier A et al. Cytokine/Jak/Stat signaling mediates regeneration and homeostasis in the Drosophila midgut. Cell 2009; 137(7):1343-1355.
64. Bischoff R. Interaction between satellite cells and skeletal muscle fibers. Development 1990; 109(4):943-952.
65. Mayack SR, Wagers AJ. Osteolineage niche cells initiate hematopoietic stem cell mobilization. Blood 2008; 112(3):519-531.
66. Dominici M, Rasini V, Bussolari R et al. Restoration and reversible expansion of the osteoblastic hematopoietic stem cell niche after marrow radioablation. Blood 2009; 114(11):2333-2343.
67. Peled A, Petit I, Kollet O et al. Dependence of human stem cell engraftment and repopulation of NOD/SCID mice on CXCR4. Science 1999; 283(5403):845-848.
68. Kortesidis A, Zannettino A, Isenmann S et al. Stromal-derived factor-1 promotes the growth, survival and development of human bone marrow stromal stem cells. Blood 2005; 105(10):3793-3801.
69. Jin DK, Shido K, Kopp HG et al. Cytokine-mediated deployment of SDF-1 induces revascularization through recruitment of CXCR4+ hemangiocytes. Nat Med 2006; 12(5):557-567.
70. Kohler A, Schmithorst V, Filippi MD et al. Altered cellular dynamics and endosteal location of aged early hematopoietic progenitor cells revealed by time-lapse intravital imaging in long bones. Blood 2009; 114(2):290-298.
71. Lewandowski D, Barroca V, Duconge F et al. In vivo cellular imaging pinpoints the role of reactive oxygen species in the early steps of adult hematopoietic reconstitution. Blood 2009.
72. Lo Celso C, Fleming HE, Wu JW et al. Live-animal tracking of individual haematopoietic stem/progenitor cells in their niche. Nature 2009; 457(7225):92-96.
73. Xie Y, Yin T, Wiegraebe W et al. Detection of functional haematopoietic stem cell niche using real-time imaging. Nature 2009; 457(7225):97-101.
74. Voog J, Jones DL. Stem Cells and the Niche: a dynamic duo. Cell Stem Cell, in press 2010; 6.
75. Biteau B, Hochmuth CE, Jasper H. JNK activity in somatic stem cells causes loss of tissue homeostasis in the aging Drosophila gut. Cell Stem Cell 2008; 3(4):442-455.
76. Janzen V, Forkert R, Fleming HE et al. Stem-cell ageing modified by the cyclin-dependent kinase inhibitor p16INK4a. Nature 2006; 443(7110):421-426.
77. Boyle M, Wong C, Rocha M et al. Decline in self-renewal factors contributes to aging of the stem cell niche in the Drosophila testis. Cell Stem Cell 2007; 1(4):470-478.
78. Hsu HJ, Drummond-Barbosa D. Insulin levels control female germline stem cell maintenance via the niche in Drosophila. Proc Natl Acad Sci USA. 2009; 106(4):1117-1121.
79. Pan L, Chen S, Weng C et al. Stem cell aging is controlled both intrinsically and extrinsically in the Drosophila ovary. Cell Stem Cell 2007; 1(4):458-469.
80. Ryu BY, Orwig KE, Oatley JM et al. Effects of aging and niche microenvironment on spermatogonial stem cell self-renewal. Stem Cells 2006; 24(6):1505-1511.
81. Liang Y, Van Zant G, Szilvassy SJ. Effects of aging on the homing and engraftment of murine hematopoietic stem and progenitor cells. Blood 2005; 106(4):1479-1487.
82. Rossi DJ, Bryder D, Zahn JM et al. Cell intrinsic alterations underlie hematopoietic stem cell aging. Proc Natl Acad Sci USA 2005; 102(26):9194-9199.
83. Ohlstein B, Spradling A. Multipotent Drosophila intestinal stem cells specify daughter cell fates by differential notch signaling. Science 2007; 315(5814):988-992.
84. Brack AS, Rando TA. Intrinsic changes and extrinsic influences of myogenic stem cell function during aging. Stem Cell Rev Fall 2007; 3(3):226-237.
85. Wagner W, Horn P, Bork S et al. Aging of hematopoietic stem cells is regulated by the stem cell niche. Exp Gerontol 2008; 43(11):974-980.
86. Paget S. The distribution of secondary growths in cancer of the breast. Lancet 1889; 1:571-573.
87. Folkman J. Tumor angiogenesis: a possible control point in tumor growth. Ann Intern Med 1975; 82(1):96-100.
88. Bergers G, Javaherian K, Lo KM et al. Effects of angiogenesis inhibitors on multistage carcinogenesis in mice. Science 1999; 284(5415):808-812.
89. Abdelrahim M, Konduri S, Basha R et al. Angiogenesis: an update and potential drug approaches (review). Int J Oncol 2010; 36(1):5-18.
90. Sneddon JB, Zhen HH, Montgomery K et al. Bone morphogenetic protein antagonist gremlin 1 is widely expressed by cancer-associated stromal cells and can promote tumor cell proliferation. Proc Natl Acad Sci USA 2006; 103(40):14842-14847.
91. Hu M, Yao J, Carroll DK et al. Regulation of in situ to invasive breast carcinoma transition. Cancer Cell 2008; 13(5):394-406.
92. Mantovani A, Romero P, Palucka AK et al. Tumour immunity: effector response to tumour and role of the microenvironment. Lancet 2008; 371(9614):771-783.
93. Raulet DH, Guerra N. Oncogenic stress sensed by the immune system: role of natural killer cell receptors. Nat Rev Immunol 2009; 9(8):568-580.

94. Walkley CR, Olsen GH, Dworkin S et al. A microenvironment-induced myeloproliferative syndrome caused by retinoic acid receptor gamma deficiency. Cell 2007; 129(6):1097-1110.
95. Walkley CR, Shea JM, Sims NA et al. Rb regulates interactions between hematopoietic stem cells and their bone marrow microenvironment. Cell 2007; 129(6):1081-1095.
96. Lane SW, Sykes SM, Shahrour F et al. The APCmin mouse has altered hematopoietic stem cell function and provides a model for cell extrinsic MPD/MDS. Blood, in press. 2010.
97. Raaijmakers MHGP, Mukherjee S, Guo S et al. Bone progenitor dysfunction induces myelodysplasia and secondary leukemia Nature, 2010.
98. Krause DS, Lazarides K, von Andrian UH et al. Requirement for CD44 in homing and engraftment of BCR-ABL-expressing leukemic stem cells. Nat Med 2006; 12(10):1175-1180.
99. Iwamoto S, Mihara K, Downing JR et al. Mesenchymal cells regulate the response of acute lymphoblastic leukemia cells to asparaginase. J Clin Invest 2007; 117(4):1049-1057.
100. Williams RT, den Besten W, Sherr CJ. Cytokine-dependent imatinib resistance in mouse BCR-ABL+, Arf-null lymphoblastic leukemia. Genes Dev 2007; 21(18):2283-2287.
101. Colmone A, Amorim M, Pontier AL et al. Leukemic cells create bone marrow niches that disrupt the behavior of normal hematopoietic progenitor cells. Science 2008; 322(5909):1861-1865.
102. Qiang YW, Chen Y, Stephens O et al. Myeloma-derived Dickkopf-1 disrupts Wnt-regulated osteoprotegerin and RANKL production by osteoblasts: a potential mechanism underlying osteolytic bone lesions in multiple myeloma. Blood 2008; 112(1):196-207.
103. Wels J, Kaplan RN, Rafii S et al. Migratory neighbors and distant invaders: tumor-associated niche cells. Genes Dev 2008; 22(5):559-574.
104. Olaso E, Salado C, Egilegor E et al. Proangiogenic role of tumor-activated hepatic stellate cells in experimental melanoma metastasis. Hepatology 2003; 37(3):674-685.
105. Olaso E, Santisteban A, Bidaurrazaga J et al. Tumor-dependent activation of rodent hepatic stellate cells during experimental melanoma metastasis. Hepatology 1997; 26(3):634-642.

第12章 成体干细胞的分化和迁移及其对疾病的影响

Ying Zhuge, Zhao-Jun Liu, Omaida C. Velazquez*

摘要： 干细胞属于非特异性祖细胞，主要定植于骨髓，对胚胎组织的生成起到重要作用。干细胞在成年期也发挥重要功能，可补充短寿命的效应细胞并修复受损的组织。干细胞有三个主要特性：自我更新、分化和体内平衡的调控。为保持干细胞库的数量，产生足够的造血细胞、基质成分和结缔组织，干细胞必须能不断补充自己的数量。干细胞还必须具有分化能力，能产生各种机能细胞。总之，干细胞能根据环境的刺激，以及体内各个器官的需求，调节和平衡自我更新及分化的能力，防止过量效应细胞的产生[1]。除了产生这些细胞，干细胞迁移的调节对器官形成、体内平衡和成体的修复都起到重要作用。干细胞需要从特定的环境进行特定输入以履行各项职能。某些类似的迁移机制在白细胞、成体干细胞、胎儿干细胞及肿瘤干细胞中都存在[1,2]。如果能让干细胞有目的地迁移，将增加干细胞靶向治疗和药物转运的效果[2]。此外，了解肿瘤干细胞归巢与迁移的相似和不同，将能进一步弄清肿瘤进展和转移的分子机制[2]。这一章我们将重点关注成体骨髓干细胞的主要类型——造血干细胞、间充质干细胞和内皮祖细胞的分化、迁移和归巢。本章中，术语"干细胞"是指"成体干细胞"，除非另有说明。

分化

造血干细胞

对于造血干细胞来说，定植于骨髓中造血干细胞的分化过程研究最多，并为大家所共识，它能逐步定向分化为循环中的成熟血细胞，包括红细胞、血小板、髓细胞和淋巴细胞（图12-1）。由于有功能的血细胞寿命短，所以在人的一生中要靠造血干细胞来补充循环中的造血成分。在受者进行适当预处理后，造血干细胞移植能够长期重建整个血液系统。

在整个脊椎动物进化中，造血过程得到很好的保存，在动物模型如小鼠和斑马鱼的研究中得到扩展，而对人的造血研究则更加广泛[3]。造血干细胞生成及定向分化主要由转录因子调节，几乎涵盖所有种类的DNA结合蛋白[4]。造血干细胞生成所必需的转录因子包括：混合系白血病基因（MLL）、Rut相关转录因子1（Runx1）、异位ets白血病/ets变异基因6（Tel/ETV6）、干细胞白血病/T细胞急性白血病1（SCL/tal1），以

---

* Corresponding Author: Omaida C. Velazquez—Division of Vascular and Endovascular Surgery, Room 3016 Holtz Center—JMH East Tower, 1611 N.W. 12th Avenue, Miami, Florida 33136, USA. Email: ovelazquez@med.miami.edu

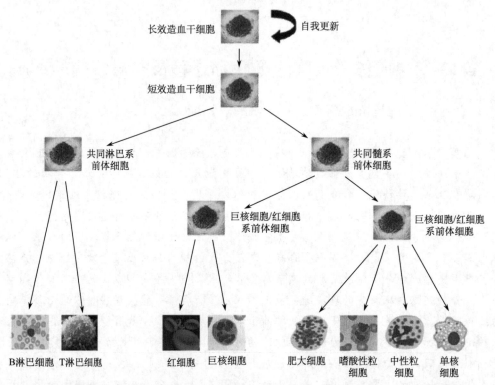

图 12-1 造血干细胞分化到各类短寿命成熟效应血细胞的能力（实心箭头）。造血干细胞具有自我更新能力（块状箭头）（改编自 Orkin SH, Zon LI. Hematopoiesis: an evolving paradigm for stem cell biology）

及 LIM 结构域 2（LMO2）[3]。基因编码的 SET 结构域含有组蛋白甲基转移酶 MLL 和 runt-domain Runx1 蛋白，两者对造血干细胞在不同部位的生成是必需的[4]。Basic-helix-loop-helix 因子 SCL/tal1 和相关蛋白 LMO2 对造血生成系统也是必需的[5]。缺乏 Runx1、SCL/tal1 或 LMO2 因子，血细胞无法生成。相反，增加 SCL/tal1 和 LMO2 的表达，能促使中胚层向造血系统转化。当造血干细胞的定向分化占主导地位时，细胞的存活和增殖可能不再继续需要这些转录因子。例如，在干细胞发育初期需要 SCL/tal1，如果在成体造血干细胞被灭活，并不影响祖细胞更新的维持[6]。同样，灭活成体 HSC 中的 Runx1，仅引起一些系的分化不平衡，但总体不影响造血干细胞的特性[7]。因此，大部分因子在干细胞发育过程只是短暂需要的，以达到对造血干细胞定向分化的调节。

除上文所提到的，还有一些转录因子参与了从多能祖细胞向定向血细胞系的分化（表 12-1）。锌指 GATA 结合因子 1（GATA-1）在巨核系和红系祖细胞中的含量很高，能产生血小板和红细胞前体细胞[8]。相反，CCAAT/促进结合蛋白 α（C/EBPα）存在于粒/髓系祖细胞中，而 PU.1 诱导髓系的发育[9]。最后，配对盒蛋白 5（Pax5）对 B 细胞的定向和分化是必需的[10]。除需要因子外，因子的浓度在发育特定时间对分化的影响也是非常重要的[4]。例如，GATA-1在低浓度时嗜酸性粒细胞生成；而在高浓度时，红系和巨核系开始发育[8]。相反，在GATA-1表达低于3倍时，导致红系祖细胞成熟障碍[11]。另外，高浓度的PU.1有利于巨核细胞的发育，而低浓度则有利于B细胞的增殖[4]。

表 12-1 造血干细胞分化的调节信号

造血干细胞类	
Mixed lineage leukemia gene (MLL)	
Runx1	
Tel/ETV6	
SCL/tal1	
LMO2	
造血干细胞分化转录因子	细胞系
GATA-1	红系、巨核细胞系
C/EBPα	髓系、粒系
PU.1	髓系
Pax5	B淋巴细胞系

虽然确定祖细胞分化子代与细胞系限制性转录因子之间——对应的关系比较简单，但对早期祖细胞的识别则是一个挑战，因为早期祖细胞可表达不同的系谱标记[12]。例如，多能祖细胞都表达GATA-1、FOG-1、Ikaros和PU.1，因而这些祖细胞有多向分化的潜能。一旦暂时达到分化条件，通过单系因子的自动上调及其他因子的拮抗，因子间产生的相互作用加强了随后分化途径的稳定[4]。例如，GATA-1和PU.1两者对它们的产物产生正反馈，分别促进红系和髓系细胞的生成[13,14]。除了在不同的细胞分化命运之间允许在分子水平上相互影响外，存在于共同祖细胞上的多系标记也证实细胞系分化启动的原理。这指的是干细胞选择向哪一系分化是一个消除选择可能性的过程，而不是凭空迫使按某一种支配途径进行[3]。这是通过早期造血干细胞保持染色质处于开放状态而完成的，伴随短暂的抑制及更持久的沉默，始终维持祖细胞的可塑性[3]。以前认为细胞的分化过程是单向的，现在有证据证明细胞可进行再编程[3]。例如，增加早期髓系祖细胞GATA-1的表达，能诱导向红系、嗜酸系和巨核系祖细胞的分化[8,15]。同样，通过增加C/EBPα的表达，能够再编程B/T淋巴细胞分化为巨嗜细胞[16,17]。细胞再编程是通过不同系谱标记的表达，这是一个循序渐进的过程[3]。

影响造血干细胞分化的主要限制性转录因子的另一个特点是：在中和其他因子的同时，能促进向同一途径分化的协同能力，提供一个分解和加强系选择的有效机制。例如，上调嗜酸系标记GATA-1，伴随着髓系标记的下调[8]。在缺乏Pax5的情况下，原定于分化成B细胞的祖细胞丧失分化特定系的能力，分化成不同的造血干细胞，如巨嗜细胞、破骨细胞和粒细胞。Pax5诱导B细胞分化的同时抑制其向他系分化[18]。这些造血生成调节因子的交叉调节也发生在蛋白质水平，例如，GATA-1和PU.1实际上是通过PU.1的氨基端和GATA-1的羧基指协同作用，阻断GATA-1识别DNA的能力[19]。同时，PU.1和GATA-1通过辅助因子的置换破坏PU.1依赖的转录[4,20]。除此之外，Pax5 对 B 细胞的发育是至关重要的，因为它抑制其他生长因子，这些因子能在无 Pax5 因子的情况下分化为 T 淋巴细胞、NK 淋巴细胞或树突状细胞、巨核细胞、粒细胞或红系的祖细胞[3]。最后，GATA-3 对从 CD4[+] T 细胞来源的 Th2 的产生是必需的，同时启动 Th1 细胞转化为 Th2 亚型[21]。反过来，转录因子 T-bet 逆转 Th2 为 Th1 细胞[22]。分化是一个持续不断的过程，需要主要调节因子的不断积极参与[4]。

几乎所有的造血转录因子直接与造血系统的肿瘤有关，如白血病。主要转录因子平衡稳定的破坏，已确定是白血病发病的一个原因[3]。这些基因突变大部分是染色体易位或分化中关键转录因子体细胞突变。在染色体易位中，不适当激活或抑制基因嵌合转录因子，都会引起不当的下游反应。例如，*Scl/tal-1*、*Lmo2*、*Tel*、*E2A* 和 *runx1* 都和染色体易位有关，结果生成融合蛋白，其机能以负显性的方式阻断定向分化因子的作用[4]。*GATA-1*、*PU.1* 和 *Ikaros* 的突变，引起转录因子表达缺陷和失调，失去了分化的控制。*GATA-1* 的突变与唐氏综合征相关的巨核细胞白血病有关[23]，*PU.1* 和 *C-EBP* 突变与髓性白血病有关[24,25]，*Pax5* 突变与 B 淋巴细胞白血病有关[26]。

间充质干细胞

MSC 是一组异质祖细胞群，具有分化成中胚层和非中胚层细胞系的潜能，包括成骨细胞、脂肪细胞、软骨细胞、肌细胞、心肌细胞、成纤维细胞、成纤维母细胞、内皮细胞和神经元（图 12-2）[27]。国际细胞治疗学会已制定判断这些细胞的三个指标：培养过程能贴壁；培养后 CD105、CD73 和 CD90 阳性率大于 95%，CD34、CD45、CD14，或 CD11b、CD79，或 CD19 和 HLD-DR 表达小于 5%；能诱导分化为成骨、软骨和脂肪细胞[28]。MSC 主要定居于骨髓，但也存在于脂肪组织、周围血、脐血、肝脏、骨膜、滑膜、滑膜液、骨骼肌、乳牙、周皮细胞、骨小梁、髌下脂体、关节软骨、胎盘、脾脏、胸腺和胎儿组织中[27]。当受到特异信号刺激时，它们会从骨髓窦释放到血循环，进入特异组织进行原位分化和组织修复[29~31]。

虽然 MSC 的组织来源不同，但具有相同的亚型特征，应用不同生长因子来诱导而表现出不同的分化潜能。具有调节作用的生长因子包括：超级家族转化生成因子 β（TGF-β）的成员、胰岛素样生长因子（IGF）、纤维母细胞生长因子（FGF）、内皮细胞生长因子（EGF）、血小板衍生生长因子（PDGF）、血管内皮生长因子（VEGF）和已知的 Wnt 生长因子家族（表 12-2）。最强的软骨诱导剂是 TGF-β 家族，包括 TGF-β1、TGF-β2 和 TGF-β3，以及骨形态生成蛋白（BMP）[32~37]。脂肪来源的 MSC 缺乏 TGF-β1 型受体的表达，与骨髓 MSC 相比，减少 BMP-2、BMP-4 和 BMP-6 的表达[27]。另外，发现典型和非典型的 Wnt 在调节干细胞增殖和成骨细胞分化过程中会相互串挠[32]。生长因子其他的调节信号通过特殊的传递途径控制下游的转录因子。当丝裂原活化蛋白激酶（MAPK）和 Smads 被激活，能诱导 Sox9、Sox5 和 Sox6 的转录因子，导致如 II 型胶原这样的细胞外基质蛋白、蛋白多糖和软骨寡聚基质蛋白的生成，最后生成软骨[27]。事实上，Sox9 是软骨表达最重要的分子，被当成软骨生成的主控开关[38]。

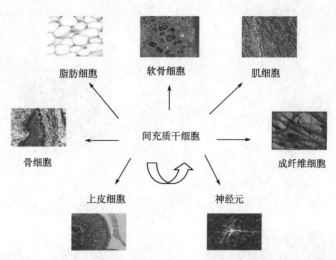

图 12-2 间充质干细胞分化到间质和非间质细胞的能力（实心箭头）。间充质干细胞还具有自我更新的能力（块状箭头）（已得到 Liu Z、Zhuge Y、Velazquez OC 的许可，转载于 Trafficking and differentiation of mensenchymal stem cells. J Cell Biochem 2009；106：984-91）

表 12-2 间充质干细胞分化调节信号

生物学信号	分化潜能
TGF-β	软骨
IGF-1	软骨
bFGF	软骨、成骨、神经
EGF	软骨
VEGF	内皮
Wnt	软骨、成骨、神经

注：改编自 Liu Z, Zhuge Y, Velazquez OC. Trafficking and differentiation of mensenchymal stem cells. J Cell Biochem 2009；106：984-91.

迁移

干细胞迁移是指细胞受特异刺激后定向和直接的运动。迁移有两种类型：归巢和间质间迁移。前者指的是干细胞/祖细胞通过血流播散到全身后，进行识别并和靶器官的微血管内皮细胞发生相互作用的过程[2]。归巢的过程是通过细胞表面的特定传感器指引的，它能感受到靶组织的趋化因子梯度，在细胞进出血流前后是一个较活跃的迁移期[2]。

间充质干细胞

招募间充质干细胞修复受损的组织是一个复杂的过程，包括细胞从远端的受损组织感受到信号,从骨髓窦释放到血循环,定植于靶组织并原位分化到成熟的机能细胞(图12-3)。

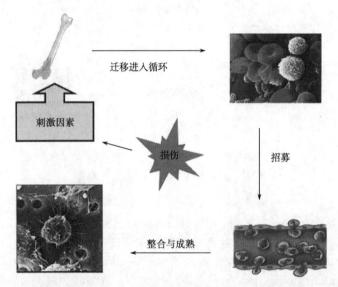

图12-3 外周损伤时释放刺激因子动员MSC从骨髓释放到血循环。在损伤部位的内皮细胞特定因子引起MSC的招募,进而从血管移居到损伤原位并成熟,融入损伤组织进行再生和愈合(已得到Liu Z、Zhuge Y、Velazquez OC的许可,转载于Trafficking and differentiation of mensenchymal stem cells. J Cell Biochem 2009;106:984-91)

一种有关动员的假设是,当组织受损时,细胞因子和趋化因子浓度增加,从远端组织释放到血循环,刺激MSC,下调使MSC留在原位的黏附分子浓度[27]。在静态情况下,祖细胞是通过与骨髓接触而保持静止状态的。动员过程依赖许多不同的分子,如基质金属蛋白酶蛋白(MMP)9和基质细胞衍生因子(SDF)-1/CXCL12,以及受体CXCR4[39]。现已证实MSC过表达CXCR4可增加心肌血管生成[40]。同样,在心梗时,通过SDF-1α信号,IGF-1在MSC的过表达可引起大量干细胞的动员,心肌血管的生成达到最高峰[41]。除SDF-1α外,高迁移率族蛋白B1(HMGB-1)是一种核蛋白,当细胞坏死通过炎症细胞因子激活细胞时,这种蛋白质在细胞外释放。在体内外,HMGB-1已作为一种趋化因子,对炎症细胞、干细胞和EPC发挥作用[42,43]。

当损害或缺血损伤后,局部定植的成纤维细胞被激活,对招募循环MSC到损伤部位起主要作用。根据目前的研究,在体外伤口愈合试验和新的三维模型中,血小板衍生生长因子B(PDGF-B)激活的成纤维细胞对调节小鼠骨髓MCS的招募、迁移和分化起到重要作用。在PDGF-B存在时,由PDGF-B激活的成纤维细胞与对照组相比,引起MSC迁移速度的明显增加,并促进MSC进入三维(3D)的胶原凝胶[27]。此外,PDGF-B激活的成纤维细胞能诱导MSC向肌纤维母细胞分化。这些作用可能是通过碱性成纤维细胞因子(bFGF)和上皮中性细胞激活肽78(ENA-78或CXCL5)来调节的,当进行蛋白阵列分析时发现这两种可溶性蛋白浓度升高。用抗体能阻断MSC的迁移和向 3D 纤维胶原的分化,当外源补充 bFGF/CXCL5 时能促进 MSC 向 3D 纤维胶原分化[37]。

MSC 定植到靶组织的机理是一种连锁反应,包括 MSC 在血管流动、黏附到毛细血管的内皮细胞表面、通过细胞外基质迁移到受损的靶组织。研究显示,这些黏附分子与机体招募白细胞到炎症部位的原理一样[44,45]。这些分子包括整合素、选择素、CAM 和趋化因子受体。悬浮的 MSC 缓慢下沉有利于随后的内皮黏附,有人提出 P 选择素与 MSC 的浮动和黏附到内皮有关。在切断流动的情况下,中和这些分子显著地减少 MSC 的浮动[46]。其他调节 MSC 相互作用的黏附因子包括:各种整合素和 CAM,如 α1、α2、α3、α4、α5、αv、β1、β3、β4、VCAM-1、ICAM-1、ICAM-3、ALCAM 和 endoglin/CD105[47,48],这些黏附分子和它们的反配体在 MSC 或内皮细胞表达。人的 MSC 近一半表达整合素非常晚期抗原(VLA)-4($\alpha 4\beta 1$, CD49d),通过体外的研究显示,在断流的情况下,人的 MSC 黏附到内皮细胞是通过 VLA-4 来调节的[46]。VLA-4 和其对应的黏附分子 VCAM-1 对人 MSC 牢固黏附到内皮细胞负责。

目前,已经证实 CD44 是骨髓来源 MSC 的定植分子,CD44 与骨髓血管中的 E 选择素相互作用,调节人 MSC 的迁移[49]。CD44 黏附分子是转膜糖蛋白的大家族,体外试验观察到在 E 选择素存在的情况下,CD44 的岩藻糖基化能产生巨大的滚动,在 NOD/SCID 小鼠模型,能大大增加静脉注射的 MSC 定植到骨髓[49]。受体和配体的特异性决定干细胞从一个器官迁移到另一个器官的方向。掌控这些定植信号能够使干细胞定位到受损的精确部位。

现在还不清楚骨髓来源的循环 MSC 与组织来源的 MSC 是否具有相同的迁移机制。在调节肌动蛋白细胞骨架动力学方面,GPT 酶的 Rho 家族起到主要作用,可能影响细胞的迁移和黏附。然而,现在一般认为通过 Rho 家族的信号影响 MSC 的迁移,但与试验结果并不一致。

在临床试验中,MSC 可通过几种途径应用于治疗,最常用的是直接部位注射和全身静脉注射。在前者,MSC 通过局部或病灶移植直接送入局部组织。在这种情况下,动员和归巢是不需要的。当应用全身静脉输入时,可避开动员的步骤,但需要归巢和招募。了解骨髓来源和循环 MSC 有关归巢的分子机理将有利于扩大 MSC 的临床应用。

内皮祖细胞

类似 MSC,当机体受损或缺血后,EPC 的动员和归巢过程由细胞因子、黏附分子和生长因子来调节并引导它们到血管壁[39]。应用细胞分选技术,目前已知功能性血管内皮生成全部来源于 $CD34^+/CD45^-$ 单核细胞,这些单核细胞 VEGFR2 阳性,CD133 阴性[50]。

SDF-1/CXCR4 轴对于动员 EPC 参与血管生成和血管新生是非常重要的[51,52]。SDF-1/CXCL12 的基因是通过转录因子缺氧诱导因子(HIF)-1 来调节的,当组织受损时,HIF-1 上调,根据缺氧的程度,通过 HIF-1 的诱导表达 SDF-1/CXCL12[35,53]。它们相互的抑制作用能部分阻断祖细胞归巢到缺血的心肌[51,54]。在后肢缺血的动物模型中,用抗 CXCR4 的中和抗体抑制 CXCR4,能显著减少由 SDF-1α/CXCL12 诱导的 EPC 黏附到成熟的内皮细胞单核层，减少体外 EPC 的迁移和体内髓性 EPC 在缺血肢体的归巢[53,55]。血管损伤可调节循环中的 EPC 短暂增加[56]。在损伤部位产生的 SDF-1/CXCL12 能诱导内皮细胞和骨髓的一氧化氮（NO）释放，动员 EPC[51,52]。

在许多情况下，EPS 的内皮再生功能对机体是有益的，如脊髓损伤的小鼠能减轻内膜的增生，帮助产后缺血组织血管的新生及治疗急慢性心肌缺血[27,51,57,58]。在载脂蛋白 E 基因敲除（ApoE$^{-/-}$）的小鼠，EPC 能减少斑块的稳定性[59]。因此，EPC 的治疗应用应该是相互关联的，如广泛动员内皮祖细胞可能减少斑块的形成，而局部刺激 EPC 可以改善动脉损伤，当缺血时可以促进血管再生[39]。

造血干细胞

造血干细胞从骨髓动员的过程是通过特殊的信号来控制的，如趋化因子、生长因子，以及骨髓和末梢的激素。例如，SDF-1α/CXCR4 轴对造血祖细胞在骨髓龛的定位起到一个举足轻重的作用，干扰这种关系会引起骨髓龛里的干细胞被迅速动员[60,61]。同样，缺血的末梢组织会引起内皮细胞 SDF-1α/CXCL12 的上调，能调节并招募 HSC 到血管损伤部位[53,62]。不同的祖细胞对趋化因子有不同的动员反应，例如，一氧化氮（NO）与骨髓来源的造血祖细胞的动员有关，但不影响低分化的干细胞[63,64]。

细胞之间黏附的调节对两个不同组织干细胞的转化是非常重要的，打破两个细胞之间的连接是细胞迁移的第一步[2]。对于 HSC，去除对龛的黏附需要蛋白水解酶的作用，如基质金属蛋白酶（MMP）-9 和半胱氨酸蛋白酶组织蛋白酶 K[65,66]。除此之外，通过 SDF-1α/CXCL12 的裂解和灭活，HSC 自身的破骨细胞和蛋白酶活性有助于骨髓 HSC 保留信号的沉默[2,66,67]。

在去黏附之后，造血干细胞离开骨髓在血液循环中播散。一旦开始血液传播，它们可能定植于靶部位，非常紧密地黏附到微血管内皮细胞。这个过程需要 HSC 辨认组织特异性微血管的特征和黏附到内皮，但必须克服血流产生的剪切力[2]。在这个过程中，附着于 HSC 和内皮细胞表面的黏附分子起到主要作用，这类似于血液中的白细胞定植于周围组织[68]。这一多步骤连续反应的第一步包括捕获和滚动，由最初的黏附分子（选择素或 α4 整合素）介导，伴随快速结合并具有很强的抗拉强度，但时间短[2]；接着，由可溶性或结合表面的趋化因子提供趋化/激活刺激；最后，由继发黏附分子、大部分的整合素（β2 或 α4）介导和免疫球蛋白超级家族内皮细胞配体的相互作用使得黏附更加牢固[2]。

在外周完成其功能后，祖细胞必须高效返回骨髓进行持续增殖补充骨髓库。大部分 HSC 表达 α4β1 整合素（VLA-4），并与骨髓基质细胞和内皮细胞上的血管细胞黏附分子（VCAM）-1 结合[69]，阻断这个轴可引起 HSC 动员的增加并丧失骨髓细胞移植到骨髓龛的能力[70~73]。其他黏附途径如选择素（E 选择素、P 选择素和 β2 选择素）也有助于祖细胞定植到骨髓[74,75]。

HSC 经血流到达周围组织，通过引流淋巴管流出。鞘氨醇-1-肌磷酸（S1P）主要控制进入淋巴管的组织造血祖细胞的流出[64]。S1P 在组织中保持低水平而在血液和淋巴中保持高水平[76,77]，由此调节从胸腺、脾和淋巴结来源的成熟淋巴细胞的运动[78,79]。可比机制也控制 HSC 从周围组织向引流淋巴管的流出[80]。例如，小鼠造血干细胞主要以 S1P1 受体依赖性的方式向 S1P 浓度高的方向迁移，阻断这些受体会减少定植组织的 HSC 流入淋巴管，抑制再循环[80]。

HSC 能在骨髓、血液、髓外组织和淋巴间进行持续的反复循环[80]。造血干细胞为保持动态平衡和补充循环细胞库的需要，在静止和自我更新之间切换[3]。因而，循环 HSC 可能有助于不断补充组织中的髓系，保持其他特异性细胞处于稳定状态[64]。除此之外，造血干细胞在组织损伤或感染时也可能参与免疫反应[64]。HSC 表达 Toll 样受体（TLR），该受体能识别外来的细菌表面分子，与其结合后促进干细胞完全进入细胞周期并引起髓样分化[81,82]。因此，迁徙的造血干细胞能探测周围组织，根据当时状况迅速作出反应，及时产生大量的免疫细胞[64]。

目前的资料显示，某些肿瘤存在肿瘤干细胞群（CSC），能维持产生不同的肿瘤细胞，这些干细胞在正常组织能引起肿瘤[1]。现已发现，这些 CSC 是通过黏附和迁移的失调播散肿瘤到全身。这些肿瘤细胞应用正常组织相似的迁移机理[2]，而特定的黏附和去黏附路径决定转移的有效性。例如，与迁移有关的 CXCR4 表达能促进某些血液和实体瘤的转移[84,85]。另外，骨髓来源的非肿瘤细胞对于形成"转移前小生境"是必需的，其引导恶性肿瘤细胞器官特异性定植模式[85]。这些小生境分泌趋化因子，如 SDF-1/CXCL12，招募转移的肿瘤干细胞在其他部位生成肿瘤[85]。因而应进行干预防止这些小生境的产生能在体内阻断转移，说明针对恶性肿瘤的定植机理，进而限制肿瘤传播的重要性[2]。

结论

干细胞在黏附、趋化和信号途径之间存在复杂的相互作用，在相互配合下实现应有的分化、迁移和归巢。这些系统往往通过胚胎的发育得以保存，存在于不同的生物体，并可延伸到肿瘤干细胞、基因载体的形成及药物传递。

MSC 可能是唯一适合于各种治疗的应用，如组织再生、各种遗传病的纠正、慢性炎症的抑制和生物药物的传递。对 MSC 分化和迁移的了解可能有助于治疗策略的发展，能促进骨髓或组织来源 MSC 的招募。以细胞为基础治疗的有效性不仅需要足够量的 MSC，而且能有效输送这些细胞到预定部位。要实现再生医学的希望有赖于弄清治疗的机理，以及控制和调节 MSC 系特异性分化和组织特异性干细胞归巢的那些分子。

同样，现已明确 HSC 归巢对临床应用的重要性，当骨髓移植临床用于各种肿瘤患者时，有赖于移植的 HSC 转运到骨髓的有效性[2]。供者的干细胞通过静脉输入受者，以便适量地扩增再定植在受者的血液。供者的 HSC 必须适当地植入受者的骨髓龛。看来，这种医学疗法利用相同的预先存在的路径，在动态平衡时通常支持造血干细胞的再循环。

最后，ESC 是一种很好的内皮细胞来源，参与修复和出生后损伤部位、缺血和肿瘤生成的血管新生，不仅在伤口愈合，而且在肢体缺血[86~88]、心梗后[89~91]、内皮的血管移植[92,93]、动脉粥样硬化[94]、视网膜和淋巴器官的血管新生[95,96]、新生儿的血管[97]和肿瘤生长[98,99]的治疗方面都起到举足轻重的作用。增加对这些细胞分化和迁移的了解,有利于对这些治疗流程进行改进。

SDF-1/CXCR4 轴广泛存在于胚胎和成体多个组织中。当发育时,胎儿小鼠 HSC 的 SDF-1/CXCR4 信号定植于胎肝和骨髓,在成人,SDF-1/CXCR4 调节人和小鼠 HSC 的动员及再进入骨髓[2,60,100,101],这个轴在癌症转移中与肿瘤形成细胞的播散有关[83],对这个轴的进一步研究将产生血液系统和实体瘤联合的新疗法。

(卓 毅 卓光生 译)

参 考 文 献

1. Dalerba P, Cho RW, Clarke MF. Cancer stem cells: models and concepts. Annu Rev Med 2007; 58:267-84.
2. Laird DJ, von Andrian UH, Wagers AJ. Stem cell trafficking in tissue development, growth and disease. Cell 2008; 132(4):612-30.
3. Orkin SH, Zon LI. Hematopoiesis: an evolving paradigm for stem cell biology. Cell 2008; 132(4):631-44.
4. Orkin SH. Diversification of haematopoietic stem cells to specific lineages. Nat Rev Genet 2000; 1(1):57-64.
5. Kim SI, Bresnick EH. Transcriptional control of erythropoiesis: emerging mechanisms and principles. Oncogene 2007; 26(47):6777-94.
6. Mikkola HK, Klintman J, Yang H et al. Haematopoietic stem cells retain long-term repopulating activity and multipotency in the absence of stem-cell leukaemia SCL/tal-1 gene. Nature 2003; 421(6922):547-51.
7. Ichikawa M, Asai T, Saito T et al. AML-1 is required for megakaryocytic maturation and lymphocytic differentiation, but not for maintenance of hematopoietic stem cells in adult hematopoiesis. Nat Med 2004; 10(3):299-304.
8. Kulessa H, Frampton J, Graf T. GATA-1 reprograms avian myelomonocytic cell lines into eosinophils, thromboblasts and erythroblasts. Genes Dev 1995; 9(10):1250-62.
9. Nerlov C, Graf T. PU.1 induces myeloid lineage commitment in multipotent hematopoietic progenitors. Genes Dev 1998; 12(15):2403-12.
10. Nutt SL, Heavey B, Rolink AG et al. Commitment to the B-lymphoid lineage depends on the transcription factor Pax5. Nature 1999; 401(6753):556-62.
11. McDevitt MA, Shivdasani RA, Fujiwara Y et al. A "knockdown" mutation created by cis-element gene targeting reveals the dependence of erythroid cell maturation on the level of transcription factor GATA-1. Proc Natl Acad Sci USA 1997; 94(13):6781-5.
12. Orkin SH. Priming the hematopoietic pump. Immunity 2003; 19(5):633-4.
13. Chen H, Ray-Gallet D, Zhang P et al. PU.1 (Spi-1) autoregulates its expression in myeloid cells. Oncogene 1995; 11(8):1549-60.
14. Tsai SF, Strauss E, Orkin SH. Functional analysis and in vivo footprinting implicate the erythroid transcription factor GATA-1 as a positive regulator of its own promoter. Genes Dev 1991; 5(6):919-31.
15. Iwasaki H, Akashi K. Myeloid lineage commitment from the hematopoietic stem cell. Immunity 2007; 26(6):726-40.
16. Laiosa CV, Stadtfeld M, Xie H et al. Reprogramming of committed T-cell progenitors to macrophages and dendritic cells by C/EBP alpha and PU.1 transcription factors. Immunity 2006; 25(5):731-44.
17. Xie H, Ye M, Feng R et al. Stepwise reprogramming of B cells into macrophages. Cell 2004; 117(5):663-76.
18. Busslinger M, Nutt SL, Rolink AG. Lineage commitment in lymphopoiesis. Curr Opin Immunol 2000; 12(2):151-8.
19. Zhang P, Zhang X, Iwama A et al. PU.1 inhibits GATA-1 function and erythroid differentiation by blocking GATA-1 DNA binding. Blood 2000; 96(8):2641-8.
20. Nerlov C, Querfurth E, Kulessa H et al. GATA-1 interacts with the myeloid PU.1 transcription factor and represses PU.1-dependent transcription. Blood 2000; 95(8):2543-51.
21. Zheng W, Flavell RA. The transcription factor GATA-3 is necessary and sufficient for Th2 cytokine gene expression in CD4 T-cells. Cell 1997; 89(4):587-96.
22. Szabo SJ, Kim ST, Costa GL et al. A novel transcription factor, T-bet, directs Th1 lineage commitment. Cell 2000; 100(6):655-69.

23. Wechsler J, Greene M, McDevitt MA et al. Acquired mutations in GATA1 in the megakaryoblastic leukemia of Down syndrome. Nat Genet 2002; 32(1):148-52.
24. Mueller BU, Pabst T, Osato M et al. Heterozygous PU.1 mutations are associated with acute myeloid leukemia. Blood 2002;100(3):998-1007.
25. Pabst T, Mueller BU, Zhang P et al. Dominant-negative mutations of CEBPA, encoding CCAAT/enhancer binding protein-alpha (C/EBPalpha), in acute myeloid leukemia. Nat Genet 2001; 27(3):263-70.
26. Mullighan CG, Goorha S, Radtke I et al. Genome-wide analysis of genetic alterations in acute lymphoblastic leukaemia. Nature 2007; 446(7137):758-64.
27. Liu ZJ, Zhuge Y, Velazquez OC. Trafficking and differentiation of mesenchymal stem cells. J Cell Biochem 2009;106(6):984-91.
28. Dominici M, Le Blanc K, Mueller I et al. Minimal criteria for defining multipotent mesenchymal stromal cells. The International Society for Cellular Therapy position statement. Cytotherapy 2006; 8(4):315-7.
29. Bianco P, Robey PG, Simmons PJ. Mesenchymal stem cells: revisiting history, concepts and assays. Cell Stem Cell 2008; 2(4):313-9.
30. Chen FH, Rousche KT, Tuan RS. Technology Insight: adult stem cells in cartilage regeneration and tissue engineering. Nat Clin Pract Rheumatol 2006; 2(7):373-82.
31. Mimeault M, Batra SK. Recent progress on normal and malignant pancreatic stem/progenitor cell research: therapeutic implications for the treatment of type 1 or 2 diabetes mellitus and aggressive pancreatic cancer. Gut 2008; 57(10):1456-68.
32. Baksh D, Tuan RS. Canonical and noncanonical Wnts differentially affect the development potential of primary isolate of human bone marrow mesenchymal stem cells. J Cell Physiol 2007; 212(3):817-26.
33. Barry F, Boynton RE, Liu B et al. Chondrogenic differentiation of mesenchymal stem cells from bone marrow: differentiation-dependent gene expression of matrix components. Exp Cell Res 2001; 268(2):189-200.
34. Ito T, Sawada R, Fujiwara Y et al. FGF-2 increases osteogenic and chondrogenic differentiation potentials of human mesenchymal stem cells by inactivation of TGF-beta signaling. Cytotechnology 2008; 56(1):1-7.
35. Kratchmarova I, Blagoev B, Haack-Sorensen M et al. Mechanism of divergent growth factor effects in mesenchymal stem cell differentiation. Science 2005; 308(5727):1472-7.
36. Matsuda C, Takagi M, Hattori T et al. Differentiation of Human Bone Marrow Mesenchymal Stem Cells to Chondrocytes for Construction of Three-dimensional Cartilage Tissue. Cytotechnology 2005; 47(1-3):11-7.
37. Nedeau AE, Bauer RJ, Gallagher K et al. A CXCL5- and bFGF-dependent effect of PDGF-B-activated fibroblasts in promoting trafficking and differentiation of bone marrow-derived mesenchymal stem cells. Exp Cell Res 2008; 314(11-12):2176-86.
38. Tsuchiya H, Kitoh H, Sugiura F et al. Chondrogenesis enhanced by overexpression of sox9 gene in mouse bone marrow-derived mesenchymal stem cells. Biochem Biophys Res Commun 2003; 301(2):338-43.
39. Hristov M, Weber C. Progenitor cell trafficking in the vascular wall. J Thromb Haemost 2009; 7 Suppl 1:31-4.
40. Zhang D, Fan GC, Zhou X et al. Over-expression of CXCR4 on mesenchymal stem cells augments myoangiogenesis in the infarcted myocardium. J Mol Cell Cardiol 2008; 44(2):281-92.
41. Haider H, Jiang S, Idris NM et al. IGF-1-overexpressing mesenchymal stem cells accelerate bone marrow stem cell mobilization via paracrine activation of SDF-1alpha/CXCR4 signaling to promote myocardial repair. Circ Res 2008; 103(11):1300-8.
42. Chavakis E, Hain A, Vinci M et al. High-mobility group box 1 activates integrin-dependent homing of endothelial progenitor cells. Circ Res 2007; 100(2):204-12.
43. Palumbo R, Bianchi ME. High mobility group box 1 protein, a cue for stem cell recruitment. Biochem Pharmacol 2004; 68(6):1165-70.
44. Ley K, Laudanna C, Cybulsky MI et al. Getting to the site of inflammation: the leukocyte adhesion cascade updated. Nat Rev Immunol 2007; 7(9):678-89.
45. Luster AD, Alon R, von Andrian UH. Immune cell migration in inflammation: present and future therapeutic targets. Nat Immunol 2005; 6(12):1182-90.
46. Ruster B, Gottig S, Ludwig RJ et al. Mesenchymal stem cells display coordinated rolling and adhesion behavior on endothelial cells. Blood 2006; 108(12):3938-44.
47. Krampera M, Pizzolo G, Aprili G et al. Mesenchymal stem cells for bone, cartilage, tendon and skeletal muscle repair. Bone 2006; 39(4):678-83.
48. Minguell JJ, Erices A, Conget P. Mesenchymal stem cells. Exp Biol Med (Maywood) 2001; 226(6):507-20.
49. Sackstein R, Merzaban JS, Cain DW et al. Ex vivo glycan engineering of CD44 programs human multipotent mesenchymal stromal cell trafficking to bone. Nat Med 2008; 14(2):181-7.
50. Timmermans F, Van Hauwermeiren F, De Smedt M et al. Endothelial outgrowth cells are not derived from CD133+ cells or CD45+ hematopoietic precursors. Arterioscler Thromb Vasc Biol 2007; 27(7):1572-9.
51. Gallagher KA, Liu ZJ, Xiao M et al. Diabetic impairments in NO-mediated endothelial progenitor cell mobilization and homing are reversed by hyperoxia and SDF-1 alpha. J Clin Invest 2007; 117(5):1249-59.

52. Petit I, Jin D, Rafii S. The SDF-1-CXCR4 signaling pathway: a molecular hub modulating neo-angiogenesis. Trends Immunol 2007; 28(7):299-307.
53. Ceradini DJ, Kulkarni AR, Callaghan MJ et al. Progenitor cell trafficking is regulated by hypoxic gradients through HIF-1 induction of SDF-1. Nat Med 2004; 10(8):858-64.
54. Abbott JD, Huang Y, Liu D et al. Stromal cell-derived factor-1alpha plays a critical role in stem cell recruitment to the heart after myocardial infarction but is not sufficient to induce homing in the absence of injury. Circulation 2004; 110(21):3300-5.
55. Walter DH, Haendeler J, Reinhold J et al. Impaired CXCR4 signaling contributes to the reduced neovascularization capacity of endothelial progenitor cells from patients with coronary artery disease. Circ Res 2005; 97(11):1142-51.
56. Hristov M, Weber C. Ambivalence of progenitor cells in vascular repair and plaque stability. Curr Opin Lipidol 2008; 19(5):491-7.
57. Hristov M, Zernecke A, Bidzhekov K et al. Importance of CXC chemokine receptor 2 in the homing of human peripheral blood endothelial progenitor cells to sites of arterial injury. Circ Res 2007; 100(4):590-7.
58. Hristov M, Zernecke A, Liehn EA et al. Regulation of endothelial progenitor cell homing after arterial injury. Thromb Haemost 2007; 98(2):274-7.
59. George J, Afek A, Abashidze A et al. Transfer of endothelial progenitor and bone marrow cells influences atherosclerotic plaque size and composition in apolipoprotein E knockout mice. Arterioscler Thromb Vasc Biol 2005; 25(12):2636-41.
60. Broxmeyer HE, Orschell CM, Clapp DW et al. Rapid mobilization of murine and human hematopoietic stem and progenitor cells with AMD3100, a CXCR4 antagonist. J Exp Med 2005; 201(8):1307-18.
61. Lapidot T, Kollet O. The essential roles of the chemokine SDF-1 and its receptor CXCR4 in human stem cell homing and repopulation of transplanted immune-deficient NOD/SCID and NOD/SCID/B2m(null) mice. Leukemia 2002; 16(10):1992-2003.
62. Ceradini DJ, Gurtner GC. Homing to hypoxia: HIF-1 as a mediator of progenitor cell recruitment to injured tissue. Trends Cardiovasc Med 2005; 15(2):57-63.
63. Aicher A, Heeschen C, Mildner-Rihm C et al. Essential role of endothelial nitric oxide synthase for mobilization of stem and progenitor cells. Nat Med 2003; 9(11):1370-6.
64. Schulz C, von Andrian UH, Massberg S. Hematopoietic stem and progenitor cells: their mobilization and homing to bone marrow and peripheral tissue. Immunol Res 2009; 44(1-3):160-8.
65. Heissig B, Hattori K, Dias S et al. Recruitment of stem and progenitor cells from the bone marrow niche requires MMP-9 mediated release of kit-ligand. Cell 2002; 109(5):625-37.
66. Kollet O, Dar A, Shivtiel S et al. Osteoclasts degrade endosteal components and promote mobilization of hematopoietic progenitor cells. Nat Med 2006; 12(6):657-64.
67. Christopherson KW, 2nd, Cooper S, Broxmeyer HE. Cell surface peptidase CD26/DPPIV mediates G-CSF mobilization of mouse progenitor cells. Blood 2003; 101(12):4680-6.
68. Springer TA. Traffic signals for lymphocyte recirculation and leukocyte emigration: the multistep paradigm. Cell 1994; 76(2):301-14.
69. Potocnik AJ, Brakebusch C, Fassler R. Fetal and adult hematopoietic stem cells require beta1 integrin function for colonizing fetal liver, spleen and bone marrow. Immunity 2000; 12(6):653-63.
70. Levesque JP, Takamatsu Y, Nilsson SK et al. Vascular cell adhesion molecule-1 (CD106) is cleaved by neutrophil proteases in the bone marrow following hematopoietic progenitor cell mobilization by granulocyte colony-stimulating factor. Blood 2001; 98(5):1289-97.
71. Papayannopoulou T, Craddock C, Nakamoto B et al. The VLA4/VCAM-1 adhesion pathway defines contrasting mechanisms of lodgement of transplanted murine hemopoietic progenitors between bone marrow and spleen. Proc Natl Acad Sci USA 1995; 92(21):9647-51.
72. Papayannopoulou T, Nakamoto B. Peripheralization of hemopoietic progenitors in primates treated with anti-VLA4 integrin. Proc Natl Acad Sci USA 1993; 90(20):9374-8.
73. Wagers AJ, Allsopp RC, Weissman IL. Changes in integrin expression are associated with altered homing properties of Lin(-/lo)Thy1.1(lo)Sca-1(+)c-kit(+) hematopoietic stem cells following mobilization by cyclophosphamide/granulocyte colony-stimulating factor. Exp Hematol 2002; 30(2):176-85.
74. Katayama Y, Hidalgo A, Furie BC et al. PSGL-1 participates in E-selectin-mediated progenitor homing to bone marrow: evidence for cooperation between E-selectin ligands and alpha 4 integrin. Blood 2003; 102(6):2060-7.
75. Mazo IB, Gutierrez-Ramos JC, Frenette PS et al von Andrian UH. Hematopoietic progenitor cell rolling in bone marrow microvessels: parallel contributions by endothelial selectins and vascular cell adhesion molecule 1. J Exp Med 1998; 188(3):465-74.
76. Hla T, Venkataraman K, Michaud J. The vascular S1P gradient-cellular sources and biological significance. Biochim Biophys Acta 2008; 1781(9):477-82.
77. Yatomi Y, Igarashi Y, Yang L et al. Sphingosine 1-phosphate, a bioactive sphingolipid abundantly stored in platelets, is a normal constituent of human plasma and serum. J Biochem 1997; 121(5):969-73.
78. Pappu R, Schwab SR, Cornelissen I et al. Promotion of lymphocyte egress into blood and lymph by distinct sources of sphingosine-1-phosphate. Science 2007; 316(5822):295-8.
79. Rosen H, Gonzalez-Cabrera P, Marsolais D et al. Modulating tone: the overture of S1P receptor immunotherapeutics. Immunol Rev 2008; 223:221-35.

80. Massberg S, Schaerli P, Knezevic-Maramica I et al. Immunosurveillance by hematopoietic progenitor cells trafficking through blood, lymph and peripheral tissues. Cell 2007; 131(5):994-1008.
81. Hoshino K, Takeuchi O, Kawai T et al. Cutting edge: Toll-like receptor 4 (TLR4)-deficient mice are hyporesponsive to lipopolysaccharide: evidence for TLR4 as the Lps gene product. J Immunol 1999; 162(7):3749-52.
82. Nagai Y, Garrett KP, Ohta S et al. Toll-like receptors on hematopoietic progenitor cells stimulate innate immune system replenishment. Immunity 2006; 24(6):801-12.
83. Balkwill F. Cancer and the chemokine network. Nat Rev Cancer 2004; 4(7):540-50.
84. Burger JA, Kipps TJ. CXCR4: a key receptor in the crosstalk between tumor cells and their microenvironment. Blood 2006; 107(5):1761-7.
85. Kaplan RN, Riba RD, Zacharoulis S et al. VEGFR1-positive haematopoietic bone marrow progenitors initiate the premetastatic niche. Nature 2005; 438(7069):820-7.
86. Kalka C, Masuda H, Takahashi T et al. Transplantation of ex vivo expanded endothelial progenitor cells for therapeutic neovascularization. Proc Natl Acad Sci USA 2000; 97(7):3422-7.
87. Majka SM, Jackson KA, Kienstra KA et al. Distinct progenitor populations in skeletal muscle are bone marrow derived and exhibit different cell fates during vascular regeneration. J Clin Invest 2003; 111(1):71-9.
88. Takahashi T, Kalka C, Masuda H et al. Ischemia- and cytokine-induced mobilization of bone marrow-derived endothelial progenitor cells for neovascularization. Nat Med 1999; 5(4):434-8.
89. Kocher AA, Schuster MD, Szabolcs MJ et al. Neovascularization of ischemic myocardium by human bone-marrow-derived angioblasts prevents cardiomyocyte apoptosis, reduces remodeling and improves cardiac function. Nat Med 2001; 7(4):430-6.
90. Orlic D, Kajstura J, Chimenti S et al. Transplanted adult bone marrow cells repair myocardial infarcts in mice. Ann N Y Acad Sci 2001; 938:221-9; discussion 29-30.
91. Orlic D, Kajstura J, Chimenti S et al. Bone marrow cells regenerate infarcted myocardium. Nature 2001; 410(6829):701-5.
92. Kaushal S, Amiel GE, Guleserian KJ et al. Functional small-diameter neovessels created using endothelial progenitor cells expanded ex vivo. Nat Med 2001; 7(9):1035-40.
93. Shi Q, Rafii S, Wu MH et al. Evidence for circulating bone marrow-derived endothelial cells. Blood 1998; 92(2):362-7.
94. Sata M, Saiura A, Kunisato A et al. Hematopoietic stem cells differentiate into vascular cells that participate in the pathogenesis of atherosclerosis. Nat Med 2002; 8(4):403-9.
95. Grant MB, May WS, Caballero S et al. Adult hematopoietic stem cells provide functional hemangioblast activity during retinal neovascularization. Nat Med 2002; 8(6):607-12.
96. Otani A, Kinder K, Ewalt K et al. Bone marrow-derived stem cells target retinal astrocytes and can promote or inhibit retinal angiogenesis. Nat Med 2002; 8(9):1004-10.
97. Young PP, Hofling AA, Sands MS. VEGF increases engraftment of bone marrow-derived endothelial progenitor cells (EPCs) into vasculature of newborn murine recipients. Proc Natl Acad Sci USA 2002; 99(18):11951-6.
98. Lyden D, Hattori K, Dias S et al. Impaired recruitment of bone-marrow-derived endothelial and hematopoietic precursor cells blocks tumor angiogenesis and growth. Nat Med 2001; 7(11):1194-201.
99. Reyes M, Dudek A, Jahagirdar B et al. Origin of endothelial progenitors in human postnatal bone marrow. J Clin Invest 2002; 109(3):337-46.
100. Peled A, Petit I, Kollet O et al. Dependence of human stem cell engraftment and repopulation of NOD/SCID mice on CXCR4. Science 1999; 283(5403):845-8.
101. Sugiyama T, Kohara H, Noda M et al. Maintenance of the hematopoietic stem cell pool by CXCL12-CXCR4 chemokine signaling in bone marrow stromal cell niches. Immunity 2006; 25(6):977-88.

第13章　脊椎动物再生模型对干细胞应用的启示

Christopher L. Antos*，Elly M. Tanaka*

摘要：一些模式动物具有再生能力，在受到严重伤害后能通过刺激局部细胞还原受损或失去的器官和附肢。这一章我们将叙述各类脊椎动物如何再生出不同的组织、器官，如中枢神经系统、心脏和附肢等，并且详细描述这些组织结构再生中详细的、特异的细胞和分子生物学特征。

脊椎动物再生模型的属性

我们对已化的细胞怎样被重编程从而具有干细胞行为的认识已经取得了重大的进展，并且很多努力集中在细胞重编程后如何能用于再生治疗的研究上[1~3]。然而，目前仅有几种治疗策略可以成功地重建一个具有多种成分的复合结构器官或附属物。目前尚不清楚将诱导的多能干细胞（induced pluripotent stem cell，iPS细胞）来源细胞移植入体内会存在什么样的长期影响（如致癌倾向）[3]。因此，有必要了解组织细胞必须在多大程度上重编程，以及这些细胞怎样被指令而再造出一个器官或附属物的复杂性。

不同的动物重建失去或损伤组织的能力有很大差别。一些动物完全用瘢痕组织替代失去的部分；另一些具有再激活发育机制的能力，重新发育（再生）出缺失的结构。残余细胞启动再生反应，重新产生原来结构而非肿瘤这一能力，要求对新生物的再生有非常严格的控制。因此，用于细胞和分子领域研究的实验动物模型为我们认识器官和附属物再生过程的基本机制提供了可能。

在脊椎动物中，哺乳类只能再生有限数量的组织，而其他一些脊椎动物包括鱼类（金鱼和斑马鱼）、两栖类（蝾螈、火蜥蜴、青蛙——在早期爬行类出现之前演化出来的动物）能够再生出一系列广泛的复杂器官，如受损的视网膜、切断的脊髓、损伤的心脏和离断的肢体。基于组织学和分子生物学观察，这种能力涉及处于休眠状态或者已分化的细胞的活化、形态学改变和转录水平变化的诱导祖细胞（progenitor cell）的产生。所以这些动物为我们理解已分化细胞重编程转变为有再生能力的前体细胞过程中的细胞和分子机制，以及了解这些分化细胞来源的前体细胞如何重建一个结构缺失部位提供了

---

* Corresponding Authors: Christopher L. Antos—DFG-Center for Regenerative Therapies Dresden, Technische Universität Dresden, Tatzberg 47/49, 01307 Dresden, Germany.
Email: christopher.antos@crt-dresden.de
Elly M. Tanaka—Max-Planck Institute for Molecular Cell Biology and Genetics, Pfotenhauerstrasse 108 and DFG-Center for Regenerative Therapies Dresden, Technische Universität Dresden, Tatzberg 47/49, 01307 Dresden, Germany. Email: elly.tanaka@crt-dresden.de

良好的研究模型。

一切组织的再生需要一个或一个以上由增生细胞群组成的细胞池,这些细胞可以来自静息的干细胞(前体细胞),或者来自回复到前体细胞状态的成熟组织细胞。有能力重建复杂组织结构的大量前体细胞的产生涉及表观遗传学和基因转录程序的改变。这些前体细胞的形成所覆盖的问题包括:①重编程必需的程度;②所涉及的机制。本章我们首先介绍特定器官系统的再生,重点关注哪种细胞机制构成了再生事件发生的基础,并提及一些分子机制;在本章的第二部分,根据我们对复杂的附属物再生系统的详细了解,重点介绍附属物的再生,并讨论从成体组织产生出一组能重建多种组织结构、具有增殖特性的前体细胞的已知细胞生物学和分子生物学机制。

组织再生中的重编程过程

各种器官系统产生具有再生特性的祖细胞策略不同,包括激活干细胞、去分化甚至转分化(图 13-1)。本章我们对这些术语进行解释。"去分化"最严格的定义是假定一个典型的有丝分裂后细胞类型失去它分化的特征,形成一个未分化的、增生的细胞。最近,"去分化"一词也被用来描述任何一种类型细胞(甚至增生的前体细胞)向更原始状态的转化。本章我们用"去分化"一词表示有丝分裂后的细胞失去其分化特征,获得增殖能力,而不论它们随后会变成哪种细胞(图 13-1B)。不同于其他研究者将"转分化"的定义限定为一种细胞类型不经过前体细胞中间状态,向另一种细胞类型的直接转化;我们对"转分化"的定义更加宽泛,我们用"转分化"来描述一种分化的细胞类型向其他分化的细胞类型的转化,而不论是否存在明显的、较低分化的中间形态。我们之所以选择这个定义,是因为有时难以确定是否存在一个较低分化的中间状态,也因为晶状体的再生中似乎形成了未分化的中间体,但过去还是被描述为转分化事件。由于这种中间体的性质和潜能尚未明确,我们觉得转分化的定义能合理地解释这个问题。因此,在我们的定义中,转分化可以通过去分化的步骤来实现。

成熟组织再生的机制

眼睛——一个转分化模型

两栖类动物眼部组织(晶状体和视网膜)的再生作为脊椎动物中两种明确的细胞类型之间转分化的例子研究得最为详尽。当切除蝾螈的晶状体后(晶状体切除术),会导致背侧的虹膜背缘的色素上皮细胞失去色素,发生增殖,转分化成晶状体组织(图 13-2)[4]。值得注意的是,虹膜色素上皮和晶状体分别来源于胚胎原肠胚形成期两种不同的发育谱系:前者来自神经外胚层,后者由非神经表面外胚层形成[5]。虹膜背缘色素上皮的转分化潜能似乎存在谱系限制性,把虹膜背缘上皮细胞移植到再生的肢体芽基中会形成晶状体。然而,由于所使用追踪方法的限制,这些实验并不能排除一些细胞可能产生了其他类型的组织。有趣的是,虹膜色素上皮移植到脑或鳍并不能引起晶状体形成,提示启动内在的晶状体形成程序需要再生许可信号[6]。

细胞的克隆培养实验确立了背侧虹膜色素上皮向晶状体转分化的必需因素。鸡胚来

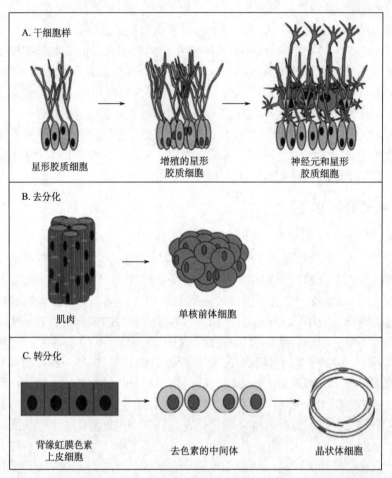

图 13-1 组织再生中观察到的细胞重编程过程。A. 一种再生出新生组织的模式是通过组织器官原位的干细胞群。损伤后，中枢神经系统的星形胶质细胞增生，一部分细胞转变成神经元，其他细胞保持原星形胶质细胞类型。B. 一部分组织经历去分化过程再生出新生组织。骨骼肌多核细胞内具有特征性的收缩装置即有横纹的肌原纤维。这类细胞去分化时分解收缩装置，断裂成单核细胞，增殖产生更多的细胞。C. 晶状体切除术诱导分化的背缘虹膜色素上皮细胞失去上皮样特征和去色素。这些无色素的细胞发生增殖并形成晶状体细胞。当前证据表明这些细胞可能是单能性的。（另见图版）

源的虹膜色素上皮克隆培养中，通过添加碱性成纤维细胞生长因子（bFGF）和褪色酶苯基硫脲（PTU），可以达到从色素上皮向晶状体的转分化[7]。在蝾螈模型中，实验表明碱性成纤维细胞生长因子能够诱导背侧虹膜色素上皮在体外培养和完整的蝾螈眼睛内形成异位的晶状体组织，而其他的生长因子（血管内皮生长因子、胰岛素样生长因子和上皮生长因子）不能引起这种转分化[8, 9]。

明确转分化事件中的这些细胞内因子——它们如何对细胞外信号产生应答和引起细胞表型转换，是重要的问题，通过对比背侧和腹侧虹膜，可以对这个转分化事件的分子

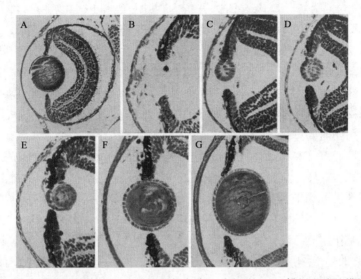

图 13-2　蝾螈晶状体再生来源于背侧色素上皮细胞。A. 蝾螈眼睛的横切面显示角膜、晶状体、背侧和腹侧虹膜区域、视网膜和视网膜色素上皮细胞。B. 手术移除晶状体后的结构。C. 晶状体切除后，背侧虹膜细胞变长，形成柱状上皮细胞并伴有色素丢失。D. 内壁来源的细胞形成囊泡并进入由于晶状体摘除形成的空隙区域。E. 囊泡细胞分化形成晶状体纤维。F. 初始的晶状体纤维从内部形成，后来在周边形成单层的纤维，完成晶状体再生。G. 晶状体纤维细胞的细胞核大部分完全丢失，只有小部分新的纤维细胞仍然有细胞核，最终，新的晶状体再生替代了失去的晶状体（本图转载和改编自 Reyer RW. Quart Rev Biol 29：1-46, 由© 1954 芝加哥大学出版社许可）

机制获得一些了解。体内完全的转分化只见于背侧虹膜色素上皮。腹侧虹膜色素上皮对损伤作出反应而再次进入细胞周期，最终并不形成晶状体结构[8]。然而，维甲酸、转录因子 sine oculis 同源基因-3（*six3*）的过表达，以及骨形态发生蛋白（BMP）信号通路的抑制，三者的组合足以诱导腹侧虹膜色素上皮向晶状体的再生[10]，提示骨形态发生蛋白信号的缺失，*six3* 基因水平和维甲酸信号通路对于腹侧虹膜生成晶状体的能力起着决定性的作用。单纯抑制 BMP 信号通路也可以重建再生能力，但是发生的频率很小，因此 BMP 信号通路的抑制是关键的限速步骤[10]。配对盒基因转录因子 Pax6 是另一个晶状体形成过程中的重要转录因子[11,12]。IPE 细胞开始向晶状体细胞分化的早期表达 Pax6[13]。但是实验证明 Pax6 并不参与到诱导晶状体再生的早期事件中[10]：*Pax6* 基因转染不能诱导初始的去分化过程，基因下调不能阻止背侧虹膜色素上皮细胞去分化（色素丢失和细胞伸长）的发生[10,14]。相反，Pax6 在去分化的细胞接下来的增生和向晶状体的分化过程是必需的[14]。

一个有趣的问题是，在哺乳类为研究对象的实验中，任何一个用来诱导多能干细胞的因子是否也能用于蝾螈的晶状体再生。早期的实验使用虹膜色素上皮细胞和已经去分化的色素上皮细胞进行差别杂交，结果表明 c-Myc 在色素上皮细胞向晶状体细胞转变的期间高度上调[15]。最近，基因表达谱分析发现 $Sox2$、Myc 和 $Klf4$ 基因随着晶状体再生过程上调[16]。除了上述转录因子，Maki等发现晶状体再生过程中，干细胞相关的核干细胞因子（nucleolar factor nucleostemin）表达并在背侧虹膜色素上皮细胞再次进入细胞周期之前几天即出现上调[17]。在晶状体再生过程中，虽然促进诱导多能干细胞形成的因子有表达，但是 Oct4 和 Nanog 没有表达，因此虹膜色素上皮细胞不会获得多能性[16]。确定 Sox2、Klf4、Myc 和核干细胞因子在虹膜色素上皮细胞向晶状体的转分化中所起的确切作用还需要进一步的功能实验。

除了蝾螈，非洲爪蟾的幼体也能再生晶状体，然而这种两栖动物形成新的晶状体是通过角膜上皮细胞的转分化完成的[18]。它们的眼睛玻璃体腔内存在由胚胎时期的视泡和幼体的神经视网膜分泌的因子，触发角膜上皮的转分化[19~22]。Gargioli等的研究表明，由角膜上皮或头上皮生成晶状体的能力是由 Pax6 给予的[23]，反映出 Pax6 在赋予眼部再生能力中的普遍作用，与其在发育过程中所扮演的角色相同[11, 12]。未来的研究需要检测介导 Pax6 活化的信号和信号通路，确定是否还有其他促进再生的因子使晶状体周围的上皮细胞获得再生能力。

数种两栖类和鱼类也能再生视网膜。两栖类的视网膜再生由视网膜色素上皮层剥离，细胞去分化，再转分化形成神经元细胞类型[24, 25]。有证据表明这个转分化过程涉及视网膜发育中没有的转录程序：去分化的视网膜色素上皮细胞（RPE）表达 CRALBP（胞内视黄醇结合蛋白），这种蛋白质在胚胎视网膜前体细胞中不表达[26]，但是在胚胎期色素上皮和米勒（Müller）神经胶质细胞中表达[26, 27]，说明视网膜色素上皮细胞的去分化过程与胚胎发育过程中视网膜形成不同。对从视网膜色素上皮细胞到神经视网膜组织的转变中详细的表达图谱分析将有助于追踪转分化过程中的基因重编程。

通过添加 bFGF 也能诱导鸟类和哺乳动物胚胎的视网膜色素上皮细胞发生转分化[28~30]。这个转分化现象与两栖类中观察到的视网膜色素上皮细胞转分化类似，但存在一些区别：鸟类和哺乳类视网膜色素上皮细胞向视网膜细胞的转分化是一种彻底的转变，不会产生另一个不同的细胞层，视网膜色素上皮细胞会全部形成视网膜层[29, 31]。另一个区别是鸟类和哺乳类眼睛发育中，视网膜色素上皮细胞对 FGF 介导的视网膜形成应答有短暂的期限[28, 29, 32]。抑制视网膜色素上皮细胞分化过程中关键的信号通路激活素信号能延迟这种应答时限[33]。鸡胚视网膜色素上皮细胞能经诱导生成视网膜神经元，新生鸡雏只能由视网膜内的米勒神经胶质细胞而非视网膜色素上皮细胞产生新的视网膜神经元细胞[34]。

与蝾螈相比，鱼类视网膜再生来自于未受损的视网膜内核层的米勒氏神经胶质细胞。感光细胞受损诱发米勒细胞增生和迁移至损伤部位[35, 36]。成熟米勒细胞重新被激活表达视网膜前体细胞标志也涉及上述机制，正常情况下，视网膜前体细胞标志物由睫状体边缘区（通常是新生的视网膜细胞产生的位置）内的视网膜前体细胞和未成熟的米勒细胞表达[37]。这些结果说明米勒胶质细胞是具有再生视网膜能力的前体细胞。

鱼类视网膜内的视杆细胞、感光细胞和多巴胺能神经元受到损害后启动再生应答导致受损细胞的更新[38]。令人惊奇的是，选择性损伤视杆细胞或视锥细胞都能造成这些细胞的再生修复，而多巴胺能神经元的选择性丢失却不能复原。只有在损伤同时涉及其他细胞类型时，多巴胺能神经元才能被再生替换[39]。其他视网膜内定居的前体细胞如定向视杆细胞的前体细胞也能在鱼类的视网膜中存在[40, 41]；但是由于视杆细胞在整个机体生命过程中都经历持续不断的更新，而且细胞替换只发生在再生的视网膜神经元进入有丝分裂后期以后，因此新的视杆细胞的形成更类似于自我稳态的更新（见参考文献42）。

神经系统——隐藏的祖细胞（sequestered progenitor cell）模型

神经系统的作用是对外界环境进行识别、理解和作出反应，神经系统的维护是动物生存的关键。令人惊奇的是，一些脊椎动物（爬行类、两栖类和鱼类）保存了神经系统的中枢与外周神经元再生的能力。随着遗传和分子生物学技术的进步，我们对两栖类和鱼类如何再生受损的脑和脊髓结构有所了解。

大脑

两栖类具有在脑组织受损后再生出新的脑组织的强大能力[42~45]。通过手术把蝾螈部分视顶盖或墨西哥蝾螈的部分端脑切除，可引起再生应答，被切除的结构有明显的恢复[45, 46]。组织切片鉴定分裂的细胞，BrdU掺入和^{3}H胸腺嘧啶核苷同位素标记实验表明，在脑部不同区域的细胞增殖与再生相关[45, 47]。在非洲爪蟾幼体进行类似的外科神经组织切除手术也能引起视顶盖和端脑的再生[44]。

为了检测特定类型神经细胞的再生能力，常使用化学药物破坏掉特定的细胞，如脑内的多巴胺能神经元细胞。对酪氨酸水解酶（用于多巴胺生物合成的酶）进行免疫组化分析发现再生的组织结构中包括多巴胺能神经元[43]。即使丧失了75％的多巴胺能神经元，30天后大部分丢失的神经元也可以被再生取代[43]。BrdU掺入、GFAP（胶质细胞标志）和Neu（分化的神经标志）的免疫组化显示，新的神经细胞是由这些区域中胶质细胞的增殖产生的[43]，说明这些胶质细胞扮演再生新神经细胞的干细胞群的角色。这些数据揭示两栖类脑内新神经元的再生来自于成熟个体脑内增生区具有干细胞行为的胶质细胞。

与两栖类相似，鱼类对脑损伤作出应答，渐进地重建分层次的组织结构[48]，并伴随着运动功能的恢复[48~50]。在未受损成体鱼类脑组织的细胞增殖实验和细胞谱系追踪实验突出显示了全脑内数个具有神经干细胞样行为的区域（图13-3）[51~55]。这些区域内细胞持续增殖，提示这些脊椎动物相对的再生能力与基本的神经发生数量相关。然而，脑受到损伤后，损伤部位附近的细胞增殖明显加快[53]。因此，目前尚不清楚这些动物所表现出的复原能力有多少来自于隐藏的前体细胞中心发生的神经再生，或者损伤周围神经组织的更广泛应答。当前的实验试图确定损伤发生后新的神经从哪里和怎样生成，以及将新生成的神经靶定到正确连接的引导信号是什么。实验追踪视神经的再生整合，发现再生后错误的神经连接的数目增多[56]。就这一点而言，金鱼的视神经再生后脑内视神经的错误连接最后消失[57]。确认引导新生的神经完成正确连接的信号还需要进一步实验研究。

脊髓

星形神经胶质细胞是一种非神经元类型细胞，在所有脊椎动物发育中的神经系统均

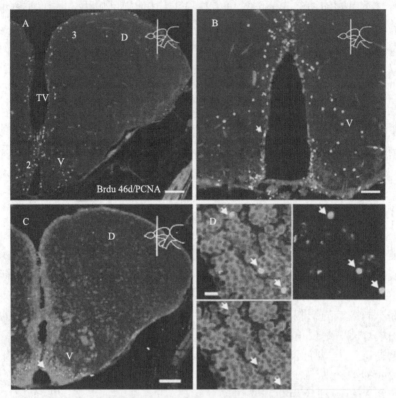

图 13-3 成年斑马鱼脑内有内源性增生区域。A. 成体斑马鱼端脑横截面，增生的细胞核抗原免疫组化染色（PCNA，红色）。BrdU 阳性细胞（绿色）从腹侧端脑（V）增生区沿着端脑室（TV）迁移。B. 高倍镜下的脑室增生区。C. 腹侧端脑增生区附近，BrdU 阳性细胞呈神经标志物 Hu 阳性（红色）。D. 如 BrdU 掺入（绿色）所示，新的神经细胞通过细胞增殖产生（本图转载和改编自 Grandel H et al. Dev Biol 295：263-277；52 2006 由 Elsevier 许可）。（另见图版）

有发现，它们可自己作为神经元前体细胞并引导新生神经元的放射状迁移来促成中枢神经系统的形成[58~60]。成体蝾螈和青蛙的幼体能依靠星形胶质细胞的再生行为，把脊髓再生作为尾再生的一部分，成功地再生出脊髓（图 13-4）[61,62]。两栖类脊髓损伤导致星形胶质细胞转录胚胎基因，经历上皮-间充质的转变，并发生增生和迁移[62,63]。这些细胞形成一个神经上皮细胞组成的管状结构（室管膜管），经过增殖，其中一些细胞会形成新的神经元[64~66]。克隆分析发现脊髓内的这些细胞具有一定程度的可塑性，能产生不保留亲本基因表达和区域组织特异性的子代细胞[63,67]。在成体动物，胚胎基因（*Shh*、*Pax7*、*Msx1*）持续表达使得沿着脊髓背-腹轴仍然保留前体细胞区域（图 13-4）[63,68]。研究表明，Shh、Wnt、Bmp 和 Notch 在尾部再生中有重要的作用。Wnt 和 Bmp 信号足以诱导脊髓和尾部再生（图 13-7A~D）[68~70]，而再生过程需要 Shh 信号，尽管提示这些因素在脊髓再生中扮演着一定的角色，但是它们对损伤周围组织直接或间接的作用尚不清楚[68]。值得一提的是,激活 Notch 信号能诱导脊髓和脊索的过生长,但是对周围肌肉组织无促生长作用,因此 Notch 信号可能对脊髓的再生有直接影响[69]。

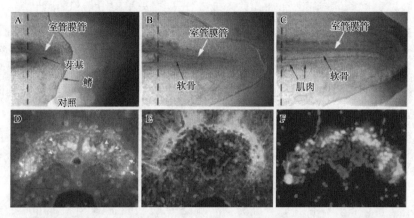

图 13-4 墨西哥蝾螈（*Ambystoma mexicanum*）尾巴和脊髓的再生涉及分化的神经细胞以外的前体细胞。A. 尾部切断 4 天以后,明显的新尾再生。再生中的尾正在形成室管膜管（将生成脊髓）、芽基（将生成尾部肌肉和软骨）、表皮鳍。B. 断尾 8 天后,新生尾有大幅度生长,可见骨骼软骨和室管膜管。C. 断尾 14 天后,再生中的尾开始沿着软骨和室管膜管分化出肌肉。D. 成体墨西哥蝾螈脊髓免疫组化染色检测 Pax6（红色）和 Pax7（绿色）,结果表明这两种参与脊髓发育的基因在成体仍然有表达。E、F. Pax7（红色）在分化的神经细胞中不表达,分化的神经细胞由Ⅲ型微管蛋白（绿色）标记,Neu 阳性（绿色）的神经细胞不表达 Pax7（A～C 转载和改编自 Schnapp E et al. Development 132：3242-3253[68], 2005；D～F 转载和改编自 Mchedlishvili L et al. Development 134：2083-2093[63], 2007；The Company of Biologist Ltd.）。（另见图版）

鱼类也有很强的脊髓轴突再生能力,使得它们能在受损害后数周内恢复运动能力[71~74]。尚不清楚鱼类神经再生是否与蝾螈一样广泛,但是一些实验提示确实发生了丢失神经元的补充。对线翎电鳗（*Sternarchus albifrons*）和黑魔鬼（*Apteronotus albifrons*）再生的脊髓做组织学分析,结果表明脊髓再生是通过室管膜管的生长介导的[75, 76]。室管膜管细胞培养表明这些细胞是新生神经元的来源[77]。最近,Reimer 等使用斑马鱼内的转基因标记示踪 Shh 信号引起的新神经元再生[78]。鱼类似乎不能再生中枢神经系统中所有的神经[79],斑马鱼幼体脊髓损伤后再生的组织缺乏毛特讷氏细胞和米勒神经元[80]。然而,当向细胞或横断的神经添加双丁酰环磷酸腺苷（cAMP）后,可以刺激毛特讷氏细胞的再生[80],说明缺失的神经细胞类型在特定信号的刺激下可以再生。

心脏——分化细胞的再生模型

两栖类和鱼类心室被切除后可以补充失去的心肌细胞再生出新的心室[81, 82]。虽然损伤后形成瘢痕组织,但是这些动物会产生新的心肌替换掉瘢痕组织从而修复心室结构（图 13-5）[82, 83]。虽然目前尚缺少确切的细胞基因学追踪,用 DNA 合成的标记物如氚化胸腺嘧啶和 BrdU 掺入法研究体内、体外培养的心肌细胞核及培养心肌细胞的分裂，发现这些细胞在分裂期间至少保留一些肌细胞特征[81,82,84,85]。一些颇有意义的实验用一种亲脂性染料标记培养的蝾螈心肌细胞，发现当处于有肢体再生的环境中时，心肌细胞可能具有进一步去分化的能力。当把荧光标记的蝾螈心肌细胞从心脏移植到再生中的肢体胚基，移植的细胞明显失去心脏特异性基因表型，最后形成骨骼肌和软骨[86]。然而，由于这些追踪实验使用细胞膜染料，不能排除染料通过细胞融合和移植细胞对其他类型细胞如结缔组织的污染使宿主细胞染色。因此需要利用基因示踪工具对上述现象再次检验以确定这些发现的真伪。

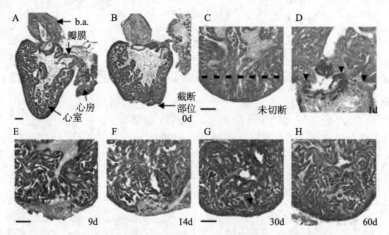

图 13-5 成体斑马鱼心脏再生心室壁。A. 成体斑马鱼心脏由一个心室、一个心房和动脉球及原始的心室流出道组成。B. 外科手术切除心室尖部、心室壁的心内膜，心肌层和心外膜层。C. 高倍镜下的心尖部，显示一道近似的截断创口（虚线）贯穿室壁。D. 切除手术后 1 天，心室组织明显缺失，创口有血凝块形成（箭头所指）。E. 切除术后 9 天，心尖部出现纤维瘢痕组织。F. 随着时间推移，瘢痕组织被心肌组织所替代。G. 瘢痕组织不断被心肌组织所替代，新的心室壁和肌肉组织形成。H. 虽然原始的损伤范围较大，大约 60 天后，心室已看不到明显的损伤迹象（转载改编自 Poss et al. Science 298: 2188-2190 2002 AAAS）。（另见图版）

最近，斑马鱼转基因报告技术实验发现已存在的心肌细胞是心肌再生的主要来源[87,87a]。一些分子学证据表明心脏发育相关的基因如 *gata4*、*hand2* 和 *nkx2.5* 能在心肌细胞内再次表达[88]。其他一些影响前体细胞的基因如 *notch 1b*、*deltaC* 和 *msxb*，在再生的心脏中表现为表达上调[89]。基因芯片分析显示心脏组织再生中有一些生长因子表达，目前的研究重点强调哪种因子诱导心肌细胞增殖。特别是，血小板衍化的生长因子（PDGF）参与心肌细胞周期的转化[90]。损伤信号怎样影响心肌细胞基因表达和细胞扩增是需要更多研究工作来解决的问题。

除了心肌层之外，损伤的心脏还需要重建心内膜和心外膜层。在两栖类和鱼类心脏再生中这些组织所经历的过程尚不清楚，但可以推测与它们在心脏发育中的作用类似。心外膜向心肌细胞提供生长信号并协助生成血管[91]。心内膜也起着同样的作用[91]。解决心外膜、心内膜和其他组织是否影响心脏再生的过程这个问题需要更多的研究工作。

附肢——从成熟组织产生祖细胞

由于附肢的实验模型容易建立，其再生过程又较复杂，因此是再生研究的一个焦点。成熟的肢体包括多种组织，在肢体再生中需要沿着前-后、背-腹和近端-远端轴发生一系列形态学模式事件。目前的研究集中在理解肢体被截断后怎样重新生成多重组织组成的复杂三维结构。组织学研究表明，一旦断肢后，伤口周围的上皮细胞迁移覆盖伤口，接着形成含有增生的前体细胞群的间充质细胞团（芽基）[92~94]。蝾螈组织嫁接和细胞谱系追踪实验表明芽基细胞首先来源于截断创口几毫米内的组织，这些组织的细胞失去了已分化的形态特征[98]。

一个重要的科学问题被提出：芽基的形成是涉及典型的有丝分裂后细胞的去分化还是干细胞池的活化？现有的证据表明，这两种过程都存在。基于超微结构的组织学分析，长期以来认为蝾螈肌肉组织缺乏在其他脊椎动物中修复横纹肌的干细胞——卫星细胞[99~102]，但是近期分子生物学实验发现蝾螈确实具有卫星细胞[103]。然而，卫星细胞在肌细胞形成芽基中似乎并不起主要的作用：一些体外培养和细胞追踪实验记录了多核肌管和肌纤维断裂成单核增生的细胞[104~108]，说明肌细胞去分化是生成芽基过程中产生祖细胞群的一种机制[98,106,109]。并且，早期移植实验使用不同的组织（可通过放射性标记跟踪或将三倍体供体组织嫁接到二倍体宿主动物），发现除肌细胞之外，皮肤成纤维细胞、软骨、施万细胞和结缔组织细胞共同参与芽基的形成[96,97,110,111]，也构成再生附肢组织结构[112~114]。

各种组织细胞失去分化特性导致芽基的形成，但是否所有的芽基细胞都去分化获得多能性，能够产生肢体所有类型的细胞，或者每种组织只能产生该种组织限制性的芽基细胞？已有的实验数据很难完整地回答这个问题，因为以往的实验手段（含氚的胸苷标记组织或在二倍体动物内的三倍体细胞进行细胞谱系追踪）产生的结果不能完全排除其他来源细胞也参与了芽基的形成。例如，Steen 早期的实验中，使用氚标记的胸腺嘧啶标记软骨细胞进行移植，产生了一小部分具有标记的肌肉组织细胞[115]，提示软骨来源的细胞可能不是高度多能性的。类似的，当缺乏分化的骨骼肌的肢体芽基移植到背鳍上皮部位，不能产生肌肉却产生软骨组织[116,117]。把芽基嫁接到眼眶也得到相似的结果[118]。只有在残肢创口的分化的肌肉组织细胞接触到移植的芽基时才会生成肌肉组织[116,117]。这些结果表明肌肉再生中需要肌组织参与，软骨形成细胞不能够产生肌肉，虽然其他的解释也是可能的。相反，肌组织移植实验表明肌组织产生的芽基细胞可以生成其他组织[119]。最近的实验接种培养的卫星细胞，明显地生成了其他类型的组织，也证实了这些结论[103,120,121]。然而，由于缺乏广泛的、长期的谱系标记和确切的组织特异性分子标记来追踪细胞身份，从而使这样的实验受到限制。

表达 GFP 的转基因技术在蝾螈细胞谱系的应用发展，以及可用分子标记的增多使得这个问题研究结果更可信，结果提示芽基细胞在分化潜能方面高度受限。Kragl 等应用胚胎移植以高纯度标记肢体主要的组织类型细胞。这个技术与直接肢体移植（嫁接）相比的优势在于能够识别胚胎阶段，每个主要组织的原基细胞，如肌肉、表皮/结缔组织等的移植不会受到其他细胞层的污染（图13-6）。以往的实验使用成体肢体组织移植，通常包含几种不同的细胞群。Pax7作为肌祖细胞分子标记的应用对于区分肌祖细胞和散在的结缔组织成纤维细胞是必需的。这种组织特异性细胞标记的结果是每种组

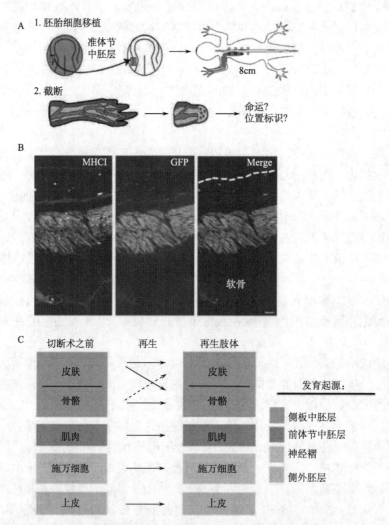

图13-6 再生肢体细胞的可塑性有限。A. 使用胚胎移植（嫁接）标记组织的方法。来源于表达GFP的胚胎组织移植导致GFP阳性细胞参与构成成体动物的组织结构。组织特异性GFP转基因的表达可以对再生组织中的这些细胞进行谱系追踪。B. 使用肌肉特异性肌球蛋白重链Ⅰ（MHCⅠ）和GFP免疫组化双标记细胞叠加观察，表明GFP标记的肌肉细胞（切除前）再生后参与形成肌肉，但是不形成皮肤或软骨。C. 在蝾螈肢体再生中细胞命运的归纳（转载改编自Kragel et al. Nature 460: 60-65; 2009）。（另见图版）

织，包括肌肉、表皮、软骨和施万细胞，产生只能向有限的组织类型分化的祖细胞[113]。因此，芽基细胞不是多能的，其分化潜能具有高度限制性。这些实验只在墨西哥蝾螈（axolotl）中进行，它是一种蝾螈家族属钝口螈科（Ambystomidae）的主要代表动物，因而在其他的蝾螈科动物内验证这个结果是比较重要的，比如许多重要的再生实验采用的蝾螈（newt）模型。

再生过程对祖细胞的引导

三维结构的再生涉及的组织提供细胞外信息和细胞内因子，后者通过这些信息的传递产生和引导祖细胞。因此，复杂结构再生的另一个重要方面就是需要维持一个促进再生的环境。当即将形成芽基的细胞呈现去分化的形态特征，这些细胞开始迁移和明显增殖，产生正确的、高度有序的、适合大小的结构，而不会产生肿瘤或癌变。关于肢体再生的研究表明上皮和周围神经调节芽基的生长，并探索了芽基沿着特定躯体轴线的行为，包括截断部位残肢上的细胞如何重编程来重新形成失去的远端结构。

表皮

伤口愈合后，残肢形成特殊类型的、增厚的上皮组织[122~125]。如果用移植的、未受损的皮肤替换这个特殊的上皮组织，再生过程将停止[126]。细胞谱系追踪实验表明表皮内的细胞在再生中只参与新表皮的形成[113,114,127,128]，提示伤口上皮细胞对芽基起调节作用但并不参与芽基的构成。最远端的伤口表皮区域形成顶外胚层帽（AEC）[122,123]，与鸡胚发生中小鸡肢体发育的顶外胚层脊（AER）有关，这个增厚的上皮结构引导肢体的发育。然而，在鸡胚 AER 和再生的两栖类 AEC 之间有一个很重要的区别：AER 在失去后不会再生，移除 AER 导致肢体截断[130]，而去除 AEC 后再生的肢体将重新形成 AEC[123,131]。这说明再生能力与 AEC 的重新形成相关。移植实验提示在 AEC 的重新形成中涉及图式形成信息的维持[131,132]。但是，再生特异性的维护程序是什么尚不清楚。

伤口部位的上皮细胞在再生中似乎起到引导的作用。将 AEC 移植到芽基后面非对称性的位置，导致相应的非对称性肢体再生，再生的肢体与残肢形成锐角[133]，而将 AEC 移植到芽基基部则形成异位生长的肢体结构[134]，这是特殊的伤口上皮细胞指导再生的结果。两栖类和鱼类的伤口上皮细胞表达数种促进相关细胞迁移、增殖和基因重编程的生长因子[135~138]。维甲酸是伤口表皮细胞产生的一个分子[136]，是维生素 A 的衍生物。维甲酸是一种强大的形态发生素，当用不同浓度的维甲酸处理发育或再生中的附肢时，可以通过把将会形成远端结构的芽基重编程为产生近端和远端结构的芽基，调整近端-远端的图式形成信息[139~141]。伤口上皮细胞内其他的生长因子包括 FGF（墨西哥蝾螈）、Wnt5（墨西哥蝾螈和斑马鱼）、shh（斑马鱼）和 $bmp2$（斑马鱼）[137,138,142,143]。在再生附肢的上皮细胞内，这些生长因子由不同的结构域表达，说明上皮细胞向芽基中基本的前体细胞提供区域性信号，调控图式形成[135]。

周围神经

周围神经在附肢再生中的重要性是在两栖类和鱼类的去神经支配实验中发现的：将

通往附肢的神经切除，则会阻止附肢的再生[144~147]。此外，重新定向两栖类肢体的周围神经到皮肤附近异常的部位造成神经尖部邻近处形成增生物[148]。这种增生物与附肢芽基具有共同的特征：在一开始表达 Msx2、Tbx5 和 Hoxa 13，并且会形成分化的组织[149,150]。尽管这种异常的神经能启动芽基的生长，但芽基的生长最终停止，被招募到芽基的细胞将分化形成异位的骨骼、肌肉和结缔组织。然而，当把对侧附肢的皮肤移植到伤口和异常的神经旁边时，会产生一个完整的异位附肢[148]。这些数据提示神经提供有助于再生的早期信号，但是只有神经不足以维持再生性生长。

在伤口部位上皮细胞和神经细胞之间的相互关系促进肢体的再生。组织学观察表明伤口上皮细胞受到神经的支配[151]，而且将神经异位植入皮肤会诱导形成一种增厚的上皮，这些上皮细胞表达 Sp9——一种在再生附肢伤口上皮细胞典型上调的转录因子[150]。它首先在施万细胞表达，接下来在维持再生的上皮内表达分泌性因子 nAg（anterior-gradient ligand）[152]。把 nAg 转染入截断后去神经支配的残肢细胞，可以重新生成周围肢体神经（图 13-7E，F）[153]。这种分泌性因子与 Prod1 相互作用，后者是一种细胞表面分子，表达在再生附肢的芽基细胞和周围神经[152,154]。这些数据表明 nAg 与 Prod1 相互作用调控再生附肢远端的生长。在悬滴培养中用抗体抑制 Prod1，将抑制远端的芽基向近端生长，Prod1 异位表达可以使本来向远端生长的芽基向近端生长，提示 nAg-Prod1 通路参与了附肢再生中近-远端的图式形成[154,155]。维甲酸可以上调 Prod1[154]，提示 Prod1 参与了维甲酸信号通路介导的图式形成活动。未来的实验需要探索 Prod1 如何将近端识别信号转入芽基细胞。

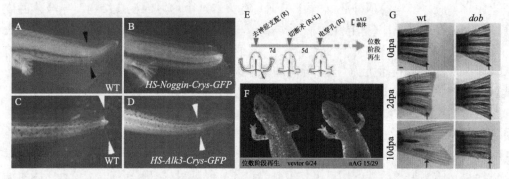

图 13-7　尾部、附肢和鳍再生所需的分子。在非洲爪蟾尾再生中需要 Bmp 信号的参与。A. 非洲爪蟾蝌蚪尾再生引起脊髓、神经管、骨骼肌和鳍组织的重新形成。B. 热激引起的 Noggin（Bmp 信号抑制剂）过表达，抑制尾部再生。C. 非洲爪蟾蝌蚪正常发育阶段的一个不应期内表现出再生减弱。D. 相反地，诱导 Bmp 受体 Alk-3 表达能重建不应期内的再生能力。去神经后附肢再生的缺失可以通过添加 nAG——一种 Prod1 跨膜受体的可能配体而解救。E. 实验设计：第 1 天去除周围神经供应，7 天后切断去神经的肢体，5 天后用 nAG（anterior-gradient ligand）基因转染附肢细胞。F. 没有 nAG 转染的附肢不能再生（载体对照），有 nAG 转染的大部分动物确实再生出远侧的肢体和足趾的组织。黄色星号提示对照组（左）和 nAG 转染组（右）截断的附肢。鳍再生中的 Dob（$fgf20a$ 突变）。G. 野生型鱼类切断尾鳍引起进行性的鳍组织再生（左侧一排）。然而，dob 突变（右）缺乏形成芽基的能力，因此不能再生鳍组织（转载和改编自 Beck et al. Dev Cell 5：429-439；Kumar et al. Science 318：722-777）

神经细胞被发现可表达其他几种能诱导芽基细胞增殖的因子：成纤维细胞生长因子2（Fgf2）、神经胶质细胞生长因子（Ggf）和转铁蛋白。Fgf2和Ggf都足以诱导去神经支配的蝾螈附肢中的细胞增殖并诱导与再生相关的基因表达，后者包括芽基标志22/18和Dlx-3（Distal-less homolog 3）[156~158]。此外，Fgf2可以在神经去除后维持附肢重建[158,159]。神经相关的转铁蛋白也参与了再生：转铁蛋白促进体外培养的芽基细胞增生，并能维持芽基细胞在去神经支配的肢体中的增生[161]。Fgf2和Ggf似乎直接参与了特定信号通路的激活，转铁蛋白参与的机制尚不清楚。这些神经相关因子的活动提示神经参与到再生的机制涉及多条分子通路。然而，这些通路之间的关系尚需进一步研究。

Fgf信号除了在神经依赖性再生中有潜在的作用外，其重要性亦在斑马鱼中用化学方法或基因重组方法破坏掉这条信号通路（图13-7）得以验证[162,163]，用一种Fgf受体信号的药理学阻断剂SU5402处理斑马鱼，会阻断芽基的形成。另外，将 *Fgf20a* 基因突变使其干扰Fgf蛋白配体与受体有效结合的能力，也能阻止正确的芽基形成和再生，但是不影响鳍的本来发育[162]，提示Fgf20a具有再生特异性的功能。其他的Fgf成员是否能取代Fgf20a的功能目前尚不清楚。

其他的信号因子

Wnt

Wnt生长因子信号在几种涉及祖细胞群及其后续分化事件维持组织发育和再生的过程中被发现。例如，基于过表达和抑制实验，Wnt2b通过阻止视网膜祖细胞区的祖细胞分化成视网膜神经元，从而参与维持视网膜祖细胞群[164,165]。在蝾螈、青蛙和鱼类的附肢再生中也有Wnt信号[137,166,167]。在非洲爪蟾附肢和尾的残端过表达Wnt信号的负调控因子（Axin-1或Dkk1）抑制经典Wnt信号通路，会减少或抑制再生[166,168]。相反，通过活化经典Wnt通路，过表达β-catenin，能在非洲爪蟾附肢的芽基没有丢失天然再生能力之前，增强其再生能力[166]。在斑马鱼中过表达dkk1、抑制β-catenin信号通路的激活也能发现抑制鳍的再生[137]。

有趣的是，Wnt-5的同源物对斑马鱼鳍和非洲爪蟾尾的再生表现出截然不同的作用。Wnt5b过表达会拮抗斑马鱼鳍的再生[137]；而在尾部被部分截断的非洲爪蟾使wnt5-a过表达，将诱导异位生成一个完整的尾部[70]。需要进一步探索的是，这种差别究竟是由于wnt-5a和wnt-5b信号通路的不同引起的，还是由于尾鳍和非洲爪蟾尾之间组织特异性的差别引起的。总体来说，似乎wnt信号在胚胎发育和组织再生中都发挥多种作用，这些作用是部分重叠的。未来定义wnt信号通路与再生特异性相关的功能显得非常重要。

Notch

Notch-delta信号通路参与数种组织中细胞命运的决定，其中包括脊椎动物中分化组织的干细胞群（综述见参考文献161）。Notch信号可能参与附肢的再生；然而，notch信号通路在附肢再生中的作用如何尚不清楚。

Tgf-β

TGF-β、激活素、抑制素、骨形态发生蛋白（Bmp）都属于转化生长因子家族。这些因子与跨膜受体一些特异的亚族发生作用，后者再与细胞内信号转导中间体的亚族相互作用，进而改变基因转录。在蝾螈前肢被截断后6h即能检测到Tgf-β的表达，这种表达持续到再生的早期阶段[169]。经一种能与Tgf-β1型受体结合的化学抑制剂（SB-431542）处理，由于细胞增殖受到抑制，早期和晚期增生阶段的再生过程被阻断[169]。在非洲爪蟾蝌蚪尾部再生过程的早期和晚期阶段也观察到需要Tgf-β[170]。特别有趣的是，SB-431542产生的抑制伤口愈合的作用是可以逆转的[170]，而这个抑制剂阻断再生的作用却是不可逆的[169,170]。这些实验中使用的这种化学抑制剂虽然选择性抑制Tgf-β1型受体，但是对（Alk4、Alk5和Alk7）也具有亲和性[171]。因此，这也能影响到这个超家族里其他成员的信号转导，如激活素（activin）。激活素在早期鳍再生过程中表达上调。原位杂交实验显示断肢损伤后，再生的新芽形成之前，在鱼类鳍上皮有激活素βA转录发生。用转录特异性吗啉代抑制激活素βA转录将阻断芽基的形成，这种阻断作用一部分是因为对鳍的基部细胞诱导增殖所致[172]。总体来说，激活素功能干预实验的结果与两栖类附肢的SB-431542抑制实验结果类似，用SB-431542处理截短的斑马鱼鳍导致产生一种锯齿状表型。再生的鳍上皮细胞覆盖伤口，但是切除边缘的鳍组织退缩到间辐条（interray）位置[172]。这个表型反映出鳍切除后激活素βA的早期表达模式[172]，提示SB-431342诱导的表型改变是由激活素信号通路受抑制引起的。所以，进一步的基因实验需要检测Tgf-β是否存在有别于活化素活化引起的功能。

Tgf-β家族生长因子成员Bmp在再生中也起着重要作用。将非洲爪蟾蝌蚪，通过转基因使其过表达一种Bmp拮抗因子Noggin，则不会有尾部再生发生[69]。类似地，过表达Bmp拮抗因子Chordin抑制Bmp信号将会减少骨的形成[173]。与此相反，在鳍的芽基中过表达bmp2导致皮鳍条骨（dermal ray bone）的产生增加[138]。除了基于配体存在与否、提供再生的诱导和抑制信号，这些结果提示再生过程是一个复杂的信号转导系统。需要确定的是这些信号怎样形成和保持，并在不同的组织中发挥再生功能。

Shh

在两栖类和鱼类的再生组织内都发现有Shh（sonic hedgehog）存在，其作用是必需的。抑制蝾螈再生的前肢和后肢中Shh信号转导导致严重的远端趾数目下降[174]，切除鱼鳍后用化学方法抑制Shh信号会阻断再生物的生长[175]。相反地，使用病毒感染的手段使再生附肢组织内Shh过表达则造成异位足趾、复趾和趾骨融合[176]，再生鱼鳍内过表达Shh导致皮鳍骨脉的错误生长[175]。切除非洲爪蟾幼蛙的附肢确实引起芽基形成，但是只形成不分支的软骨组成的刺状附肢[177,178]。不能再生出所有足趾的原因可以归于幼蛙附肢芽基中Shh缺乏再次表达[177,178]。提供Shh信号可以在再生的刺突内生长出分支的软骨结构，但是不形成有指的附肢[179]，提示正确的图式形成需要其他因素参与。基于基因表达分析和现有的功能实验数据，Shh信号在再生中发挥的作用似乎与其在发育中发挥的作用相同。迄今为止，两个关于Shh和其他hedgehog家族成员的基本问题是：

①Shh在组织再生中扮演的角色是否有别于其在胚胎发育中扮演的角色？②再生过程中激活细胞内 Shh 表达的因素是什么？

生长促进信号的细胞内翻译

芽基细胞分化状态的改变和周期中细胞的增加是同时发生的，随后发生的细胞分裂增加是芽基形成所必需的。一旦达到一个特定的身体-器官比例关系以后，再生的组织会停止生长。随着芽基细胞增生，芽基近端区域的细胞分化并不留痕迹地与残余未损伤的分化组织整合[180~182]。细胞分化和增生的状态必须同时伴有引导的机制，调控组织再建以重新形成失去部分的正确结构。因此，关于这个调控机制有两个问题：是什么机制激活静息的、终末分化的细胞进入细胞周期；当达到正确的组织比例时，是什么机制使细胞增殖停止。

细胞周期控制斑马鱼鳍再生

观测再生鱼鳍生长速率的实验发现，对于再生新生物的生长速率有一个位置元件——当再生的鳍新生物接触到远端的结构时生长速率下降。这个位置信息与成纤维母细胞生长因子信号的水平相关，因为当生长速率降低时，下游 Fgf 转录靶点的表达下降[183]。因此，人们可以推测生长中断的大致机制是源于 Fgfs 和其他促进新生物形成的再生基因下调引起的。

脊椎动物基因 fam53b/simplet（smp）分子似乎把细胞增殖和组织图式形成联系在一起。在再生的鳍，smp 敲除通过抑制细胞增殖和上调 msx、shh 的表达，减少再生新生物生长，并引起异位骨形成[184]。这个基因编码一种细胞内蛋白，包括两个保守的区域，其功能未知[185]。因此，探索这个细胞内因子如何将细胞外再生前信号传递到细胞增殖的调控和组织图式形成中将是一个有趣的事情。

两栖类附肢的细胞周期调控

自然状态下，细胞周期调控因子参与细胞暴发增殖，管理芽基的形成，但是似乎在再生中特定的细胞周期调控因子发挥作用，或被调控过程存在着独特的特点。p53 是一个肿瘤抑制基因，当细胞处于应激状态或者受到 DNA 损伤时能抑制细胞分裂或者诱导凋亡[186]。有趣的是，p53 化学抑制剂能损害附肢再生[187]。假设所用的药物具有特异性，这些抑制性结果提示需要发生显著细胞增殖的过程的再生新生物的生长需要 p53 的活性。长期培养的两栖类（蝾螈）芽基细胞持续增殖，不发生任何明显的改变或衰老[188]。p53 和芽基细胞潜在的永生化特征的必要性提示 p53 可能在这里发挥一个有别于其抑制细胞分裂和促进凋亡的作用。未来需要通过实验研究哪些细胞需要 p53 的活性、是否 p53 在再生中发挥除了中止细胞周期之外的功能。除了抑癌基因 *p53* 之外，在成年蝾螈的心肌和横纹肌中，细胞周期抑制剂 Rb（retinoblastoma）蛋白的调控与哺乳动物不同。分化的蝾螈肌肉保留磷酸化 Rb 的能力，这个能力与蝾螈肌纤维在血清刺激后重新进入细胞周期相关[187,189]。相比之下，血清刺激的分化的哺乳类肌小管不能进入细胞周期[190,191]，除非去除 Rb 蛋白：实验发现培养的、来源于 *Rb* 基因敲除小鼠的肌小管可以被诱导进入细胞周期,并呈血清依赖性[192]。这些结果提示使 Rb 介导的细胞周期抑制解除的调控机制在哺乳类肌肉中缺如,但在蝾螈肌纤维中存在。这些机制的具体细节是未来需要继续研究的。

祖细胞和芽基形成所涉及的分子因素

祖细胞群的形成,如芽基内的祖细胞,很可能涉及几种细胞外和细胞内的分子机制,将成熟附肢组织细胞转变为芽基细胞的分子因素是再生学研究的一个主要焦点。已发现有一些因子,如 Msx1、Twist、Pax7 等,参与了附肢再生中祖细胞状态的转变或调控,关于诱导分化的组织细胞变成芽基细胞的表观遗传学和转录调控机制还有很多内容尚不清楚。

Msx1

近些年来一些颇吸引人的实验结果指出 Msx1 在肌细胞去分化中的潜在功能。Msx1 首先被确认为一个在增生的、未分化的、发育中的附肢胚芽尖部表达的因子[193~195],能够阻止体外培养的成肌细胞向肌小管的形态学分化[196,197]。Msx1 在再生过程中重新表达这一发现提出了另一个问题:Msx1 是否可以推动肌小管向成肌细胞的去分化[198~200]。Odelberg 发现在体外形成的小鼠肌小管内强制表达 Msx1,一小部分细胞似乎发生了去分化[121]。一项检测从蝾螈分离的肌纤维的实验提示 Msx1 表达可能与肌细胞断裂有关[201],因此可能参与了肌小管向成肌细胞的转变过程[121,202]。现在已经清楚非肌细胞中的 Msx1 在再生过程中也有上调,斑马鱼中 Msx 的同源基因在再生的尾鳍内也被重新激活[198],再生的尾鳍缺乏肌细胞,但是仍然需要 Msx 活性:基因敲除再生鱼鳍中 Msx 的鱼类同源基因($msxb$),能够抑制再生新生物的生长[203]。在未来的实验中,条件性激活和抑制芽基形成期间及其后的 Msx1 表达,需要直接验证究竟是一部分程序在芽基形成中驱动着去分化,还是一旦芽基细胞形成后即通过某种机制来阻止它们在未成熟状态的再次分化。

Twist

Twist 是另一个参与肌组织去分化的分子。Twist 所属的基因家族参与数种重要的中胚层-外胚层来源的组织分化过程[204]。原始的 $Twist$ 基因在中胚层的形成中是必需的[205,206],它的功能与成肌细胞的发生有关[207,208]。使用腺病毒转移方式使 Twist 在培养的肌小管内过表达能阻止培养的成肌细胞形成肌小管,并且似乎也能诱导已形成的肌小管断裂成碎片,同时伴有肌组织相关基因转录的下调[209]。未来的实验需要验证这个基因是否在再生的附肢表达、对于再生是否是必需的,以及这个基因是否足以促进肌组织或其他组织再生中的去分化。

另一种哺乳类 $Twist$ 基因、蝾螈 AmTwist 同源物——Twist-2/Dermo-1,被发现在芽基近端区域表达[210]。基于 AmTwist 在再生附肢有表达,它在再生过程中可能具有与哺乳类 Twist-2 在附肢发育中类似的功能。Twist-2 在表皮分化和骨骼元件发育的早期阶段在外胚层下间质表达[211,212]。将蝾螈肢体内周围神经转移到皮肤附近的异常位置,导致表皮成纤维细胞增殖产生一种与早期切除附肢的芽基类似的结构[148]。AmTwist沿着新愈合的伤口的整个表皮表达，这种表达受近端周围边缘的限制，并且在神经诱导的芽基末梢中心区域没有表达[210]。在再生的附肢，芽基近端区域可以观察到组织分化；因此，近端AmTwist的表达提示其可能参与了表皮细胞的分化或成熟过程。未来还需要进一步研究AmTwist在再生新生物近端区域的功能。

Pax7

配对盒（paired-box）转录因子在胚胎发生的不同阶段表达于不同的组织，但是一些成员在成人祖细胞内仍然有表达[120,213~216]。例如，发育过程中骨骼肌祖细胞的行为受到Pax3和Pax7的调控[217]。在成体鱼类、蝾螈、青蛙和哺乳动物，Pax7表达在静息的、新激活的骨骼肌干细胞，并对骨骼肌干细胞的保持具有重要意义[217]。Pax7足以引起原位的肌肉干细胞（卫星细胞）和CD45/Sca1阳性细胞的肌原性分化[218]，后者是肌组织中一个成年干细胞群的侧群细胞，来源于造血细胞[219~221]。

非洲爪蟾实验也提示Pax7在横纹肌再生中的重要性。免疫组化和电镜可以鉴别出横纹肌中的Pax7阳性细胞，这些细胞具有卫星细胞的形态学特征：小细胞由在肌纤维周围的细胞核支配[222]。非洲爪蟾幼虫截断尾部后，Pax7表达明显上调，在再生的尾部所有细胞内过表达一个主要的阴性Pax7构建体，将阻断肌肉的再生，提示Pax7对于肌肉再生是必需的[222]。在哺乳类细胞培养中，保持肌干细胞群的部分机理是Pax7通过招募组蛋白甲基转移酶MLL2到肌原性调控因子Myf5[223]，Myf5是一个参与向成肌细胞转化的基因。这种分子机制是否参与了两栖类附肢肌肉再生尚未明了。

Meis 和 Hox 基因

切除后的附肢近端的组织产生新的远端附肢组织，因此，再生能力涉及上臂的细胞重编程形成下臂的组织结构。移植实验已经表明皮肤成纤维细胞含有再生肢体图式形成必要的信息[113,224~232]。因此，成纤维细胞被认为参与了再生中位置信息的再次确定。成纤维细胞中传递图式形成的信息是什么？

Meis转录因子家族已经被确认能在再生附肢内传递位置识别信息。Meis1和Meis2在发育中和再生中的附肢近端区域表达[233~235]。在再生的附肢中过表达Meis，Meis阳性细胞不能参与形成远端的肢体组织，而是形成近端的结构[233]。这个结果与Prod1的近端化功能类似[155]。Prod1和Meis基因都能被维甲酸激活[154,236]，提示这些基因传递维甲酸信号的近端化活性。未来的工作需要研究Prod1和Meis是通过同一条信号转导机制还是各自的通路来诱导细胞向近端结构细胞的转化。

发育研究已经表明Hox基因对于沿着几个躯体轴线的图式形成有重要意义[237]，因为Hox基因突变导致沿着前-后或近侧-远侧轴线产生分段的结构[238]。表达数据显示Hox基因在再生的附肢再次表达[239,240]，Hox基因对于再生是必需的[241]。另外，Meis是Hox基因的转录辅因子[242]，提示Meis通过与Hox基因相互作用调控图式形成。需要阐明的是，是否图式形成信息只有在再生中的图式形成过程中存在，或者图式形成基因具有Meis和Hox的类似功能促进了再生过程。

有趣的是，来自于胚胎和成体组织的人成纤维细胞表达分析提示体内的成纤维细胞有位置-依赖性的 *Hox* 基因转录谱[243]。从躯体分化的区域分离的成纤维细胞，在标准的培养基内培养数代后，仍然保持与原躯体特定区域相关的特异性 *Hox* 基因表达谱[244]。这些结果提示图式形成信息在损伤之前就已经存在，因此又引发两个有趣的问题：①如果人类细胞保持区域识别身份，这个信息是否能用于器官和附肢的再生；②相反地，这种区域特异性信息在人类成纤维细胞的保持是否会妨碍再生？这个区域特异性表达谱有多么灵活、是否保持特异性 *Hox* 基因表达谱就能阻止再生过程仍需要进一步实验确定。

MicroRNA

最近几年发现 microRNA 在组织形成和模式上有重要的作用，近期组织再生实验也表明这些转录子参与了鱼类和蝾螈的组织再生[245~247]。microRNA-133 在未受损的成体鱼鳍表达，但是在再生的鱼鳍内表达下调[248]。斑马鱼鳍内缺失 microRNA-133 能重建再生能力，即使缺少再生所需的 Fgf 信号[248]。microRNA-133 通过抑制细胞周期调控因子 mps1 参与抑制细胞增生[248]，后者对于鱼类鳍芽基细胞增殖是必需的[249]。microRNA-133 也能拮抗肌组织的分化[250]。其能够拮抗鱼鳍再生过程中的细胞增殖，同时在发育中的肌组织能拮抗细胞增殖抑制剂的作用，提示即使同一个 microRNA 也有不同的活性，这些活性取决于组织特异性，或者受其他的调控组件的影响。当前，这些芽基细胞的增殖和分化状态是如何联系在一起的尚不清楚。因此，阐明 microRNA-133 是否部分参与细胞增殖与细胞分化相互协调的分子机制具有重要意义。

microRNA-196 在蝾螈的尾部再生过程中，表达在尾部芽基、芽基近侧的脊髓背侧和侧面的细胞[251]。抑制 miR-196 导致形成异常短小的尾部并有脊髓缺陷，而过表达一个 miR-196 类似物能增加再生尾部的长度[251]。进一步研究 miR-196 和其他 miRNA 如何调控组织再生的详细机制，能够为分化的细胞怎样成为芽基细胞和芽基细胞如何再生出新的组织提供另一个层面调控的认识。

结论

这是再生生物学上一个激动人心的时刻，因为一个多世纪以来关于细胞生物学的有价值的观察结果逐渐可以用分子生物学来解释。虽然仍在早期阶段，目前新的分子生物学工具的发展，包括鱼类、两栖类和其他动物的再生模型，为探索祖细胞如何有助于损伤或失去复杂结构组织的重建提供了必需的技术支持。这些研究正在阐明终末分化的细胞怎样获得适当的可塑性、什么机制控制这些细胞的增殖和分化状态来重新创造与原来非常类似相的结构。这个成果与哺乳类干细胞生物学的进展一起将使哺乳类和再生动物模型之间更具体的比较成为可能。这些比较可能不仅为重编程成体细胞成为干细胞开辟新的途径，而且为指导重编程的细胞再生出功能性的器官和附属器官提供了重要的信息。

（周曼倩　李宗金　译）

参 考 文 献

1. Lewitzky M, Yamanaka S. Reprogramming somatic cells torwards pluripotency by defined factors. Curr Opin Biotechnol 2007; 18:467-473.
2. Graf T, Enver T. Forcing cells to change lineages. Nature 2009; 462:587-594.
3. Yamanaka S. A fresh look at ips cells. Cell 2009; 137:13-17.
4. Okada TS. Transdifferentiation: flexibility in cell differentiation. Oxford: Oxford University Press; 1991.
5. Graw J. Genetic aspects of embryonic eye development in vertebrates. Dev Genet 1996; 18:181-197.
6. Reyer RW, Wooliftt RA, Withersty LT. Stimulation of lens regeneration from the newt dorsal iris when implanted into the blastema of the regenerating limb. Dev Biol 1973; 32:258-281.
7. Hyaga M, Kodama R, Eguchi G. Basic fibroblast growth factor as one of the essential factors regulating lens transdifferentiation of pigment epithelial cells. Int J Dev Biol 1993; 37:319-326.
8. Hayashi T, Mizuno N, Ueda Y et al. Fgf2 triggers iris-derived lens regeneration in newt eye. Mech Dev 2004; 121:519-526.
9. Hayashi T, Mizuno N, Owaribe K et al. Regulated lens regeneration from isolated pigment epithelial cells of newt iris in culture in response to fgf2/4. Differentiation 2002; 70:101-108.
10. Grogg MW, Call MK, Okamoto M et al. Bmp-inhibition-driven regulation of six-3 underlies induction of newt lens regeneration. Nature 2005; 438:858-862.
11. Halder G, Callaerts P, Gehring WJ. Induction of ectopic eyes by targeted expression of the eyeless gene in drosophila. Science 1995; 267:1788-1792.
12. Altmann CR, Chow RL, Lang RA et al. Lens induction by pax-6 in Xenopus laevis. Dev Biol 1997; 185:119-123.
13. Rio-Tsonis KD, Washabaugh CH, Tsonis PA. Expression of pax-6 during urodele eye development and lens regeneration. Proc Natl Acad Sci USA 1995; 92:5092-5096.
14. Madhavan M, Haynes TL, Frisch NC et al. The role of pax-6 in lens regeneration. Proc Natl Acad Sci USA 2006; 103:14848-14853.
15. Agata K, Kobayashi H, Itoh Y et al. Genetic charaterization of the multipotent dedifferentiated state of pigmented epithelial cells in vitro. Development 1993; 118:1993.
16. Maki N, Suetsugu-Maki R, Tarui H et al. Expression of stem cell pluripotency factors during regeneration in newts. Dev Dyn 2009; 238:1613-1616.
17. Maki N, Takechi K, Sano S et al. Rapid accumulation of nucleostemin in nucleolus during newt regeneration. Dev Dyn 2007; 236:941-950.
18. Freeman G. Lens regeneration from cornea in Xenopus laevis. J Exp Zool 963; 154:39-66.
19. Filoni S, Bosco L, Cioni C. The role of the neural retina in lens regeneration from cornea in larval Xenopus leavis. Acta Embryol Morphol Exp 1982; 3:15-28.
20. Filoni S, Bernardini S, Cannata SM. Experimental analysis of lens-forming capacity in Xenopus borealis larvae. J Exp Zool2006; 305A:538-550.
21. Bosco L, Filoni S. Relationships between presence of the eye cup and maintenance of lens-forming capacity in larval xenopus laevis. Dev Growth Differ 1992; 34:619-625.
22. Grainger RM, Henry JJ, Henderson R. Reinvestigation of the role of the optic vesicle in embryonic lens induction. Development 1988; 102:517-526.
23. Gargioli C, Giambra V, Santoni S et al. The lens-regenerating competence in the outer cornea and epidermis of larval Xenopus laevis is related to pax6 expression. J Anat 2008; 212:612-620.
24. Stone LS. The role of retinal pigment cells in regenerating neural retinae of adult salamander eyes. J Exp Zoo 1950; 113:9-31.
25. Bonnet C. Sur les Reproductions des Salamandres. Oeuvres d'Histoire naturelle et de Philosophie. Neutchatel 1781; 5:356.
26. Sakami S, Histomi O, Sakakibara S et al. Downregulation of otx2 in the dedifferentiated RPE cells of regenerating newt retina. Developmental Brain Research 2005; 155:49-59.
27. De Leeuw AM, Gaur VP, Saari JC et al. Immunolocalization of cellular retinol-, retinaldhyde- and retinoic acid-binding proteins in rat retina during pre- and postnatal development. J Neurocytol 1990; 19:253-264.
28. Park CM, Hollenberg MJ. Basic fibroblast growth factor induces retinal regeneration in vivo. Dev Biol 1989; 134:201-205.
29. Zhao S, Thornquist SC, Barnstable CJ. In vitro transdifferentiation of embryonic rat retinal pigment epithelium to neural retina. Brain Res 1995; 677:300-310.
30. Stroeva OG, Mitashov VI. Retinal pigment epithelium: proliferation and differentiation during development and regeneration. Int Rev Cytol 1983; 83:211-293.
31. Park CM, Hollenberg MJ. Induction of retinal regeneration in vivo by growth factors. Dev Biol 1991; 148:322-333.
32. Reh TA, Levine EM. Multipotent stem cells and progenitors in the vertebrate retina.J Neurobiol 1998; 80:206-220.
33. Sakami S, Etter P, Reh TA. Activin signaling limits the competence for retinal regeneration from the pigment epithelium. Mech Dev 2008; 125:106-116.

34. Fischer AJ, Reh TA. Müller glia are a potential source of neural regeneration in the postnatal chicken retina. Nat Neurosci 2001; 4:247-252.
35. Thummel R, Kassen SC, Enright JM et al. Characterization of Müeller glia and neuronal progenitors during adult zebrafish retinal regeneration. Exp Eye Res 2008; 87:433-444.
36. Bernardos RL, Barthel LK, Meyers JR et al. Late-stage neuronal progenitors in the retina are radial Müeller glia that function as retinal stem cells. J Neurosci 2007; 27:7028-7040.
37. Raymond PA, Barthel LK, Bernardos RL et al. Molecular characterization of retinal stem cells and their niches in adult zebrafish. BMC Dev Biol 2006; 6:36-53.
38. Braisted JE, Essman TF, Raymond PA. Selective regeneration of photoreceptors in goldfish retina. Development 1994; 120:2409-2419.
39. Braisted JE, Raymond PA. Regeneration of dopaminergic neurons in goldfish retina. Development 1992; 114:913-919.
40. Raymond PA. The unique origin of rod photoreceptors in the teleost retina. Trends Neurosci 1985; 8:12-17.
41. Raymond PA, Reifler MJ, Rivlin PK. Regeneration of goldfish retina: rod precursors are a likely source of regenerated cells. J Neurobiol 1988; 19:431-463.
42. Kirsche W, Kirsche K. Experimentelle untersuchungen zur frage der regeneration und funktion des tectums opticum von carassius carassuis. Z Mikrosk-Anat Forsch 1961; 67:140-182.
43. Parish CL, Beljajeva A, Arenas E et al. Midbrain dopaminergic neurogenesis and behavioural recovery in a salamander lesion-induced regeneration model. Development 2007; 134:2881-2887.
44. Endo T, Yoshino J, Kado K et al. Brain regeneration in anuran amphibians. Dev Growth Differ 2007; 49:121-129.
45. Okamoto M, Ohsawa H, Hayashi T et al. Regeneration of retinotectal projections after optic tectum removal in adult newts. Mol Vis 2007; 13:2112-2118.
46. Kirsche K, Kirsche W. Regenerative vorgänge im telencephalon von Ambystoma mexicanum. Journal für Hirnforschung 1963/1964; 6:421-436.
47. Richter W, Kranz D. Autoradiographische Untersuchungen der postnatalen Proliferationsaktivität in den Matrixzonen des Telencephalons und des Diencephalons beim Axolotl (Ambystoma mexicanum), unter Berücksichtigung der Proliferation im olfactorischen Organ. Zeitschrift Mikrosk-Anat Forschung 1981; 95:883-904.
48. Kirsche W. The significance of matrix zones for brain regeneration and brain transplantation with special consideration of lower vertebrates. In: Wallace RB, Das GD, eds. Neural Tissue Transplantation Research. New York: Springer-Verlag, 1983:65-104.
49. Stevenson JA, Yoon MG. Regeneration of optic nerve fibers enhances cell proliferation in the goldfish optic tectum. Brain Res 1978; 153:345-351.
50. Stevenson JA, Yoon MG. Kenetics of cell proliferation in the halved tectum of adult goldfish. Brain Res 1980; 184:11-22.
51. Kaslin J, Ganz J, Geffarth M et al. Stem cells in the adult zebrafish cerebellum: initiation and maintenance of a novel stem cell niche. J Neurosci 2009; 29:6142-6153.
52. Grandel H, Kaslin J, Ganz J et al. Neural stem cells and neurogenesis in the adult zebrafish brain: origin, proliferation dynamics, migration and cell fate. Dev Biol 2006; 295:263-277.
53. Zupanc GKH, Ott R. Cell proliferation after lesions in the cerebellum of adult teleost fish: time course, origin and type of new cells produced. Exp Neurol 1999; 160:78-87.
54. Zupanc GKH, Hinsch K, Gage FH. Proliferation, migration, differentiation and long-term survival of new cells in the adult zebrafish brain. J Comp Neurol 2005; 488:290-319.
55. Ekström P, Johnsson C-M, Ohlin L-M. Ventricular proliferation zones in the brain of an adult teleost fish and their relation to neuromeres and migration (secondary matrix) zones. J Comp Neurol 2001; 436:92-110.
56. Becker CG, Becker T. Gradients of ephrin-a2 and ephrin-a5b mrna during retinotopic regeneration of the optic projection is adult zebrafish. J Comp Neurol 2000; 427:469-483.
57. Springer AD. Normal and abnormal retinal projections following the crush of one optic nerve in goldfish (Carrassius auratus). J Comp Neurol 1981; 199:87-95.
58. Hatten ME. Central nervous system neuronal migration. Annu Rev Neurosci 1999; 22:511-539.
59. Campbell K, Goetz M. Radial glia: multi-purpose cells for vertebrate brain development. Trends Neurosci 2002; 25:235-238.
60. Noctor SC, Flint AC, Weissman TA et al. Neurons derived from radial glial cells establish radial units in neocortex. Nature 2001; 409:714-720.
61. Holder N, Clarke JDW, Kamalati T, et al. Heterogeneity in spinal radial glia demonstrated by intermediate filament expression and hrp labelling. J Neurocytol 1990; 19:915-928.
62. O'Hara CM, Egar MW, Chernoff EAG. Reorganization of the ependyma during axolotl spinal cord regeneration: changes in intermediate filament and fibronectin expression. Dev Dyn 1992; 193:103-115.
63. Mchedlishvili L, Epperlein HH, Telzerow A et al. A clonal analysis of neural progenitors during axolotl spinal cord regeneration reveals evidence for both spatially restricted and multipotent progenitors. Development 2007; 134:2083-2093.

64. Butler EG, Ward MB. Reconstitution of the spinal cord after ablation in adult triturus. Dev Biol 1967; 15:454-486.
65. Egar MW, Singer M. The role of ependyma in spinal cord after ablation in adult triturus. Exp Neurol 1972; 37:422-430.
66. Nordlander RH, Singer M. The role of ependyma in regeneration of the spinal cord in the urodele amphibian tail. J Comp Neurol 1978; 180:349-374.
67. Echeverri K, Tanaka EM. Ectoderm to mesoderm lineage switching during axolotl tail regeneration. Science 2002; 298:1993-1996.
68. Schnapp E, Kragl M, Rubin L et al. Hedgehog signaling controls dorsoventral patterning, blastema cell proliferation and cartilage induction during axolotl tail regeneration. Development 2005; 132:3243-3253.
69. Beck CW, Christen B, Slack JMW. Molecular pathways needed for regeneration of spinal cord and muscle in a vertebrate. Dev Cell 2003; 5:429-439.
70. Sugiura T, Tazaki A, Ueno N et al. Xenopus wnt-5a induces an ectopic larval tail at injured site, suggesting a crucial role for noncanonical wnt signal in tail regeneration. Mech Dev 2009; 126:56-67.
71. Becker T, Wullimann MF, Becker CG et al. Axonal regrowth after spinal cord transection in adult zebrafish. J Comp Neurol 1997; 337:577-595.
72. Bernstein JJ. Relation of spinal cord regeneration to age in adult goldfish. Exp Neurol 1964; 9:161-174.
73. Reimer MM, Soerensen I, Kuscha V et al. Motor neuron regeneration in adult zebrafish. J Neurosci 2008; 28:8510-8516.
74. Becker T, Lieberoth BC, Becker CB et al. Differences in the regenerative response of neuronal cell populations and indications for plasticity in intraspinal neurons after spinal cord transection in adult zebrafish. Mol Cell Neurosci 2005; 30:265-278.
75. Anderson MJ, Waxman SG, Laufer M. Fine structure of regenerated ependyma and spinal cord in Sternarchus albifrons. Anat Rec 1983; 205:73-83.
76. Anderson MJ, Rossettl DL, Lorenz LA. Neuronal differentiation in vitro from precursor cells of regenerating spinal cord of the adult teleost Apteronotus albifrons. Cell Tissue Res 1994; 278:243-248.
77. Anderson MJ, Waxman SG, Fong HL. Explant cultures of teleost spinal cord: source of neurite outgrowth. Dev Biol 1987; 119:601-604.
78. Reimer MM, Kuscha V, Wyatt C et al. Sonic hedgehog is a polarized signal for motor neuron regeneration in adult zebrafish. J Neurosci 2009; 29(48):15073-15082.
79. Bernstein JJ, Gelderd JB. Regeneration of the long spinal tracts in the goldfish. Brain Res 1970; 20:33-38.
80. Bhatt DH, Otto SJ, Depoister B et al. Cyclic amp-induced repair of zebrafish spinal circuits. Science 2004; 305:254-258.
81. Oberpriller JO, Oberpriller JC. Response of the adult newt ventricle to injury. J Exp Zoo 1974; 187:249-260.
82. Poss KD, Wilson LG, Keating MT. Heart regeneration in zebrafish. Science 2002; 298:2188-2190.
83. Bader D, Oberpriller JO. Repair and reorganization of minced cardiac muscle in the adult newt (Notophthalmus viridescens. J Morphol 1978; 155:349-358.
84. Betencourt-Dias M, Mittnacht S, Brockes JP. Heterogeneous proliferative potential in regenerative adult newt cardiomyocytes. J Cell Biol 2003; 116:4001-4009.
85. Flink IL. Cell cycle reentry of ventricular and atrial cardiomyocytes and cells within the epicardium following amputation of the ventricular apex in the axolotl, Amblystoma mexicanum. Anat Embryol 2002; 205:235-244.
86. Laube F, Heister M, Scholz C et al. Re-programming of newt cardiomyocytes is induced by tissue regeneration. J of Cell Science 2006; 119:4719-4729.
87. Kikuchi K, Holdaway JE, Werdich AA et al. Primary contribution to zebrafish heart regeneration by gata4+ cardiomyocytes. Nature 2010; 464:601-605.
87a. Joplin C, Sleep E, Raya M et al. Zebrafish heart regeneration occurs by cardiomyocyte dedifferentiation and proliferation. Nature 2010; 464:606-609.
88. Lepilina A, Coon AN, Kikuchi K et al. A dynamic epicardial injury response supports progenitor cell activity during zebrafish heart regeneration. Cell 2006; 127:607-619.
89. Raya A, Koth CM, Buescher D et al. Activation of notch signaling pathway precedes heart regeneration in zebrafish. Proc Natl Acad Sci USA 2003; 100:11889-11895.
90. Lien CL, Schebesta M, Makino S et al. Gene expression analysis of zebrafish heart regeneration. PLoS Biology 2006; 4:1386-1396.
91. Lavine KJ, Yu K, White AC et al. Endocardial and epicardial derived fgf signals regulate myocardial proliferation and differentiation in vivo. Dev Cell 2005; 8:85-95.
92. Brockes JP, Kumar A. Appendage regeneration in adult vertebrates and implications for regenerative medicine. Science 2005; 310:1919-1923.
93. Poss KD, Keating MT, Nechiporuk A. Tales of regeneration in zebrafish. Dev Dyn 2003; 226:202-210.
94. Straube WL, Tanaka EM. Reversibility of the differentiated state: regeneration in amphibians. Artif Organs 2006; 30:743-755.
95. Muneoka K, Fox WF, Bryant S. Cellular contribution from dermis and cartilage to the regenerating llimb blastema in axolotls. Dev Biol 1986; 116:256-260.

96. Gardiner DM, Muneoka K, Bryant SV. The migration of dermal cells during blastema formation in axolotls. Dev Biol 1986; 118:488-493.
97. Bryant SV, Endo T, Gardiner DM. Vertebrate limb regeneration and the origin of the limb stem cells. Int J Dev Biol 2002; 46:887-896.
98. Hay ED. Electron microscopic observations of muscle dedifferentiation in regenerating Amblystoma limbs. Dev Biol 1959; 1:555-585.
99. Hay ED. The fine structure of differentiating muscle in the salamander tail. Zeitschrift für Zellforschung 1963; 59:6-34.
100. Mauro A. Satellite cell of skeletal muscle fibers. J Biophys Biochem Cytol 1961; 9:493-495.
101. Collins CA, Olsen I, Zammit PS et al. Stem cell function, self-renewal and behavioral heterogeneity of cells from the adult muscle satellite cell niche. Cell 2005; 122:289-301.
102. Montarras D, Morgan J, Collins C et al. Direct isolation of satellite cells for skeletal muscle regeneration. Science 2005; 309:2064-2067.
103. Morrison JI, Loeoef S, He P et al. Salamander limb regeneration involves the activation of a multipotent skeletal muscle satellite cell population. J Cell Biol 2006; 172:433-440.
104. Thornton CS. The histogenesis of muscle in the regenerating fore limb of larval Amblyostoma punctatum. J Morphol 1938; 62:17-47.
105. Kintner CR, Brockes JP. Monoclonal antibodies identify blastemal cells derived from dedifferentiating muscle in newt limb regeneration. Nature 1984; 308:67-69.
106. Lo DC, Allen F, Brockes JP. Reversal of muscle differentiation during urodele limb regeneration. Proc Natl Acad Sci USA 1993; 90:7230-7234.
107. Kumar A, Velloso CP, Imokawa Y et al. Plasticity of retrovirus-labelled myotubes in the newt limb regeneration blastema. Dev Biol 2000; 218:125-136.
108. Echeverri K, Clarke JDW, Tanaka EM. In vivo imaging indicates muscle fiber dedifferentiation is a major contributor to the regenerating tail blastema. Dev Biol 2001; 236:151-164.
109. Tanaka EM, Drechsel DN, Brockes JP. Thrombin regulates S-phase re-entry by cultured newt myotubes. Current Biology 1999; 9:792-799.
110. Rollman-Dinsmore C, Bryant SV. The distrubution of marked dermal cells from small localized implants in limb regenerates. Dev Biol 1984; 106:275-281.
111. Muneoka K, Fox WF, Bryant SV. Cellular contibution from dermis and cartilage to the regenerating limb blastema in axololts. Dev Biol 1986; 116:256-260.
112. Chalkey DT. A quantitative histological analysis of forelimb regeneration in Triturus viridescens. J Morphol 1954; 94:21-70.
113. Kragl M, Knapp D, Nacu E et al. Cells keep a memory fo their tissue origin during axolotl limb regeneration. Nature 2009; 460:60-65.
114. Namenwirth M. The inheritance of cell differentiation during limb regeneration in the axolotl. Dev Biol 1974; 41:42-56.
115. Steen TP. Stability of chondrocyte differentiation and contribution of muscle to cartilage during limb regeneration in the axolotl (Siredon mexicanum). J Exp Zoo 1968; 167:49-77.
116. Holtzer H, Avery G, Holtzer S. Some properties of the regenerating limb blastema cells of salamanders. Biological Bulletin 1954; 107:313.
117. Pietsch P. Differentiation in regeneration: i. The development of muscle and cartilage following deplantation of regenerating limb blastemata of amblystoma larvae. Dev Biol 1961; 3:255-264.
118. Pietsch P. The effect of heterotropic musculature on myogenesis during limb regeneration in Amblystoma larvae. The Anat Rec 1961; 141:295-303.
119. Steen TP. Stability of chondrocyte differentiation and contribution of muscle to cartilage during limb regeneration in the axolotl (Siredon mexicanum). J Exp Zoo 1968; 167:49-78.
120. Morrison JI, Borg P, Simon A. Plasticity and recovery of skeletal muscle satellite cells during limb regeneration. FASEB Journal 2009; 24:Nov 6. [Epub ahead of print].
121. Odelberg SJ, Kollhoff A, Keating MT. Dedifferentiation of mammalian myotubes induced by msx1. Cell 2000; 103:1099-1109.
122. Mescher AL. Effects of adult newt limb regeneration of partial and complete skin flaps over the amputation surface. J Exp Zoo 1976; 195:117-128.
123. Thornton CS. The effect of apical cap removal on limb regeneration in amblystoma larvae. J Exp Zoo 1957; 134:357-381.
124. Tassava RA, Garling DJ. Regenerative responses in larval axolotl limbs with skin grafts over the amputation surface. J Exp Zoo 1979; 208:97-110.
125. Polezhaev LV, Faworina WN. Über die Rolle des Epithels in den Anfänglichen Entwicklungstadien einer Regenerationsanlage der Extremität beim Axolotl. Wilhelm Roux' Archiv Entwicklungsmech Org 1935; 133:701-727.
126. Tank PW. Skin of non-limb origin blocks regeneration of the newt forlimb. Prog Clin Biol Res 1983; 110:565-575.

127. Riddiford LM. Autoradiographic studies of tritiated thymidine infused into the blastema of the early regenerate in the adult newt, triturus. J Exp Zoo 1960; 144:25-31.
128. Hay ED, Fischman DA. Origin of the blastema in regenerating limbs of the newt triturus viridescens. Dev Biol 1961; 3:26-59.
129. Saunders JW, Gasseling MT, Errick JE. Inductive activity and enduring cellular constitution of a supernumerary apical ectodermal ridge grafted to the limb bud of the chick embryo. Dev Biol 1976; 50:16-25.
130. Saunders JW. The proximo-distal sequence of origin of the parts of the chick wing and the role of the ectoderm. J Exp Zoo 1948; 108:363-403.
131. Carlson BM. Morphogenetic interactions between rotated skin duffs and underlying stump tissues in regenerating axolotl forelimbs. Dev Biol 1974; 39:263-285.
132. Bryant SV, Iten LE. Supernumerary limbs in amphibians: experimental production in Notophthalmus viridescens and a new interpretation of their formation. Dev Biol 1976; 50:212-234.
133. Thornton CS. Influence of an eccetric epidermal cap on limb regeneration in Ambystoma larvae. Dev Biol 1960; 2:551-569.
134. Thornton CS, Thornton MT. The regeneration of accessory limb parts following epidermal cap transplantation in urodeles. Experimentia 1965; 21:146-148.
135. Lee Y, Hami D, Val SD et al. Maintenance of blastemal proliferation by functionally diverse epidermis in regenerating zebrafish fins. Dev Biol 2009; 331:270-280.
136. Viviano CM, Horton CE, Maden M et al. Synthesis and release of 9-cis retinoic acid by the urodele wound epidermis. Development 1995; 121:3753-3762.
137. Stoick-Cooper CL, Weidinger G, Riehle KJ et al. Distinct wnt signaling pathways have opposing roles in appendage regeneration. Development 2007; 134:479-489.
138. Laforest L, Brown CW, Poleo G et al. Involvement of the sonic hedgehog, patched1 and bmp2 genes in patterning of the zebrafish dermal fin rays. Development 1998; 125:4175-4184.
139. Maden M. Vitamin a and pattern formation in the regenerating limb. Nature 1982; 295:672-675.
140. White JA, Boffa MB, Jones B et al. A zebrafish retinoic acid receptor expressed in the regenerating caudal fin. Development 1994; 120:1861-1872.
141. Saxena S, Niazi IA. Effect of vitamin A excess on hind limb regeneration in tadpoles of the toad, bufo andersonii (boulenger). Indian J Exp Biol 1977; 15:435-439.
142. Ghosh S, Roy S, Seguin C et al. Analysis of the expression and function of wnt-5a and wnt-5b in developing and regenerating axolotl (Ambystoma mexicanum) limbs. Dev Growth Differ 2000; 50:289-297.
143. Christensen RN, Weinstein M, Tassava RA. Expression of fibroblast growth factors 4, 8 and 10 in limbs, flanks and blastemas of ambystoma. Dev Dyn 2002; 223:193-203.
144. Geraudie J, Singer M. Necessity of an adequate nerve supply for regeneration of the amputated pectoral fin in the teleost fundulus. J Exp Zoo 1985; 234:367-374.
145. Singer M. The influence of the nerve in regeneration of the amphibian extremity. Q Rev Biol 1952; 27:169-200.
146. Singer M. The nervous system and regeneration of the forelimb of adult triturus. J Exp Zoo 1942; 90:377-399.
147. Goss RJ, Stagg MW. The regeneration of fins and fin rays in Fundulus heteroclitus. J Exp Zoo 1957; 136:487-507.
148. Endo T, Bryant SV, Gardiner DM. A stepwise model system for limb regeneration. Dev Biol 2004; 270:135-145.
149. Satoh A, Gardiner DM, Bryant SV et al. Nerve-induced ectopic limb blastemas in the axolotl are equivalent to amputation-induced blastemas. Dev Biol 2007; 312:231-244.
150. Satoh A, Graham GMC, Bryant SV et al. Neurotrophic regulation of epidermal dedifferentiation during wound healing and limb regeneration in the axolotl (Ambystoma mexicanum). Dev Biol 2008; 319:321-335.
151. Singer M, Inoue S. The nerve and the epidermal apical cap in regeneration of the forelimb of adult triturus. J Exp Zoo 964; 115:105-116.
152. Kumar A, Gates PB, Brockes JP. Positional indentity of adult stem cells in salamander limb regeneration. Comptes Rendus Biologies 2007; 330:485-490.
153. Kumar A, Godwin JW, Gates PB et al. Molecular basis for the nerve dependence of limb regeneration in an adult vertebrate. Science 2007; 318:772-777.
154. Silva SMd, Gates PB, Brockes JP. The newt orthology of CD59 is implicated in proximodistal identity during amphibian limb regeneration. Dev Cell 2002; 3:547-555.
155. Echeverri K, Tanaka EM. Proximodistal patterning during limb regeneration. Dev Biol 2005; 279:391-401.
156. Brockes JP, Kintner CR. Glial growth factor and nerve-dependent proliferation in the regeneration blastema of urodele amphibians. Cell 1986; 45:301-306.
157. Albert P, Boilly B, Courty J et al. Stimulation in cell culture of mesenchymal cells of newt limb blastemas by edgfi and ii (basic or acidic fgf). Cell Differ 1987; 21:63-68.
158. Wang L, Marchionni MA, Tassava RA. Cloning and neuronal expression of a type iii newt neuregulin and rescue of denerated, nerve-dependent newt limb blastemas. J Neurobiol 2000; 43:150-158.

159. Mullen LM, Bryant SV, Torok MA et al. Nerve dependency of regeneration: the role of distal-less and fgf signaling in amphibian limb regeneration. Development 1996; 122:3487-3497.
160. Kuehn LC, Schulman HM, Ponka P. Iron-transferrin requirements and transferrin receptor expression in proliferating cells. In: Ponka P, Schulman HM, Woodworth RC, eds. Iron Transport and Storage. Boca Raton: CRC Press, 1990:149-191.
161. Mescher AL, Connell E, Hsu C et al. Transferrin is necessary and sufficient for the neural effect on growth in amphibian limb regeneration blastemas. Dev Growth Differ 1997; 39:677-684.
162. Whitehead GG, Makino S, Lien CL et al. Fgf20 is essential for initiating zebrafish fin regeneration. Science 2005; 310:1957-1960.
163. Poss KD, Shen J, Nechiporuk A et al. Roles for fgf signaling during zebrafish fin regeneration. Dev Biol 2000; 222:347-358.
164. Kubo F, Takeichi M, Nakagawa S. Wnt2b controls retinal cell differentiation at the ciliary marginal zone. Development 2003; 130:587-598.
165. Kubo F, Takeichi M, Nakagawa S. Wnt2b inhibits differentiation of retinal progenitor cells in the absence of notch activity by downregulating the expression of proneural genes. Development 2005; 132:2759-2770.
166. Kawakami Y, Esteban CR, Raya M et al. Wnt/β-catenin signaling regulates vertebrate limb regeneration. Genes and Development 2006; 20:3232-3237.
167. Yokoyama H, Ogino H, Stoick-Cooper CL et al. Wnt/β-catenin signaling has an essential role in the initiation of limb regeneration. Dev Biol 2007; 306:170-178.
168. Lin G, Slack JMW. Requirement for wnt and fgf signaling in Xenopus tadpole tail regeneration. Dev Biol 2008; 316:323-335.
169. Levesque M, Gatien S, Finnson K et al. Transforming growth factor: β signaling is essential for limb regeneration in axolotls. PLoS One 2007; 11:2-14.
170. Ho DM, Whitman M. Tgf-B signaling is required for multiple processes during Xenopus tail regeneration. Dev Biol 2008; 315:203-216.
171. Inman GJ, Nicolas FJ, Callahan JF et al. Sb-431542 is a potent and specific inhibitor of transforming growth factor-B superfamily type I activin receptor-like kinase (alk) receptors alk4, alk5 and alk7. Molecular Pharmacology 2002; 62:65-74.
172. Jazwinska A, Badakov R, Keating MT. Activin-βa signaling is required for zebrafish fin regeneration. Current Biology 2007; 17:1390-1395.
173. Smith A, Avaron F, Guay D et al. Inhibition of bmp signaling during zebrafish fin regeneration disrupts fin growth and scleroblast differentiation and function. Dev Biol 2006; 299:438-454.
174. Roy S, Gardiner DM. Cyclopamine induces digit loss in regenerating axolotl limbs. J Exp Zool 2002; 293:186-190.
175. Quint E, Smith A, Avaron F et al. Bone patterning is altered in the regenerating zebrafish caudal fin after ectopic expression of sonic hedgehog and bmp2b or exposure to cyclopamine. Proc Natl Acad Sci USA 2002; 99:8713-8718.
176. Roy S, Gardiner DM, Bryant SB. Vaccinia as a tool for functional analysis in regenerating limbs: ectopic expression of shh. Dev Biol 2000; 218:199-205.
177. Endo T, Tamura K, Ide H. Analysis of gene expression during Xenopus forelimb regeneration. Dev Biol 2000; 220:296-306.
178. Yakushiji N, Suzuki M, Satoh A et al. Correlation between shh expression and DNA methylation status of the limb-specific shh enhancer region during limb regenertion in amphibians. Dev Biol 2007; 312:171-182.
179. Yakushiji N, Suzuki M, Satoh A et al. Effects of activation of hedgehog signaling on patterning, growth and differentiation in Xenopus froglet limb regeneration. Dev Dyn 2007; 238:1887-1896.
180. Dent JN. Limb regeneration is larvae and metamorphosing individuals of the south african clawed toad. J Morphol 1962; 110:61-77.
181. Endo T, Yokoyama H, Tamura K et al. Shh expression in developing and regenerating limb buds of xenopus laevis. Dev Dyn 1997; 209:227-232.
182. Moneoka K, Holler-Dinsmore G, Bryant SV. Intrinsic control of regenerative loss in Xenopus laevis limb. J Exp Zoo 986; 240:47-54.
183. Lee Y, Grill S, Sanchez A et al. Fgf signaling instructs position-dependent growth rate during zebrafish fin regeneration. Development 2005; 132:5173-5183.
184. Kizil C, Otto GW, Geisler R et al. Simplet controls cell proliferation and gene transcription during zebrafish caudal fin regeneration. Dev Biol 2009; 325:329-340.
185. Thermes V, Candal E, Alunni A et al. Medaka simplet (fam53b) belongs to a family of novel vertebrate genes controlling cell proliferation. Development 2006; 133:1881-1890.
186. Vogelstein B, Lane D, Levine AJ. Surfing the p53 network. Nature 2000; 408:307-310.
187. Villiard E, Brinkmann H, Moiseeva O et al. Urodele p53 tolerates amino acid changes found in p53 varients linked to human cancer. BMC Evolutionary Biology 2007; 7:180-194.
188. Ferretti P, Brockes JP. Culture of newt cells from different tissues and their expression of a regeneration-associated antigen. J Exp Zoo 1988; 247:77-91.
189. Tanaka EM, Gann AAF, Gates PB et al. Newt myotubes reenter the cell cycle by phosphorylation of the retinoblastoma protein. J Cell Biol 1997; 136:155-165.

190. Florini JR, Ewton DZ, Magri KA. Hormones, growth factors and myogenic differentiation. Annual Review Physiology 1991; 53:201-216.
191. Olson EN. Proto-oncogenes in the regulatory circuit for myogenesis. Seminars in Cell Biology 1992; 3:127-136.
192. Schneider JW, Gu W, Zhu L et al. Reversal of terminal differentiation mediates by p107 in Rb-/- muscle cells. Science 1994; 264(1467-1471).
193. Davidson DR, Crawley A, Hill RE et al. Position-dependent expression of two related homeobox genes in developing vertebrate limbs. Nature 1991; 352:429-431.
194. Summerbell D, Lewis JH, Wolpert L. Positional information in chick limb morphogenesis. Nature 1973; 244:492-496.
195. Robert B, Sassoon D, Jacq B et al. Hox-7, a mouse homeobox gene with a novel pattern of expression during embryogenesis. EMBO Journal 1989; 8:91-100.
196. Song K, Wang Y, Sassoon D. Expression of hox-7.1 in myoblasts inhibits terminal differentiation and induces cell transformation. Nature 1992; 360:477-481.
197. Woloshin P, Song K, Degnin C et al. Msx1 inhibits myod expression in fibroblast x 10t1/2 cell hybrids. Cell 1995; 82:611-620.
198. Akimenko MA, Johnson SL, Westerfield M et al. Differential induction of four msx homeobox genes during fin development and regeneration in zebrafish. Development 1995; 121:347-357.
199. Carlson MRJ, Bryant SV, Gardiner DM. Expression of msx-2 during development, regeneration and wound healing in axolotl limbs. J Exp Zoo 1998; 282:715-723.
200. Crews L, Gates PB, Brown R et al. Expression and activity of the newt msx-1 gene in relation to limb regeneration. Proc R Soc Lond B 1995; 259:161-171.
201. Kumar A, Velloso CP, Imokawa Y et al. The regenerative plasticity of isolated urodele myofibers and its dependence on msx1. PLoS Biology 2004; 2:1168-1176.
202. Hu G, Lee H, Price SM et al. Msx homeobox genes inhibit differentiation through upregulation of cyclin D1. Development 2001; 128:2373-2384.
203. Thummel R, Bai S, Michael P et al. Inhibition of zebrafish fin regeneration using in vivo electroporation of morpholinos against fgfr1 and msxb. Dev Dyn 2006; 235:335-346.
204. Barnes RM, Firulli AB. A twist of insight-the role of twist-family bHLH factors in development. Int J Dev Biol 2009; 53:909-924.
205. Nüsslein-Volhard C, Wieschaus E, Kluding H. Mutations affecting the pattern of the larval cuticle in drosophila melanogaster. Roux's Archives of Dev Biol 1984; 193:267-282.
206. Simpson P. Maternal-zygotic gene interactions during formation of the dorsoventral pattern in Drosophila embryos. Genetics 1983; 105:615-632.
207. Bate M, Rushton E, Currie DA. Cells with persistant twist expression are the embryonic precursors of adult muscles in Drosophila. Development 1991; 113:79-89.
208. Currie DA, Bate M. The development of adult abdominal muscles in drosophila: myoblasts express twist and are associated with nerves. Development 1991; 113:19-101.
209. Hjiantoniou E, Anayasa M, Nicolaou P et al. Twist induces reversal of myotubes formation. Differentiation 2007; 76:182-192.
210. Satoh A, Bryant SV, Gardiner DM. Regulation of dermal fibroblast dedifferentiation and redifferentiation during wound healining and limb regeneration in the axolotl. Dev Growth Differ 2008; 50:743-754.
211. Scaal M, Fürchtbauer EM, Brand-Saberi B. cDermo-1 expression indicates a role in avian skin development. Anat Embryol 2001; 203:1-7.
212. Li L, Cserjesi P, Olson EN. Dermo-1: a novel twist-related bHLH protein expressed in the developing dermis. Dev Biol 1995; 172:280-292.
213. Seale P, Sabourin LA, Girgis-Gabardo A et al. Pax7 is required for the specification of myogenic satellite cells. Cell 2000; 102:777-786.
214. Seale P, Polesskaya A, Rudnicki MA. Adult stem cell specification by wnt signaling in muscle regeneration. Cell Cycle 2003; 2:418-419.
215. Maekawa M, Takashima N, Arai Y et al. Pax6 is required for production and maintenance of progenitor cells in postnatal hippocampal neurogenesis. Genes Cells 2005; 10:1001-1014.
216. Lang D, LM M, Huang L et al. Pax3 functions at a nodal point in melanocyte stem cell differentiation. Nature 2005; 433:884-887.
217. Lagha M, Sato T, Bajard L et al. Regulation of skeletal muscle stem cell behavior by pax3 and pax7. Cold Spring Harbor Symposia of Quantitative Biology 2008; 73:307-315.
218. Seale P, Ishibashi J, Scime A et al. Pax7 is necessary and sufficient for the myogenic specification of cd45+:sca1+ stem cells from injured muscle. PLoS Biology 2004; 2:E130.
219. Asakura A, Seale P, Girgis-Gabardo A et al. Myogenic specification of side population cells in skeletal muscle. J Cell Biol 2002; 159:123-134.
220. Jackson KA, Mi T, Goodell MA. Hematopoeitic potential of stem cells isolated from murine skeletal muscle. Proc Natl Acad Sci USA 1999; 96:14482-14486.

221. Gussoni E, Soneoka Y, Strickland CD et al. Dystrophin expression in the mdx mouse restored by stem cell transplantation. Nature 1999; 401:390-394.
222. Chen Y, Lin G, Slack JMW. Control of muscle regeneration in the Xenopus tadpole tail by pax7. Development 2006; 133:2303-2313.
223. McKinnell IW, Ishibashi J, Grand FL et al. Pax7 activates myogenic genes by recruitment of a histone methyltransferase complex. Nat Cell Biol 2008; 10:77-84.
224. Slack JMW. Morphogenetic properties of the skin in axolotl limb regeneration. J Embryol Exp Morphol 1980; 58:265-288.
225. Slack JMW. Positional information in the forelimb of the axolotl: properties of the posterior skin. J Embryol Exp Morphol 1983; 73:233-247.
226. Rollman-Dinsmore C, Bryant SV. Pattern regulation between hind- and forelimbs after blastema exchanges and skin grafts in Notophthalmus viridescens. J Exp Zoo 1982; 223:51-56.
227. Tank PW. The ability of localized implants of whole or minced dermis to disrupt pattern formation in the regeneration forelimb of the axolotl. Am J Anat 1981; 162:315-326.
228. Dunis DA, Namenwirth M. The role of grafted skin in the regeneration of x-irradiated axolotl limbs. Dev Biol 1977; 56:97-109.
229. Holder N. Organization of connective tissue patterns by dermal fibroblasts in the regenerating axolotl limb. Development 1989; 105:585-593.
230. Mescher AL. The cellular basis of limb regeneration in urodeles. Int J Dev Biol 1996; 40:785-795.
231. Bryant SV, Gardiner DM. Limb development and regeneration. Am Zool 1987; 27:675-696.
232. Carlson BM. Multiple regeneration from axolotl limb stumps bearing cross-transplanted minced muscle regenerates. Dev Biol 1975; 45:203-208.
233. Mercader N, Tanaka EM, Torres M. Proximodistal identity during vertebrate limb regeneration is regulated by meis homeodomain proteins. Development 2005; 132:4131-4142.
234. Mercader N, Selleri L, Criado LM et al. Ectopic meis1 expression in the mouse limb bud alters P-Dpatterning in a pbx1-independent manner. Int J Dev Biol 2008; 53:1483-1494.
235. Capdevila J, Tsukui T, Esteban CR et al. Control of vertebrate limb outgrowth by the proximal factor meis2 and distal antagonism of bmp by gremlin. Molecular Cell 1999; 4:839-849.
236. Mercader N, Leonardo E, Piedra ME et al. Opposing RA and fgf signals control proximodistal vertebrate limb development through regulation of meis genes. Development 2000; 127:3961-3970.
237. Izpisua-Belmonte JC, Tickle C, Dolle P et al. Expression of the homeobox hox-4 genes and the specification of position in chick wing development. Nature 1991; 350:585-589.
238. Krumlauf R. Hox genes in vertebrate development. Cell 1994; 78:191-201.
239. Gardiner DM, Blumberg B, Komine Y et al. Regulation of hoxa expression in developing and regenerating axolotl limbs. Development 1995; 121:1731-1741.
240. Geraudie J, Birraux VB. Posterior hoxa genes expression during zebrafish bony ray development and regeneration suggests their involvement in scleroblast differentiation. Dev Genes Evol 2003; 213:182-186.
241. Thummel R, Ju M, Michael P et al. Both hoxc13 orthologs are functinally important for zebrafish tail fin regeneration. Dev Genes Evol 2007; 217:413-420.
242. Moens CB, Selleri L. Hox cofactors in vertebrate development. Dev Biol 2006; 291:193-206.
243. Rinn JL, Bondre C, Gladstone HB et al. Anatomic demarcation by positional variation in fibroblast gene expression programs. PLoS Genetics 2006; 2:10841096.
244. Chang HY, Chi JT, Dudoit S et al. Diversity, topographic differentiation and positional memory in human fibroblasts. Proc Natl Acad Sci USA 2002; 99:12877-12882.
245. Makarev E, Spence JR, Rio-Tsonis KD et al. Identification of microRNAs and other small RNAs from the adult newt eye. Mol Vis 2006; 12:1386-1391.
246. Tsonis PA, Call MK, Grogg MW et al. MicroRNAs and regeneration: let-7 members as potential regulators of dedifferentiation in lens and inner ear hair cell regeneration of the adult newt. Biochem Biophys Res Commun 2007; 362:940-945.
247. Thatcher EJ, Paydar I, Anderson KK et al. Regulation of zebrafish fin regeneration by microRNAs. Proc Natl Acad Sci USA 2008; 105:18384-18389.
248. Yin VP, Thomson JM, Thummel R et al. Fgf-dependent depletion of microRNA-133 promotes appendage regeneration in zebrafish. Genes Dev 2008; 22:728-733.
249. Poss KD, Nechiporuk A, Hillam AM et al. Msp1 defines a proximal blastemal proliferative compartment essential for zebrafish fin regeneration. Development 2002; 129:5141-5149.
250. Chen JF, Mandel EM, Thomson JM et al. The role of microRNA-1 and microRNA-133 in skeletal muscle proliferation and differentiation. Nat Genet 2006; 38:228-233.
251. Sehm T, Sachse C, Frenzel C et al. Mir-196 is an essential early-stage regulator of tail regeneration, upstream of key spinal cord patterning events. Dev Biol 2009; 334:468-480.

第14章　成体细胞重编程获得多能性

Masato Nakagawa*，Shinya Yamanaka

摘要：2006年，研究人员通过引入Sox2、Oct3/4、Klf4和c-Myc四个转录因子成功地使体细胞重编程为多能干细胞。这些诱导多能干细胞［induced pluripotent stem (iPS) cell］能够避免胚胎干细胞在应用中所面临的伦理学问题及固有免疫反应，为再生医学的发展提供了新的希望。然而，目前细胞重编程的潜在分子机制仍不清楚。本章我们将回顾从两栖动物到哺乳动物细胞的重编程研究历史，并讨论目前对重编程分子机制的认识，以及在再生医学中利用重编程细胞的可能性。

引言

许多研究人员进行了大量的实验希望能够解决"重编程"这个具有重要意义且困难重重的科学问题。在发育过程中，不同的干细胞群体分化为不同胚系的成体细胞，而干细胞微环境（niche），如因子分泌、细胞间或细胞与基质的相互作用、机械力以及其他外界刺激，调控着分化过程有序进行。虽然有关细胞分化的研究数据在不断增加，但是逆向研究及去分化作用相关的研究相对滞后。成体细胞重编程（细胞核重编程）的研究最初是在青蛙体内进行的，相关成果不但促成了克隆羊"多莉"的诞生，而且是成功获得iPS细胞的理论基础。自此，曾经被认为只能是单向的分化过程，如今可以通过实验重新开始。

青蛙成体细胞核重编程研究

研究人员利用核转移技术（SCNT）成功地将非洲爪蟾体细胞重编程，并将这种重编程细胞培育成为可育成体。SCNT已经成为重编程研究的常规技术。同一生物体内的分化细胞和干细胞具有相同的DNA序列，然而不同细胞中DNA序列的读取方式存在非常大的差异，从而导致了不同的基因表达谱。基因表达由几种染色质修饰进行调控，如DNA甲基化及组蛋白修饰。SCNT技术是将体细胞的核植入到无核的卵母细胞中，植入的细胞核在新的环境中进行了DNA和染色质修饰，从而被重新编程。尽管效率较低，但是包含有重编程体细胞核的胚胎能够按照胚胎发育阶段进行正常发育。有一个重要的实验是将完全分化的肠上皮细胞核分离并转移至卵母细胞，最终转入的细胞核进行

---

* Corresponding Author: Masato Nakagawa—Center for iPS Research and Application, Kyoto University, 53 Kawahara-cho, Shogoin, Sakyo-ku, Kyoto 606-8507, Japan. Email: nakagawa@cira.kyoto-u.ac.jp

了重编程。通过这项实验，研究者获得了具有繁殖能力的雄性和雌性青蛙[1]。这项研究及其他相关的研究[2~4]表明体细胞的分化过程是可逆的，卵母细胞具有使体细胞核重编程的能力，也就是说卵细胞中存在着重编程的因子。

克隆羊"多莉"的诞生

在进行青蛙的体细胞核重编程研究大约 30 年之后，哺乳动物细胞重编程研究获得了成功[5]。研究人员利用三种成体细胞的单一细胞核进行核转移并成功获得转核后细胞，这三种细胞分别取自小羊的乳腺上皮细胞、羊胎儿成纤维细胞和羊胚胎来源细胞。这项研究是对哺乳动物克隆的首次报道。虽然通过核重编程成功培育了几只克隆羊，然而实验成功的效率似乎取决于供体细胞类型。胎儿成纤维细胞或胚胎来源细胞的体细胞核转移获得的细胞可成功产生子代的效率要高于乳腺上皮细胞来源的细胞核。这个结果表明来源于发育早期阶段的哺乳动物细胞核更容易被重编程。

改变细胞命运的因子 MyoD（成肌分化抗原）

在青蛙和绵羊的重编程实验中，研究人员使用的卵母细胞包含许多蛋白及因子。由于体细胞重编程涉及很多的分子机制，使得进一步明确运用 SCNT 进行的重编程机制存在困难。5-氮胞苷（5-aza）是一种去甲基化试剂，在培养系统中使用它可以促使小鼠胚胎细胞 C3H/10T1/2CL8 和 Swiss 3T3 细胞转化为成肌、成脂和成软骨细胞。这种细胞型的变化被认为是"转化"而不是"重编程"[6]，然而使用 5-aza 是否可以导致完全的转化还是未知的。另有研究表明，单基因可以使某些细胞转化为其他类型的细胞。这个基因为 MyoD，是一个碱性/螺旋-环-螺旋转录因子[7]，它是调节肌细胞发生的一个主基因[8]，只单一引入 MyoD 就可以使小鼠成纤维细胞、成脂细胞或猴肾脏细胞转化成肌细胞。这个研究表明仅通过异常表达诱导因子即可实现细胞向其他细胞系转化，类似的细胞命运转化的研究亦有报道[9]。例如，单核前体细胞可以通过 GATA1 转化成红系-巨核细胞系细胞[10]，B 细胞可以通过 C/EBPα 转化为巨噬细胞[11]，抑制 Pax5[12] 可使 B 细胞转化为 T 细胞，而成纤维细胞可以通过 PU.1 和 C/EBPα/β 转化为巨噬细胞样细胞[13]。以上研究证明单一的因子对细胞的命运及特化具有巨大的影响。

通过细胞融合来重编程体细胞

体外细胞融合实验同样可以实现体细胞核重编程。小鼠成体胸腺细胞与胚胎干细胞（ES）融合后可再次获得核重编程能力，这个结论可以通过 T 细胞受体的 V-（D）-J DNA 重组的重编程来证实[14]。在成熟胸腺细胞中 Oct3/4-GFP 报告基因没有被激活，而是在细胞融合 48h 后激活，融合后的杂交细胞具有多能性，并具有早期小鼠发育的能力。这项研究表明在 ES 细胞中存在着重编程因子。最近，有报道称 AID（活化诱导的胞嘧啶核苷脱氨酶）使 DNA 去甲基化是细胞融合重编程实验的必需步骤[15]。在这项研究中，小鼠的 ES 细胞和人的成纤维细胞融合生成种间异核体。在重编程的起始阶段，敲减 AID 表达可以抑制 Oct4 启动子区域的去甲基化作用并导致 Oct4 和其他多能基因的表达抑制。这些结论表明 DNA 去甲基化作用对重编程很重要。

转染 Sox2、Oct3/4、Klf4 和 c-Myc 从而产生诱导多能干细胞

我们身体所有细胞均由来自于囊胚的外胚层祖细胞所分化的细胞构成，这些祖细胞具有多能性[16~18]。ES 细胞是由这些细胞在体外派生而得，具有与囊胚外胚层祖细胞相同的多能性及体外无限增殖能力，因此 ES 细胞对再生医学来说是非常有用的手段。

虽然人 ES 细胞理论上可以用于细胞移植治疗，但是它们在应用中面临着反对使用人类胚胎的伦理学障碍。针对这个问题有一个解决方案，即利用直接来源于成体细胞的多能细胞而避免使用植入前的胚胎。SCNT 和细胞融合研究的相关分子机制还不清楚，但是结果表明成体细胞是能够被重编程为多能性的。然而由于这些技术效率低且重复性差，造成了它们无法成为常规技术手段。为了解决这些问题，研究人员开展了利用已有的因子进行直接体细胞核重编程的研究。

最初，我们假设在维持 ES 细胞多能性中起重要作用的因子可以作为重编程因子。然后在芯片表达数据库中分析查询了 ES 细胞特异性高表达且成体细胞不表达的基因，这样我们获得了很多候选的重编程因子，并将它们命名为 ECAT（胚胎干细胞相关转录物）[19]。Nanog 是其中一个，它对维持 ES 细胞的多能性具有非常重要的作用[19]。Nanog 敲除（Nanog-KO）的 ES 细胞失去多能性并分化为胚外内胚层细胞。也有报道表明可以建立 Nanog-KO 的 ES 细胞[20]。这些 Nanog-KO 的 ES 细胞有助于嵌合体小鼠研究而非生殖细胞。Nanog-KO 的 ES 细胞自我更新能力较野生型 ES 细胞要低，此外它们还易分化成为原始内胚层样细胞。这些结果表明 Nanog 对 ES 细胞的自我更新能力不是必需的，它的主要功能是抑制分化，也就是维持 ES 细胞的状态。ECAT 还包括对 ES 细胞多能性维持起重要作用的基因，如 Sox2 和 Oct3/4[21,22]。最终，我们的关注点主要集中在 24 个潜在的重编程相关因子[23]。

我们把这些因子分为三类。第一类是对维持 ES 细胞多能性起重要作用的因子，包括 Nanog、Sox2 和 Oct3/4；第二类是 Tcl1、Stat3 和 c-Myc 等肿瘤相关基因；第三类是 ECAT1、Esg1 和 Klf4 等最初由我们实验室发现有可能在 ES 细胞中具有特定功能的基因。我们使用逆转录病毒将所有的 24 个基因转染小鼠胚胎成纤维细胞，获得了 ES 细胞样细胞[23]。随后，我们采用每次去掉一个因子的方法进行实验，最终将 24 个因子精炼到 4 个关键的因子，包括 Sox2、Oct3/4、Klf4 和 c-Myc（图 14-1）。通过转染这 4 种因子可以获得 ES 细胞样的细胞，这种细胞具有与 ES 细胞相似的多能性，以及与肿瘤细胞相似的无限增殖能力。我们将这些细胞命名为诱导多能性干细胞，即 iPS 细胞。第一代 iPS 细胞（Fbx-iPSC）是利用 *Fbx15* 报告基因对 G418 具有抗性的特点筛选出来的，*Fbx15* 是 ES 细胞的特有基因，因此它在成纤维细胞中不表达而在 ES 细胞中高表达。Fbx-iPSC 能够产生嵌合胚胎但是无法成功繁育嵌合小鼠，这个结果表明 Fbx-iPSC 并不完全等同于 ES 细胞，它是非完全的重编程细胞。

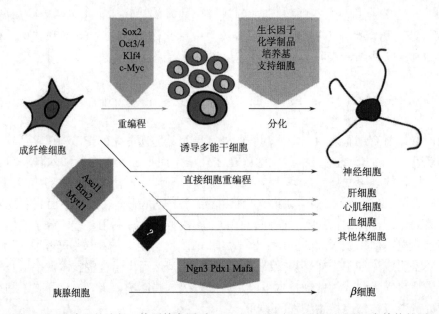

图 14-1 重编程的途径。使用特定因子（Sox2、Oct3/4、Klf4、c-Myc 或其他的因子组合）将成纤维细胞重编程为 iPS 细胞。iPS 细胞在特定条件下培养分化为几种不同的成体细胞。直接由成纤维细胞生成的神经元细胞。体内胰腺外分泌细胞也可以直接生成 β-细胞。灰线标明了未确认的重编程过程

为了获得可以繁育成熟嵌合小鼠的 iPS 细胞，我们及其他研究团队构建了具有嘌呤霉素抗性的 Nanog 报告基因，用以筛选高质量的 iPS 细胞[24]。利用第二代 iPS 细胞（Nanog-iPSC），我们获得了可繁育的成熟嵌合小鼠[24~26]，通过这项研究我们发现除了报告基因外，药物筛选时机也是非常重要的。在我们的研究中，鼠尾来源的成纤维细胞[23]、干细胞[27]和胃上皮细胞[27]等完全分化的细胞都被这 4 种因子成功地重编程为 iPS 细胞，这些结果表明体细胞能够被特定的因子重编程。许多实验室相继报道使用不同来源细胞建立了小鼠的 iPS 细胞系。

随后使用相同的 4 种修饰后因子成功诱导获得了人的 iPS 细胞[28]。同时，其他研究组还通过 SOX2、OCT3/4、NANOG 和 LIN28 这 4 种因子成功获得了人 iPS 细胞[29]。人 iPS 细胞在形态、增殖能力及分化能力方面与人 ES 细胞极其相似。

iPS 细胞诱导的方法

最初，iPS 细胞来源于用逆转录病毒转染 4 个转录因子的成纤维细胞。这种诱导 iPS 的方法效率很高且可重复。使用逆转录病毒进行转染，会把用于重编程的基因的 DNA 和逆转录病毒载体的一小部分一起整合到宿主基因中，这增加了由 iPS 细胞发育而来的嵌合小鼠患肿瘤的风险。在这些肿瘤中，我们检测到转染的 c-Myc 基因的活化[24]，表明在构建 iPS 细胞时应该避免使用逆转录病毒转染 c-Myc。基于以上结果，我

们尝试不使用带有 c-Myc 基因的逆转录病毒进行转染，虽然获得 iPS 的效率很低，还是得到了更安全的 iPS 细胞产物（3F-iPSC）[30,31]。3F-iPSC 的形态、增殖效率和分化能力与 4 因子诱导的 iPS 细胞（4F-iPSC）相当。而由 3F-iPSC 细胞发育而来的嵌合小鼠的肿瘤发生效率明显低于由 4F-iPSC 细胞发育而来的嵌合小鼠。这些结果表明诱导 iPS 细胞时 c-Myc 不是必需的，如果想要建立安全的 iPS 细胞系，应该避免使用带有 c-Myc 的逆转录病毒。根据我们最近的研究，3F-iPSC 的种系传递效率要低于 4F-iPSC（Aoi 等，未发表结果），这个结果意味着仅使用 3 种因子的重编程是不完全的，寻找新的因子以替代 c-Myc 对建立安全和高质量的 iPS 细胞来说是必要的。最近有研究报道，Tbx3 可以替代 c-Myc 提高小鼠 iPS 细胞的种系能力[32]。3 因子和 Tbx3 组合可以提高 iPSC 产生的效率，但是不能提高 iPSC 克隆的数量。另外，虽然转染 Tbx3 可以增加种系传递的效率，但是获得的嵌合小鼠肿瘤发生的情况目前还没有定论。相关的实验结果还需要进一步的观察。

几种不同的将重编程因子引入成纤维细胞的方法已经被报道，这些研究都是为了获得更安全的 iPS 细胞。我们先前报道过使用普通转染试剂进行的质粒载体转染可以获得小鼠 iPS 细胞，而且这些细胞可以进行种系传递[33]。其他研究团队也相继报道了获得 iPS 细胞的替代方法[34~48]。已经有综述对 iPS 细胞研究的进展情况进行了很好的总结[49]，但目前还不清楚这些方法中哪个是最佳方案。

iPS 细胞产生的分子机制

虽然将 4 因子转导到成纤维细胞中能够将成体细胞转变成多能细胞，但是相关的分子机制还不清楚。已知 Sox2 和 Oct3/4 在维持 ES 细胞多能性和功能方面发挥着重要的作用，在 iPS 细胞诱导过程中它们具有同样的机制。c-Myc 是重要的原癌基因，并且在许多肿瘤中都高表达。c-Myc 的功能主要是能够改变染色质的结构并激活几种基因。大约有 3 万个 DNA 结合区域被带有或不带有接头蛋白的 c-Myc 所识别。Klf4 则同时具有原癌基因和肿瘤抑制基因两种功能[50,51]。在 iPS 细胞诱导过程中，Klf4 有可能会抑制 c-Myc 的功能从而阻碍 iPS 细胞的产生[52,53]。虽然每种因子的功能都有详尽的研究，但是究竟为什么将特定的 3 种或 4 种因子转染到体细胞中可将其重编程为多能细胞还是未知的。

直接重编程：来源于胰腺细胞的 β 细胞

再生医学需要特定的健康体细胞，运用 iPS 技术我们可以利用已经建立的 iPS 细胞系在体外分化获得几种不同类型的成体细胞。然而，从 iPS 细胞诱导获得细胞的效率、活性及纯度都很低，不足以在再生医学特别是移植治疗中使用。在 iPS 细胞技术快速发展的同时，胰腺外分泌细胞体内转变成 β 细胞的相关研究已被报道（图 14-1）[54]。这个体内重编程过程是通过腺病毒系统将特定的因子直接转入胰腺细胞而实现的。这项研究表明小鼠细胞有可能具有无需 iPS 过程而直接重编程其他类型细胞的能力。

直接重编程：来源于成纤维细胞的神经元细胞

最近的研究表明成体细胞可以在体外被重编程为多能干细胞，也可以在体内被直接重编程为其他类型的成体细胞。在体内直接细胞重编程被报道之后，体外诱导成体细胞也获得成功[55]。有研究团队使小鼠成纤维细胞表达特定的因子并将它们转化成神经元细胞，这些特定因子仅为 3 个神经元系统特殊的转录因子，而这些细胞被命名为诱导神经元（iN）细胞（图 14-1）。iN 细胞表达神经元特异基因，并且产生具有功能的神经元突触。这项技术对建立疾病发展过程中的特定异常细胞的细胞系非常有帮助。

可用于临床的疾病 iPS 细胞系

人的 iPS 细胞系建立将促进再生医学的发展（图 14-2）。对细胞移植治疗来说，ES 细胞曾被认为是良好的种子细胞，然而对于大多数患者来说，拥有自身来源的 ES 细胞几乎是不可能的。使用患者自身的成纤维细胞（或其他细胞）产生的 iPS 细胞可以避免 ES 细胞的免疫排斥问题，这有可能促进细胞移植治疗的快速发展。重要的是，人 iPS 细胞可以直接由患者的那些含有致病遗传缺陷的成体细胞产生（disease-iPS），disease-iPSC 同样具有这些遗传缺陷。在 disease-iPS 细胞分化为特定细胞的过程中观察它们，

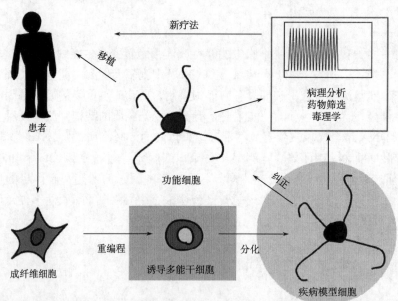

图 14-2　再生医学中 iPS 细胞的应用。收集患者的成纤维细胞（或其他细胞），将这些细胞重编程为 iPS 细胞并在体外分化获得疾病模型细胞。为了进行移植治疗，来源于患者的疾病细胞被校正为具有功能的正常细胞并被移植到他们体内。疾病模型细胞也可以用作病理学分析、药物筛选和毒理分析。通过这些研究，可以找到新的治疗方法和药物并使患者受益

有可能重现并了解患者疾病发生、发展的过程。这些细胞可用于病理学分析、药物筛选、毒理分析及药物副作用研究,在 iPS 细胞诞生之前这些研究过程是很困难的。

结论

长期以来,成体细胞重编程对许多研究者来说只是一个梦想。重编程实验起始于青蛙,随之是克隆动物研究,最终通过直接重编程产生了 iPS 细胞。重编程或转化一般是通过激活分化细胞的关键因子实现的,而这些关键因子的功能在分化的细胞中被一些精巧的机制所抑制,如基因组 DNA 的甲基化作用或染色质修饰。一旦这些因子被激活,细胞的特性将发生改变并被重编程为其他类型细胞。重编程方法比以前认为的更加容易,但是精确的重编程因子组合对确定的细胞类型产生很重要,在未来的探索中,研究者将确定更多的因子组合以产生不同类型的细胞(图 14-1)。定向的细胞重编程同时有希望成为获得疾病特有异常细胞的技术,这使在短期内从患者的皮肤成纤维细胞中直接产生神经系统疾病的模型细胞成为可能。目前这些细胞还不能用于临床应用,还需要进一步的研究。

基础研究,特别是小鼠和人 ES 细胞的研究为成功获得 iPS 细胞奠定了基础。许多科学研究者对 iPS 细胞技术进行了完善,它的研究将会提高再生医学的水平。目前,iPS 细胞的质量似乎已经符合应用需求,然而 iPS 细胞技术还未完全成熟,仍存在着一些问题,如 iPS 细胞克隆的选择和特性、构建方法的发展(病毒、质粒、蛋白质或化学方法)、培养条件的优化化,以及验证 iPS 细胞安全性的方法。

(池 颖 译)

参 考 文 献

1. Gurdon JB, Uehlinger V. "Fertile" intestine nuclei. Nature 1966; 210(5042):1240-1241.
2. Briggs R, King TJ. Transplantation of Living Nuclei From Blastula Cells into Enucleated Frogs' Eggs. Proc Natl Acad Sci USA 1952; 38(5):455-463.
3. King TJ, Briggs R. Changes in the Nuclei of Differentiating Gastrula Cells, as Demonstrated by Nuclear Transplantation. Proc Natl Acad Sci USA 1955; 41(5):321-325.
4. Gurdon JB. The developmental capacity of nuclei taken from intestinal epithelium cells of feeding tadpoles. J Embryol Exp Morphol 1962; 10:622-640.
5. Wilmut I, Schnieke AE, McWhir J et al. Viable offspring derived from fetal and adult mammalian cells. Nature 1997; 385(6619):810-813.
6. Taylor SM, Jones PA. Multiple new phenotypes induced in 10T1/2 and 3T3 cells treated with 5-azacytidine. Cell 1979; 17(4):771-779.
7. Tapscott SJ. The circuitry of a master switch: Myod and the regulation of skeletal muscle gene transcription. Development 2005; 132(12):2685-2695.
8. Davis RL, Weintraub H, Lassar AB. Expression of a single transfected cDNA converts fibroblasts to myoblasts. Cell 1987; 51(6):987-1000.
9. Graf T, Enver T. Forcing cells to change lineages. Nature 2009; 462(7273):587-594.
10. Kulessa H, Frampton J, Graf T. GATA-1 reprograms avian myelomonocytic cell lines into eosinophils, thromboblasts and erythroblasts. Genes Dev 1995; 9(10):1250-1262.
11. Xie H, Ye M, Feng R et al reprogramming of B cells into macrophages. Cell 2004; 117(5):663-676.
12. Nutt SL, Heavey B, Rolink AG et al. Commitment to the B-lymphoid lineage depends on the transcription factor Pax5. Nature 1999; 401(6753):556-562.
13. Feng R, Desbordes SC, Xie H et al. PU.1 and C/EBPalpha/beta convert fibroblasts into macrophage-like cells. Proc Natl Acad Sci USA 2008; 105(16):6057-6062.
14. Tada M, Takahama Y, Abe K et al. Nuclear reprogramming of somatic cells by in vitro hybridization with ES cells. Curr Biol 2001; 11(19):1553-1558.

15. Bhutani N, Brady JJ, Damian M et al. Reprogramming towards pluripotency requires AID-dependent DNA demethylation. Nature 2010; 463(7284):1042-7.
16. Evans MJ, Kaufman MH. Establishment in culture of pluripotential cells from mouse embryos. Nature 1981; 292(5819):154-156.
17. Martin GR. Isolation of a pluripotent cell line from early mouse embryos cultured in medium conditioned by teratocarcinoma stem cells. Proc Natl Acad Sci USA 1981; 78(12):7634-7638.
18. Thomson JA, Itskovitz-Eldor J, Shapiro SS et al. Embryonic stem cell lines derived from human blastocysts. Science 1998; 282(5391):1145-1147.
19. Mitsui K, Tokuzawa Y, Itoh H et al. The Homeoprotein Nanog Is Required for Maintenance of Pluripotency in Mouse Epiblast and ES Cells. Cell 2003; 113(5):631-642.
20. Chambers I, Silva J, Colby D et al. Nanog safeguards pluripotency and mediates germline development. Nature 2007; 450(7173):1230-1234.
21. Boyer LA, Lee TI, Cole MF et al. Core Transcriptional Regulatory Circuitry in Human Embryonic Stem Cells. Cell 2005; 122:947-956.
22. Wang J, Rao S, Chu J et al. A protein interaction network for pluripotency of embryonic stem cells. Nature 2006; 444(7117):364-368.
23. Takahashi K, Yamanaka S. Induction of pluripotent stem cells from mouse embryonic and adult fibroblast cultures by defined factors. Cell 2006; 126(4):663-676.
24. Okita K, Ichisaka T, Yamanaka S. Generation of germ-line competent induced pluripotent stem cells. Nature 2007; 448:313-317.
25. Maherali N, Sridharan R, Xie W et al. Directly reprogrammed fibroblasts show global epigenetic remodelling and widespread tissue contribution. Cell Stem Cell 2007; 1(1):55-70.
26. Wernig M, Meissner A, Foreman R et al. In vitro reprogramming of fibroblasts into a pluripotent ES cell-like state. Nature 2007; 448:318-324.
27. Aoi T, Yae K, Nakagawa M et al. Generation of Pluripotent Stem Cells from Adult Mouse Liver and Stomach Cells Science 2008; 321:699-702.
28. Takahashi K, Tanabe K, Ohnuki M et al. Induction of pluripotent stem cells from adult human fibroblasts by defined factors. Cell 2007; 131(5):861-872.
29. Yu J, Vodyanik MA, Smuga-Otto K et al. Induced pluripotent stem cell lines derived from human somatic cells. Science 2007; 318(5858):1917-1920.
30. Nakagawa M, Koyanagi M, Tanabe K et al. Generation of induced pluripotent stem cells without Myc from mouse and human fibroblasts. Nat Biotechnol 2008; 26(1):101-106.
31. Wernig M, Meissner A, Cassady JP et al. c-Myc is dispensable for direct reprogramming of mouse fibroblasts. Cell Stem Cell 2008; 2(1):10-12.
32. Han J, Yuan P, Yang H et al. Tbx3 improves the germ-line competency of induced pluripotent stem cells. Nature 463(7284):1096-1100.
33. Okita K, Nakagawa M, Hyenjong H et al. Generation of Mouse Induced Pluripotent Stem Cells Without Viral Vectors. Science 2008; 322(5903):949-953.
34. Carey BW, Markoulaki S, Hanna J et al. Reprogramming of murine and human somatic cells using a single polycistronic vector. Proc Natl Acad Sci USA 2008.
35. Sommer CA, Stadtfeld M, Murphy GJ et al. iPS Cell Generation Using a Single Lentiviral Stem Cell Cassette. Stem Cells. 2009; 27(3):543-9.
36. Stadtfeld M, Nagaya M, Utikal J et al. Induced Pluripotent Stem Cells Generated Without Viral Integration. Science 2008; 322(5903):945-949.
37. Gonzalez F, Barragan Monasterio M, Tiscornia G et al. Generation of mouse-induced pluripotent stem cells by transient expression of a single nonviral polycistronic vector. Proc Natl Acad Sci USA 2009.
38. Hotta A, Cheung AYL, Farra N et al. Isolation of human iPS cells using EOS lentiviral vectors to select for pluripotency. Nature Methods. 2009; advance online publication.
39. Kaji K, Norrby K, Paca A et al. Virus-free induction of pluripotency and subsequent excision of reprogramming factors. Nature 2009.
40. Lyssiotis CA, Foreman RK, Staerk J et al. Reprogramming of murine fibroblasts to induced pluripotent stem cells with chemical complementation of Klf4. Proc Natl Acad Sci USA 2009.
41. Shao L, Feng W, Sun Y et al. Generation of iPS cells using defined factors linked via the self-cleaving 2A sequences in a single open reading frame. Cell Research. 2009; advance online publication.
42. Soldner F, Hockemeyer D, Beard C et al. Parkinson's Disease Patient-Derived Induced Pluripotent Stem Cells Free of Viral Reprogramming Factors. Cell 2009; 136:964-977.
43. Sommer CA, Stadtfeld M, Murphy GJ et al. Induced pluripotent stem cell generation using a single lentiviral stem cell cassette. Stem Cells 2009; 27(3):543-549.
44. Woltjen K, Michael IP, Mohseni P et al. piggyBac transposition reprograms fibroblasts to induced pluripotent stem cells. Nature 2009.
45. Zhou W, Freed CR. Adenoviral Gene Delivery Can Reprogram Human Fibroblasts to Induced Pluripotent Stem Cells. Stem Cells 2009.
46. Kim D, Kim CH, Moon JI et al. Generation of human induced pluripotent stem cells by direct delivery of reprogramming proteins. Cell Stem Cell 2009; 4(6):472-476.

47. Zhou H, Wu S, Joo JY et al. Generation of induced pluripotent stem cells using recombinant proteins. Cell Stem Cell 2009; 4(5):381-384.
48. Jia F, Wilson KD, Sun N et al. A nonviral minicircle vector for deriving human iPS cells. Nat Methods; 7(3):197-199.
49. Chen L, Liu L. Current progress and prospects of induced pluripotent stem cells. Sci China C Life Sci 2009; 52(7):622-636.
50. Zhao W, Hisamuddin IM, Nandan MO et al. Identification of Kruppel-like factor 4 as a potential tumor suppressor gene in colorectal cancer. Oncogene 2004; 23(2):395-402.
51. Rowland BD, Bernards R, Peeper DS. The KLF4 tumour suppressor is a transcriptional repressor of p53 that acts as a context-dependent oncogene. Nat Cell Biol 2005; 7(11):1074-1082.
52. Cartwright P, McLean C, Sheppard A et al. LIF/STAT3 controls ES cell self-renewal and pluripotency by a Myc-dependent mechanism. Development 2005; 132(5):885-896.
53. Sumi T, Tsuneyoshi N, Nakatsuji N et al. Apoptosis and differentiation of human embryonic stem cells induced by sustained activation of c-Myc. Oncogene 2007; 26(38):5564-5576.
54. Zhou Q, Brown J, Kanarek A et al. In vivo reprogramming of adult pancreatic exocrine cells to beta-cells. Nature 2008; 455(7213):627-632.
55. Vierbuchen T, Ostermeier A, Pang ZP et al. Direct conversion of fibroblasts to functional neurons by defined factors. Nature 2010; 463(7284):1035-41.